AF726476

FACTS and HVDC

Other related titles:

You may also like

- PBPO0300 | Song | Flexible AC Transmission Systems (FACTS) | 1999
- PBPO0290 | Arrillaga | High Voltage Direct Current Transmission, 2nd Edition | 1998
- PBPO116 | Han | Modelling and Simulation of HVDC Transmission | 2020

We also publish a wide range of books on the following topics:
Computing and Networks
Control, Robotics and Sensors
Electrical Regulations
Electromagnetics and Radar
Energy Engineering
Healthcare Technologies
History and Management of Technology
IET Codes and Guidance
Materials, Circuits and Devices
Model Forms
Nanomaterials and Nanotechnologies
Optics, Photonics and Lasers
Production, Design and Manufacturing
Security
Telecommunications
Transportation

All books are available in print via https://shop.theiet.org or as eBooks via our Digital Library https://digital-library.theiet.org.

IET ENERY ENGINEERING SERIES 173

FACTS and HVDC

Advances in technology, modelling, and regulatory requirements

Edited by
Enrique Acha, Claudio R. Fuerte-Esquivel and
Francisco De La Rosa

The Institution of Engineering and Technology

About the IET

This book is published by the Institution of Engineering and Technology (The IET).

We inspire, inform and influence the global engineering community to engineer a better world. As a diverse home across engineering and technology, we share knowledge that helps make better sense of the world, to accelerate innovation and solve the global challenges that matter.

The IET is a not-for-profit organisation. The surplus we make from our books is used to support activities and products for the engineering community and promote the positive role of science, engineering and technology in the world. This includes education resources and outreach, scholarships and awards, events and courses, publications, professional development and mentoring, and advocacy to governments.

To discover more about the IET please visit https://www.theiet.org/

About IET books

The IET publishes books across many engineering and technology disciplines. Our authors and editors offer fresh perspectives from universities and industry. Within our subject areas, we have several book series steered by editorial boards made up of leading subject experts.

We peer review each book at the proposal stage to ensure the quality and relevance of our publications.

Get involved

If you are interested in becoming an author, editor, series advisor, or peer reviewer please visit https://www.theiet.org/publishing/publishing-with-iet-books/ or contact author_support@theiet.org.

Discovering our electronic content

All of our books are available online via the IET's Digital Library. Our Digital Library is the home of technical documents, eBooks, conference publications, real-life case studies and journal articles. To find out more, please visit https://digital-library.theiet.org.

In collaboration with the United Nations and the International Publishers Association, the IET is a Signatory member of the SDG Publishers Compact. The Compact aims to accelerate progress to achieve the Sustainable Development Goals (SDGs) by 2030. Signatories aspire to develop sustainable practices and act as champions of the SDGs during the Decade of Action (2020–30), publishing books and journals that will help inform, develop, and inspire action in that direction.

In line with our sustainable goals, our UK printing partner has FSC accreditation, which is reducing our environmental impact to the planet. We use a print-on-demand model to further reduce our carbon footprint.

Published by The Institution of Engineering and Technology, London, United Kingdom

The Institution of Engineering and Technology (the "**Publisher**") is registered as a Charity in England & Wales (no. 211014) and Scotland (no. SC038698).

Copyright © The Institution of Engineering and Technology and its licensors 2025

First published 2025

All intellectual property rights (including copyright) in and to this publication are owned by the Publisher and/or its licensors. All such rights are hereby reserved by their owners and are protected under the Copyright, Designs and Patents Act 1988 ("**CDPA**"), the Berne Convention and the Universal Copyright Convention.

With the exception of:

(i) any use of the publication solely to the extent as permitted under:

a. the CDPA (including fair dealing for the purposes of research, private study, criticism or review); or
b. the terms of a licence granted by the Copyright Licensing Agency ("**CLA**") (only applicable where the publication is represented by the CLA); and/or

(ii) any use of those parts of the publication which are identified within this publication as being reproduced by the Publisher under a Creative Commons licence, Open Government Licence or other open source licence (if any) in accordance with the terms of such licence, no part of this publication, including any article, illustration, trade mark or other content whatsoever, may be used, reproduced, stored in a retrieval system, distributed or transmitted in any form or by any means (including electronically) without the prior permission in writing of the Publisher and/or its licensors (as applicable).

The commission of any unauthorised activity may give rise to civil or criminal liability.

Please visit https://digital-library.theiet.org/copyrights-and-permissions for information regarding seeking permission to reuse material from this and/or other publications published by the Publisher. Enquiries relating to the use, including any distribution, of this publication (or any part thereof) should be sent to the Publisher at the address below:

The Institution of Engineering and Technology
Futures Place,
Kings Way, Stevenage,
Herts, SG1 2UA
United Kingdom

www.theiet.org

Whilst the Publisher and/or its licensors believe that the information and guidance given in this publication is correct, an individual must rely upon their own skill and judgement when performing any action or omitting to perform any action as a result of any statement, opinion or view expressed in the publication and neither the Publisher nor its licensors assume and hereby expressly disclaim any and all liability to anyone for any loss or damage caused by any action or omission of an action made in reliance on the publication and/or any error or omission in the publication, whether or not such an error or omission is the result of negligence or any other cause. Without limiting or otherwise affecting the generality of this statement and the disclaimer, whilst all URLs cited in the publication are correct at the time of press, the Publisher has no responsibility for the persistence or accuracy of URLs for external or third-party internet websites and does not guarantee that any content on such websites is, or will remain, accurate or appropriate.

Whilst every reasonable effort has been undertaken by the Publisher and its licensors to acknowledge copyright on material reproduced, if there has been an oversight, please contact the Publisher and we will endeavour to correct this upon a reprint.

Trade mark notice: Product or corporate names referred to within this publication may be trade marks or registered trade marks and are used only for identification and explanation without intent to infringe.

Where an author and/or contributor is identified in this publication by name, such author and/or contributor asserts their moral right under the CPDA to be identified as the author and/or contributor of this work.

British Library Cataloguing in Publication Data
A catalogue record for this product is available from the British Library

ISBN 978-1-83953-137-8 (hardback)
ISBN 978-1-83953-138-5 (PDF)

Typeset in India by MPS Limited

Cover image: Photo courtesy of Hitachi Energy

Contents

Preface

Electronic semiconductor valves, converters and control methods continue to evolve and transform the fabric of electrical power systems, enabling a more reliable, energy-efficient and environmentally friendly global power supply industry. Such developments started to become established from the turn of the millennium, but their widespread uptake is now set on an irreversible and accelerated upward trend. Early developments have been penned by leading academics, such as J Arrillaga, NG Hingorani, L Gyugyi and TJE Miller, in books which have become classics. However, the inexorable march of power electronics permeating all sectors of the electric power industry, namely, generation, transmission and distribution, has continued, and it may now be a good time to sum up the surge of developments that have taken place over the last two decades in power electronics applications in electrical power systems, and to consider what the face of the future electrical power system would look like.

Vast hydropower resources are being transported in electrical form, at GW levels, using HVDC transmission, over distances spanning thousands of kilometres in China and Brazil to meet the energy needs of large industrial complexes and conurbation areas lacking clean, primary energy resources in their neighbourhoods, to continue fuelling their economic prosperity and clean way of life of their dwellers. In a different corner of the world, the North Sea has been housing one of the most magnificent accelerations of sustained development and innovation that the electrical power industry has ever witnessed, namely, the deployment of a large number of offshore wind farms made up of variable speed wind turbines rated at 10 MW and higher, producing electricity from a clean, free, renewable primary energy resource and feeding it to hundreds of millions of European homes. Leading-edge power electronics-based submarine transmission systems are the enabler technologies that links those awesome offshore wind farms to the household, industrial and commercial loads located hundreds of kilometres from where electricity is produced. To put these developments into context, the 443 MW Viking wind farm, comprising 103 turbines on the Shetland Islands north of Scotland, meets the annual power needs of 26 million homes and cut carbon emissions by more than 35 million tonnes a year. The Viking wind farm is starting operations at the time of writing this (28 August 2024), bringing the total wind power operational capacity of the UK to just above the historic mark of 30 GW. The 260 km submarine HVDC cable linking the Viking windfarm and a super DC node in the northern tip of the Scottish mainland, in Caithness/Spittal, has given rise to what is the first multi-terminal VSC-HVDC system in Europe. In its present form, it is a three-terminal,

Y-connected DC link, with a potential capability for connecting two future DC branches to receive offshore wind power coming from the North Sea and hydropower from Norway.

These are mammoth achievement, moving in the right direction, but much more is urgently required to attain the net-zero-carbon emission by 2050 that international bodies have set to be our Rubicon if we are to put the brakes on irreversible damage caused to our delicately balanced climate system by CO_2 emissions. To achieve this, the IEA estimates that every year, 630 GW of solar PV and 390 GW of wind power would need to be installed, starting as soon as possible but certainly no later than 2030. That is four times as much as our current annual rate.

What is called for is an even more radical change in the use of technology that we already have, which needs perfecting and making less expensive. To tackle the variability of renewable energy resources, suitable energy storage systems would need to be embedded into the structure of wind and solar PV power generators, just as power electronics converters are now an integral part of these systems. Furthermore, the power electronics converters should be fitted with the so-called *grid-forming* control capabilities. This would enable the existence of power grids dominated by inverter-based generation and energy storage, where the 'fuel' is clean, renewable and free, and current operational teething issues, such as intermittence and loss of grid inertia, are a thing of the recent past. Nonetheless, new issues may arise such as the appearance of inter-harmonics and supra-harmonics. For example, protection practices may need adjusting given decreasing fault current levels in power systems with a higher share of renewable generation connected to the bulk power system through power electronics converters. These new operational problems will need to be addressed, and their solution reflected in updates to the latest grid codes.

Inverter-dominated power grids already exist, and their resilience has successfully been put through its paces under different operating scenarios. This is well exemplified by the South Australian power grid, having such high levels of solar PV power generation that at solar irradiation peaks, solar power generation regularly meets most of the power demand. This game changer is being achieved with Li-ion batteries supplied by Tesla and connected to the grid through GFM inverters. The current largest battery system in Australia is a 500 MW/1,000 MWh with GFM inverters at Liddell, New South Wales, installed at the site of a decommissioned coal plant. Several other large GFM systems are undergoing construction in Australia. The peak demand of South Wales is around 3 GW, and to scale up similar developments to larger power grids, such as the Texan power grid with a peak demand of 85 GW, will require more advanced and diverse technologies than what we already have. For instance, Li-ion batteries may become a bottleneck and nanoparticle-fluid batteries, with their higher energy densities, small footprint, low heat dissipation and no fire-catching issues, low maintenance and low cost seem the ideal energy storage media for grid power applications and a suitable partner for GFM systems. Perhaps such striking attributes would find immediate use in offshore wind farms in the rough North Sea waters. Companies like Influit Energy, an Illinois-based firm, may be ready to launch this product in 2025. Technology like this will strengthen the emissions-free expansion of the UK power grid, with wind forming the backbone of its future energy system. The current wind power contribution to the energy mix stands at more than 1/3, with new offshore wind farms under construction. The UK government's target is to reach 40 GW offshore wind capacity by 2030 and, together with the installed onshore capacity, to attain an overall figure of 50 GW. This could go even higher as the development of floating wind turbines consolidates further, potentially achieving the decarbonisation of its electricity supply sector. This would come along with the formation of a meshed DC power grid in the North Sea, as the dozens of point-to-point VSC-HVDC links already built, start to be joined together, as exemplified by the Shetland-Caithness-Moray HVDC link in northern Scotland.

We perceive that electricity generation that is completely carbon-free in its operation is around the corner, but much more R&D work is needed, including developing new materials, devices, equipment, control methods, computer algorithms and scaling up the all-important but early developments that we have reached heretofore. This book aims to contribute to this global endeavour by focusing on how the power grid could be made more flexible and controllable to enable an ever larger penetration of converter-interfaced power generation resources and storage systems.

IEEE Life Fellow Enrique Acha, Bristol, England, UK.
IEEE Fellow Claudio R. Fuerte-Esquivel, Morelia, Michoacán, México.
IEEE Life Senior Member Francisco De La Rosa, San Antonio, Texas, USA.

About the editors

Enrique Acha has pursued an international career in power electronics applications in electrical power systems. This has taken him from his native Mexico in the early 1980s to pursue R&D work in academic institutions and industry in the UK, New Zealand, Canada, France, Finland and Mexico. As a member of the IEEE Distinguished Lecturers Program, he has lectured globally. He has written five textbooks and more than 200 papers. He is an IEEE Life Fellow.

Claudio R. Fuerte-Esquivel is a professor at the Universidad Michoacana de San Nicolás de Hidalgo, Mexico. His research focuses on modelling, analysis and control of electrical systems. He is an independent advisor of the EPS-II Generation Company of the Federal Electricity Commission. He has served as a consultant supporting the operation of the Mexican National Center for Energy Control. He is a regular member of the Mexican Academy of Sciences and Fellow of the IEEE.

Francisco De La Rosa is an independent senior power systems consultant in San Antonio, Texas, USA, focusing on the operational planning of transmission systems incorporating renewable power. He has extensive experience in R&D, assessing and integrating innovative technologies in T&D systems in North America and other countries. He has authored several power systems books and is an IEEE Life Senior Member, an Associate Member of ESIG and a member of several NERC Task Forces.

Chapter 1
Introduction

Enrique Acha[1]

1.1 Background

The global power industry continues to experience unprecedented growth and technological innovation. Some of the factors driving these developments are the unquenching global demand for electricity, the urgent need for the decarbonisation of the electricity sector to ameliorate global warming, the increasing liberalisation of the energy markets, and the appeal of "green jobs" opportunities [1]. The grid integration of large-scale renewable power installations, such as photovoltaics (PV) and wind power, brings about new challenges to the operation of the otherwise conventional power grid. For one, the variable nature of the solar and wind primary energy resources is mainly responsible for the intermittent operation of PV solar power plants and wind farms. Fully-fledged, MW (megawatt)-level battery storage is available today from several vendors to reign on the intermittence of the power output and voltage of these kinds of renewable power plants. Economics may not yet be in favour of the widespread use of such a solution. However, energy storage technology is developing very fast, and the teething problems caused by intermittence could soon become a lesser concern. This issue is of paramount importance to the long-term success of the decarbonisation of the electrical power industry, and a great many efforts are being made globally to develop the next generation of batteries that will overcome the limitations that the current Li-ion battery technology exhibits: high costs, high weights, highly susceptible to catching fire when subjected to stress of any kind, scalability issues when reaching the MW-level and the all-important issue of waste disposal. One battery option that is developing fast is the nano-electro-fuel batteries, which is a new take on the old redox flow batteries. It outperforms the Li-ion battery in all counts, and it is to be able to scale up to 100 MW with no perceived difficulty. Unlike the conventional flow batteries that pack very little energy into a given volume and mass, the new breed of flow batteries using nanofluids in suspension [2] seem to be an excellent match for onshore and offshore wind farms.

[1]Electrical Energy Consultant, Power Systems Laboratory, Bristol, United Kingdom

Arguably, this trend of the rapid evolution of the electricity supply industry seems to have started in the mid-1980s, which was spearheaded by the privatisation of several of the largest electrical power utilities in the world. Still, it has gradually quickened its pace since the start of the current millennium, when the power output of wind power generators showed their potential for replacing that of the conventional, large synchronous generators, which were nearing the end of their useful lives. This was particularly the case when the first MW-level wind farms made up of variable speed, doubly-fed induction generators started to be installed – first inland and then offshore [3]. A few years later, after the wind power technology had been firmly established, a striking decrease in the price of photovoltaic panels – largely due to the entrance of China into the photovoltaic market – enabled the photovoltaic solar power industry to take a quantum leap in its capacity to expand and started installing MW-level photovoltaic solar power systems, which today rival wind farms in terms of economics and operational flexibility. Here-to-fore, wind power and photovoltaic solar power are the two most advanced forms of renewable power generation, which are fronting the global effort for the decarbonisation of the electricity supply industry, together with energy-saving equipment, advanced planning methodologies, and smart operating practices [4]. The global installed capacity of solar PV power and wind power is around the 1,000 GW mark each.

Power electronics valves, converters, and switching control techniques may be considered the *nervous system* of the new breed of electrical power equipment and networks [5]. A case in point is Scotland's next-generation power transmission network based on what has become the first European multiterminal VSC-HVDC (Voltage Source Converter High Voltage Direct Current (VSC-HVDC)) system, operated at $\pm$320 kV. The first phase of this far-reaching project was commissioned in 2018. It is a 160 km long, point-to-point VSC-HVDC running undersea between Caithness/Spittal and Moray/Blackhillock in Northern Scotland. The second phase is a 260 km long submarine cable linking the main island of Shetland to the Scottish mainland. The second phase, built to evacuate the power output of Shetland's Viking wind farm, which commenced operation on 28 August 2024, has been designed to form with the Spittal-Blackhillok DC cable, a Y-connected, three-terminal VSC-HVDC system, as illustrated in Figure 1.1. The first phase is rated at 1,200 MW, and the second at 600 MW. The first phase was built to evacuate the output of renewable energy sources in the north of Scotland, e.g., Beatrice (588 MW) and Dorenell (177 MW) wind farms. The immediate use of the Shetland's submarine cable will be to evacuate the power output of the 443 MW Viking wind farm in the Shetlands to the Scottish mainland. Up to about 10% of this power will be used to meet the local needs of the Shetlands, and the rest will travel south to the mainland via the DC submarine power cable. Provisions are being made for the future expansion of two more terminals, arriving at the central point of the Y, to pick up offshore wind power from the North Sea, as illustrated in Figure 1.1 [6].

Elsewhere in Europe, plans are well advanced for constructing multiterminal VSC-HVDC systems. Germany, Holland, Belgium, and England have ambitious programs for offshore wind power developments in the North Sea. To support these

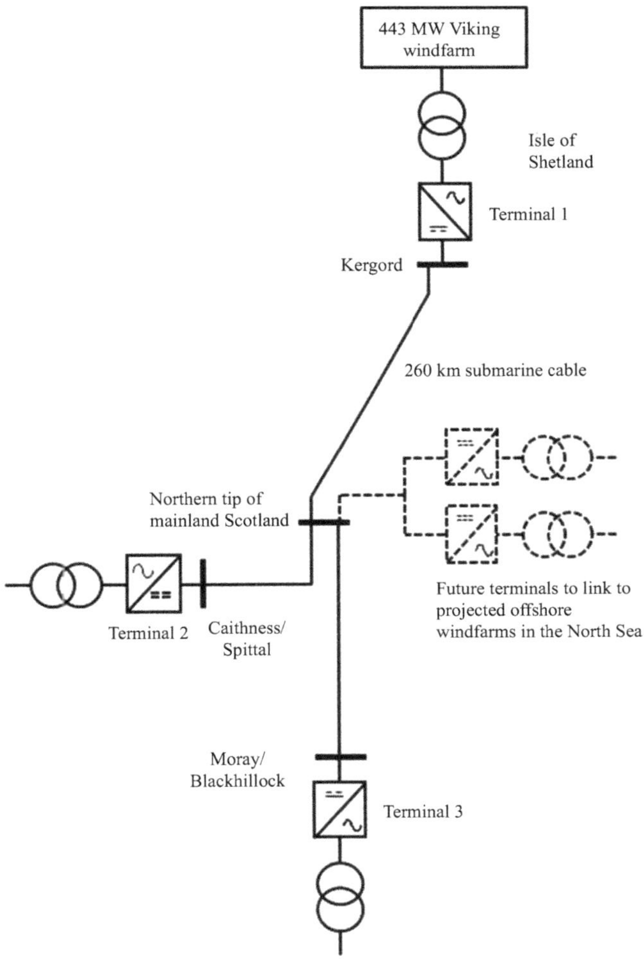

Figure 1.1 Schematic diagram of a three-terminal VSC-HVDC system built in Scotland to evacuate 443 MW of wind power in the Island of Shetland, using submarine power transmission, adapted from [6]

developments, a power grid made up of DC submarine cables and VSC-HVDC multilevel converter stations is needed. The next natural step would be to start joining the dozens of point-to-point VSC HVDC links already built in the North Sea, following in the footsteps of the Shetland-Spittal-Blackhillok HVDC system described above. In 2020, China State Grid started the construction of a 500 kV, four-terminal DC power ring near Beijing, with DC circuit breakers being a major part of this ambitious undertaking. All the major manufacturers of HVDC equipment are reportedly working on DC circuit breakers. Units of 350 kV developed by Hitachi Energy have been verified and are being considered for planned offshore projects [6].

Some of the largest wind, solar, and hydropower resources in the world are being transported in electricity form, using point-to-point, LCC (Line-Commutated Converter)-HVDC transmission systems, over distances spanning thousands of kilometres in China and Brazil, to supply their largest conurbation areas with clean electricity, in a bold drive to decarbonise their electricity supply sectors. Six of the ten largest HVDC systems in the world are in China, three in Brazil, and one in India. The largest is the Changji-Guquan HVDC system rated at 12 GW, operating at ±1,100 kV and running from the Xinjiang region at the tip of the northwest of China, to the Anhui province in the eastern coastal area, spanning 3,300 km. The DC link was commissioned in 2019.

On the other side of the scale in terms of size, but with the same commitment to decarbonization, the westernmost Hawaiian island – Kauai, home to 100,000 residents at any one time, produces 70% of its energy needs from renewable sources [7]. Its peak demand is about 80 MW. During the day, 90% of the demand may be met from generation coming only from solar PV power sources. This small, islanded grid can operate almost trouble-free owing to the backing of Li-ion battery storage systems strategically distributed on the grid and, crucially important, the use of two grid-forming (GFM) inverters connected to two of the battery systems. The Kapaia substation houses a 13 MW PV plant, a 52 MWh battery storage, and a GFM inverter. The other GFM inverter is in the Barking Sands substation, which has a 14 MW PV plant and a 70 MWh battery storage. The third battery storage, rated at 100 MWh, is in the Aepo substation, where there is a 20 MW PV plant. Note that there is no GFM inverter at Aepo. At present, two GFM inverters are the optimum number in Kauai's power grid, with the rest being GFL inverters. The operating experience attained in this renewable energy-dominated power grid is serving as a laboratory for the world.

1.2 Power electronics applications in electrical power systems

It should be brought to the attention of the unaware reader that the use of power electronics in electrical power systems did not start with its application in electrical power generation *circa* the year 2000. Early efforts were made to rectify the AC three-phase waveforms dating back to the prewar years in the 1930s, using mercury arc bulbs, i.e., *vapour* power electronics [8]. By the early 1970s, solid-state power electronics technology was mature enough to be employed in applications of long-distance and submarine HVDC electrical power transmission, using power thyristor valves, phase control, and Graetz bridges. We now term this flavour of the technology *classic HVDC transmission* [9]. This is to differentiate it from the new breed of HVDC power transmission, which uses self-commutated power electronic valves, multilevel converters, and pulse-width-modulation switching control methods of various kinds. It is remarkable that this new form of rectification of the electrical waveform, which enables DC power transmission, only appeared in the year 1997 and has already revolutionized the fabric of the electrical power industry – encompassing the generation, transmission, and distribution sectors [10].

Indeed, throughout the 1980s, when the global electricity demand was beginning to pick up following the oil crisis of 1979, the expansion programs of many power utilities were being put on hold by a variety of well-founded environmental, land-use, and regulatory pressures, which severely delayed the licensing and building of new transmission lines and electricity generating plants. At the time, an in-depth analysis of the options available for maximizing the use of existing transmission equipment and installations, while ensuring high levels of reliability and stability, pointed towards power electronics. The view of the working group who carried out the analysis was that the power electronics equipment and techniques available at the time were potential substitutes for the conventional solutions employed [11].

Perhaps the most far-reaching program of power electronics applications into electrical power systems has been the so-called FACTS initiative in the mid-1980s. It is true that many of the ideas upon which the foundation of FACTS rests evolved over decades in university laboratories and research institutes worldwide. Nonetheless, FACTS, as an integrated philosophy, was a novel concept brought to fruition at EPRI in Palo Alto, California, USA [12]. This was a mammoth effort led by Dr. N.G. Hingorani. *The acronym FACTS stands for Flexible AC Transmission Systems*. Initially, FACTS looked at ways of capitalising on the many breakthroughs that were taking place in the high-voltage and high-current power electronics industry, aiming at increasing the control of power flows and voltages in the high-voltage side of the AC power network during both steady-state and transient conditions. Soon, the new reality of making the power network electronically controllable began to alter how power plant equipment was designed and built, as well as the thinking and procedures that go into the planning and operation of transmission and distribution networks. Owing to the many economic and technical benefits that FACTS promised, it received the enthusiastic backing of electrical equipment manufacturers, electrical power utilities, and research organisations around the world [13]. The rapid success of the FACTS initiative, intended for applications in high-voltage, high-power transmission, soon motivated the use of new equipment and concepts in low-voltage, low-power distribution applications, such as power quality enhancement and reliability of supply. This initiative in electrical power distribution systems is also credited to Dr. N.G. Hingorani, who called it *Custom Power* [14]. However, the name *Low-Voltage FACTS* is more widely used today. Such a pioneering initiative has been expanded to put the new generation of power electronics valves, converter structures, and control methods to the service of the whole low-voltage distribution system, rendering it more sophisticated and resilient by enabling the incorporation of in-situ distributed generation and switching centres with battery recharging facilities. This has transformed the otherwise passive distribution system into an active one and paved the way for the development of AC microgrids and looking forward to the future: the consolidation of DC microgrids and Smart Grids [15].

The wide range of equipment emanating from the FACTS initiative may be divided into first- and second-generation FACTS equipment. In general terms, the first generation corresponds to the equipment that was conceptualised and built in the early days of the FACTS initiative and employed thyristor valves: the Static Var Compensator (SVC), the Thyristor-Controlled Series Compensator (TCSC), and the solid-state tap changer and phase shifter are the most prominent equipment in this category [11,16]. Figure 1.2 shows the SVC and the TCSC in symbol and diagrammatic form. There is an ample range of possible power electronics topologies that would add great operational flexibility to the classic tap-changing and phase-shifting transformers, which are based on slow-acting electromechanical servomechanisms. Examples of such topologies are presented in Chapter 2.

It should be noted that the Static Var Compensator (SVC) existed before the FACTS initiative had been conceived [17]. The SVC comprises a Thyristor-Controlled Reactor (TCR) in parallel with a Thyristor-Switched Capacitor (TSC), working in a coordinated fashion to enable a finely tuned voltage magnitude regulation at the high-voltage side of the interfacing transformer, say node k in Figure 1.2(a) [18]. The SVC is covered in Chapters 2 and 4. The TCSC is connected in series with a long-distance transmission line to vary on an almost instantaneous basis the series impedance of the compensated transmission line. The range of impedance variation takes place in the capacitive and inductive ranges according to operational requirements up to the rated capacitive/inductive values of the TCSC [19]. As illustrated in Figure 1.2(b), the TCSC is connected in series with the compensated transmission line and comprises a TCR in parallel with a capacitor. Various aspects of the TCSC are addressed in detail in Chapters 2, 4, and 5.

The arrival of the STATCOM, which is essentially a Voltage Source Converter (VSC) using forced commutation valves and PWM (Pulse Width Modulation) control, handling MW-level powers, represented a game changer in the development of newer and more advanced FACTS equipment, spearheading the second generation of FACTS equipment [20,21]. The Unified Power Flow Controller (UPFC) definitively belongs to this category [22]. In the argot of electrical power engineering, the VSC connected in shunt is termed STATCOM, and the VSC connected in series is termed Solid-State Series Compensator (SSSC). The development of the original STATCOM, SSSC, and UPFC are largely credited to Dr. L. Gyugyi of Westinghouse, now incorporated into Siemens Energy. The converter bridges used GTO valves switching at just over 1 kHz.

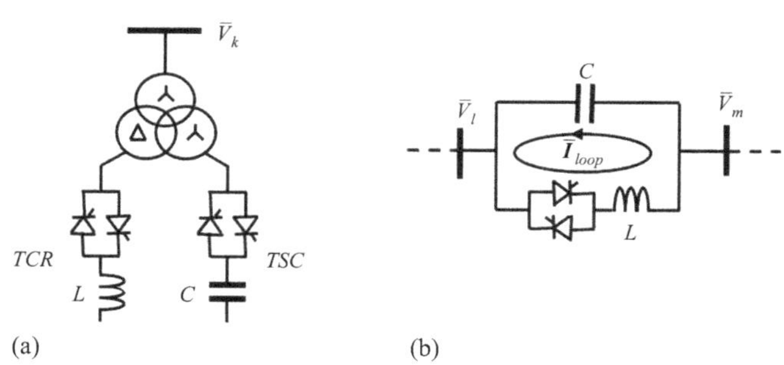

Figure 1.2 Schematic representations of (a) the SVC and (b) the TCSC

However, interest in the application of forced commutation valves in high-power applications was developing globally. For instance, a low-power STATCOM using GTO valves was developed and commissioned in Japan in 1981, and it was built at Mitsubishi's industrial premises [23]. Present-day STATCOMs and UPFCs use IGBT valves. The various VSC topologies and their switching controls are addressed in depth in Chapter 4. Their symbol/diagrammatic forms are shown in Figure 1.3(a) and (b) for the case of a VSC connected in shunt and in series, respectively. The UPFC is addressed in detail in Chapters 2 and 5.

The STATCOM is a three-phase, multilevel VSC connected to the secondary side of a conventional two-winding, three-phase power transformer whose primary winding is shunt-connected to the high-voltage bus of the electrical power grid – it is primarily used for tight voltage magnitude control at the high-voltage point of connection with the AC power grid. The SSSC is a three-phase VSC connected to the secondary side of a conventional two-winding, three-phase power transformer whose primary winding is series-connected to the high-voltage bus of the electrical power grid – it provides a means for the instantaneous regulation of the active power flowing through its series-connected winding by the tight control of the phase angle of the voltage across the series-connected winding.

Present-day VSC designs employed in high voltage, high power applications use IGBT valves as opposed to GTO valves owing to their better switching and controllability performance. These characteristics are addressed in Chapter 3. The structures used in STATCOMs have progressed from their early two-level bridges to today's multilevel bridges using modular designs, IGBT valves switching just over the fundamental frequency and with it, a drastic reduction in switching power losses and an overall improvement in the quality of the sinusoidal nature of the voltage waveform [15]. The VSC has resulted in a rather versatile device; it cannot only be connected either in shunt or in series with the electrical power system but also two or more VSCs may be connected in series or in parallel to achieve compound structures that suit specific power systems applications [5]. The main VSC-based compound equipment is shown diagrammatically in Figure 1.4.

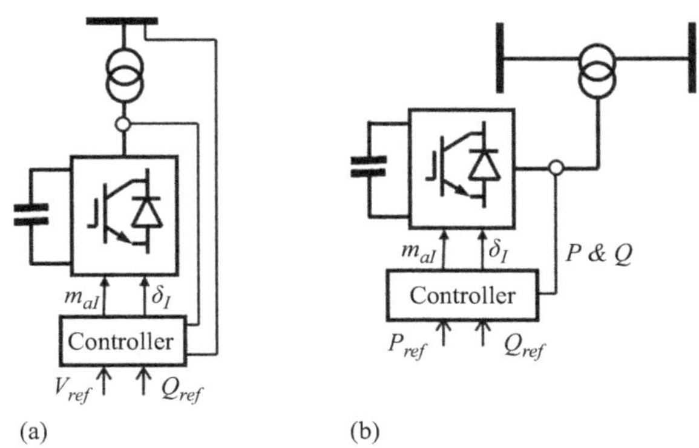

Figure 1.3 Basic VSC-based equipment: (a) STATCOM and (b) SSSC

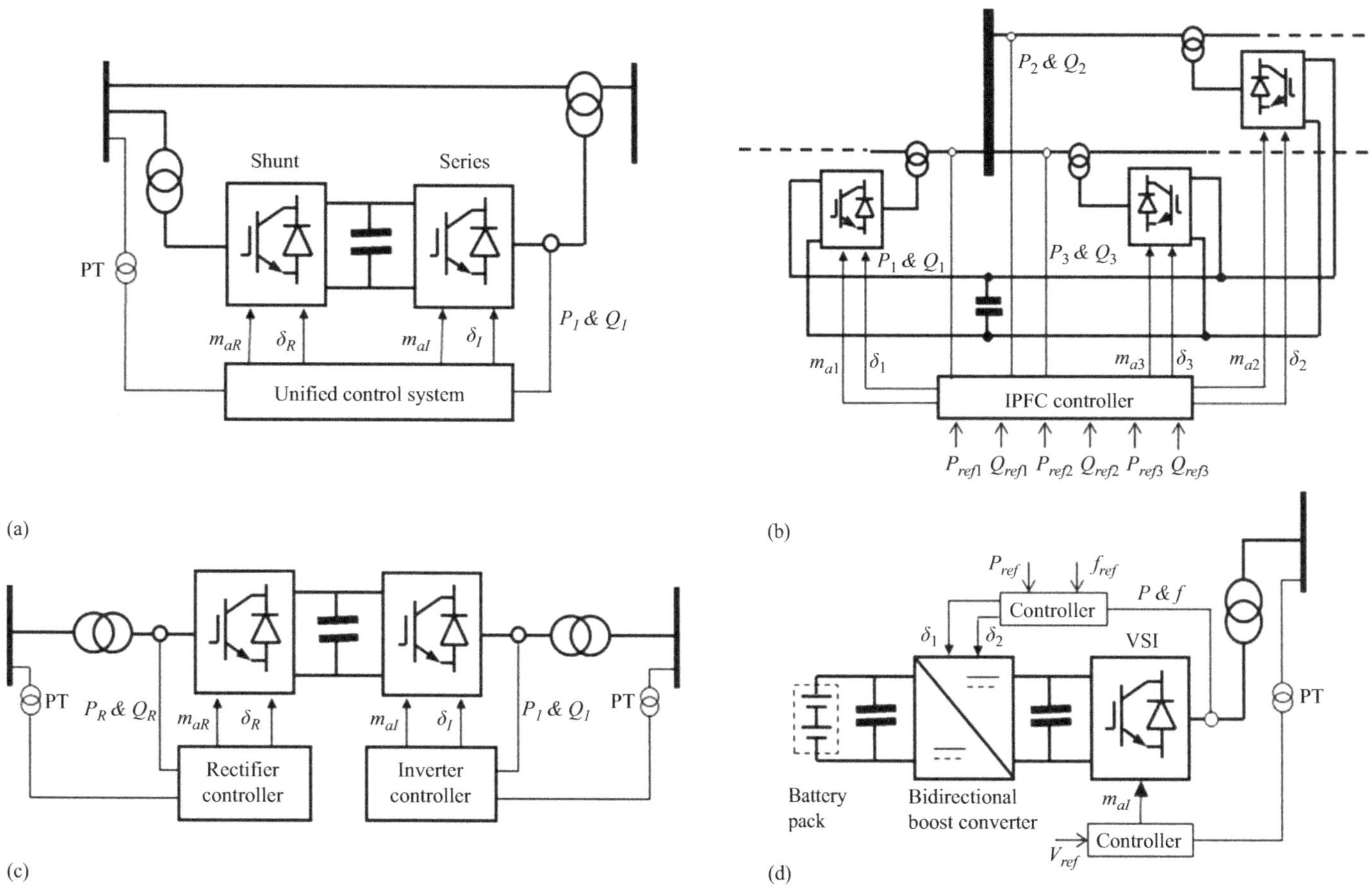

Figure 1.4 Main VSC-based compound equipment: (a) UPFC; (b) IPFC; (c) back-to-back VSC-HVDC; and (d) BESS

As illustrated in Figure 1.4(a), the UPFC combines a STATCOM and an SSSC connected in a back-to-back fashion on their DC sides and employs a unified control system to incorporate the tight voltage regulation capabilities of the STATCOM and the active power control exerted by the SSSC. As a result of this coordinated control, the UPFC can also control the amount of reactive power injected at the opposite node where the voltage magnitude is regulated. Hence, the UPFC can control one or more of these three parameters to suit specific operational demands. The ratings of the shunt and series converters limit their upper and lower regulating bounds. The UPFC plays the role, in any combination, of a STATCOM, a TCSC, an SSSC, a phase-shifting transformer, and a tap-changing transformer simultaneously and in any combination. These unique characteristics of the UPFC are explained in Chapters 2 and 5.

The IPFC comprises a coordinated array of SSSCs embedded in a bundle of transmission lines – one SSSC per transmission line – designed to control the amount of active power flowing through each one of the SSSC-fitted transmission lines. This SSSC arrangement is exemplified in Figure 1.4(b), where the use of a common DC bus is clear. The SSSCs exchange power through the DC bus to meet scheduled power in each transmission line. Note that the IPFC is, in fact, a back-to-back, multiterminal VSC-HVDC system.

As shown in Figure 1.4(c), the back-to-back VSC-HVDC comprises two STATCOMs sharing a DC capacitor. Note that such an assembly of VSCs is a natural frequency converter. It enables the transfer of active power of a given frequency into a power network exhibiting a different power frequency, say, from a 50 Hz power system into a 60 Hz power system. The frequency conversion would be 50–0–60 Hz in this case. In practice, this rather specialised requirement does exist. A case in point is the Sakuma back-to-back HVDC link, which has allowed the interchange of power between Japan's 50 and 60 Hz AC networks since 1965. It started using mercury-arc valves, which were replaced in 1993 by thyristor valves. Work has commenced to replace the link with a VSC-HVDC link to be commissioned in 2028. It has a capacity of 300 MW. Three other back-to-back CSC-HVDC links exist in Japan: Shin-Shinano, Minami-Fukumitsu, Higashi-Shimuzu. However, the main application of back-to-back HVDC links is the asynchronous interconnection of power grids of nominally equal frequencies, 50 Hz or 60 Hz, but different electrical strengths, which prevent the use of conventional AC tie-lines. This topic is covered in Chapter 8. The power converter technology employed for this application has used mercury-arc valves, which over time were replaced by Si thyristor valves and are now moving decisively into the use of back-to-back, VSC-HVDC links using Si IGBT valves. Refer to Chapter 2 for a list of this kind of installation.

By comparing Figure 1.4(a) and (c), the likeness of the UPFC and the back-to-back VSC-HVDC power electronics structures is self-evident. As seen from their respective AC sending ends, both structures have their rectifier bridges connected in shunt. However, the UPFC uses the second converter connected in series with the AC system, whereas in the latter structure, the second converter is connected in shunt. In both cases, the two converters share the DC capacitor since they are connected back-to-back. In the case of the UPFC, the shunt converter is rated at full carrying-current capacity, whereas the series converter is rated at only a fraction of its full capacity. Both converters have an equal rating in the case of the back-to-back VSC-HVDC. Both power electronics structures exhibit similar operational flexibility. Nonetheless, the great advantage of the VSC-HVDC structure is that the two VSC stations are not limited to being connected in a back-to-back fashion. Instead, the two converters may be linked by an overhead transmission line or an underground cable. This is known as point-to-point VSC-HVDC. Its schematic representation is shown in Figure 1.5 for the case of a bipolar VSC-HVDC system, which enables the transmission of large blocks of electrical power in DC form over considerable distances, including submarine electrical power transmission. This new form of transmitting electrical power in DC form, which only started in the year 1997, was initially termed HVDC Light [24]. It is generally termed VSC-HVDC to distinguish it from the classical HVDC transmission, which uses thyristor-based bridges and phase control [8,9]. The schematic representation of the bipolar HVDC link based on classical thyristor bridges is shown in Figure 1.6.

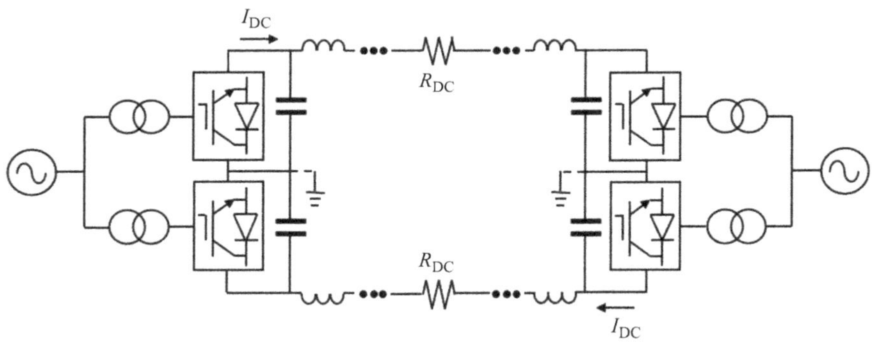

Figure 1.5 Schematic representation of a bipolar VSC-HVDC system

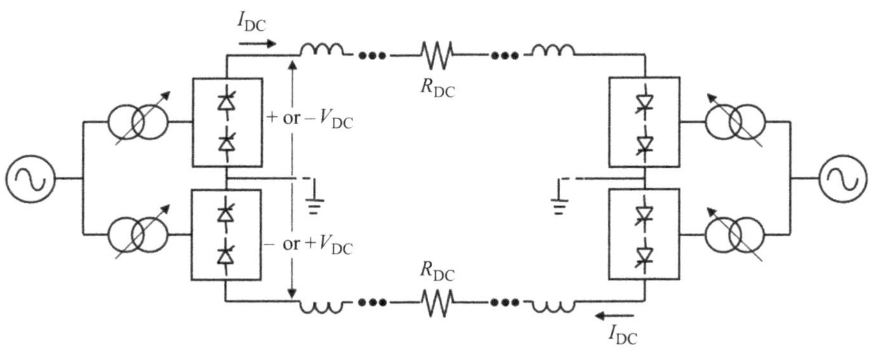

Figure 1.6 Schematic representation of a classic, bipolar HVDC system – it uses thyristor bridges and phase control

Three of the largest vendors of power electronics equipment, Hitachi-ABB, Siemens, and Alstom, have proprietary VSC-HVDC equipment termed HVDC Light, HVDC Plus, and MonSin, respectively. More recently, a new equipment manufacturer has entered this highly specialized market: NR from China. NR has recently supplied proprietary VSC converters and DC circuit breakers for the Kangbao–Zhangbei–Fengning–Changping DC system; this being the first meshed DC power grid in the world [25].

It should be noted that the original boundaries between HVDC and FACTS were drawn along the type of power systems applications and solid-state converters and their controls, but these boundaries have become blurred with the arrival of newer technology [5]. A case in point is the STATCOM, which is essentially a VSC applied in power grid applications and considered a product of the FACTS technology designed to provide reactive power support [20]. It has been reported that the first use of VSCs in high-power applications occurred circa 1980 [23] and 1996 [20], as STATCOMs. This installation falls squarely within the realm of the FACTS technology. The first VSC application in HVDC transmission took place in the year 1997 – an application comprising two STATCOMs connected back-to-back or through a DC cable [24]. This has resulted in a rather versatile and efficient means for transporting electrical power in DC form with utmost controllability. It is unassailable in submarine power transmission for evacuating large blocks of off-shore wind electrical power [26]. The present status of the VSC-HVDC technology is in point-to-point applications and moving decisively into the arena of multi-terminal VSC-HVDC systems [6,25]. Hence, it may be argued that the artificial lines that were drawn along FACTS and HVDC ought to be removed and that, instead, the focus should be on Flexible Transmission Systems (FTS), a unifying concept bridging the FACTS and HVDC technologies – the aim is to enable the best of breed solutions underpinning the new power-carrying structures that the Smart Grid demands [4].

Furthermore, the use of the VSC working as an inverter (VSI) has been instrumental in incorporating into the AC power grid equipment, which in natural form operates with DC voltages and currents, such as solar PV power generators and various forms of energy storage equipment, i.e., Battery Energy Storage System (BESS) [27], Superconducting Magnetic Energy Storage (SMES) [28], etc. For the purposes of energy efficiency management and voltage and power flow control, the use of a DC/DC converter is used in tandem with the DC/AC inverter (VSI), making up a hybrid, back-to-back, HVDC system. In schematic form, this is illustrated in Figure 1.4(d) for the case of the BESS. In this application, the DC/DC converter is a bidirectional converter that enables the charging and discharging cycles of the battery pack [27].

Solar PV power generation technology is classified as renewable power generation, along with wind and ocean wave power generation. Until now, the first two power generation technologies have reached technical and economic success, whereas the third one has yet to deliver economic viability. The solar PV power generators are static, i.e., have no moving parts, and produce electrical power in DC form. In cases of grid-connected PV installations, a VSC working as an inverter is required to enable conversion of the DC voltages and currents into AC voltages and currents at the nominal grid power frequency. Furthermore, a DC/DC power converter enables the optimum handling of the available solar energy resource [29]. As shown in Figure 1.7, the two power converters are connected in tandem, forming a hybrid, back-to-back, HVDC link. This power electronics structure is almost identical to that of the BESS, but in this application, the DC/DC converter is unidirectional instead of bidirectional. It also has the task of maximising the production of electrical power commensurate with a given amount of solar resource by means of an embedded controller termed MPPT, i.e., maximum power point tracker. Solar PV power parks have grown over the last decade. The Midog Solar Park in China, with a capacity of 3,500 MW, is the largest single PV station in the world. It was commissioned in June 2024. The Badhla Solar Park in northwest India, with a capacity of 2,245 MW, started operating in 2018, but more panels are being added to eventually reach the 3,500 MW capacity. Several more solar PV parks that have surpassed the 2,000 MW mark exist in China, India, and the Arab Emirates. To put this in context, the peak demand of South Australia is *circa* 3,000 MW, and solar power routinely supplies most of the demand during the middle of the day. Scaling up these figures to achieve similar success in larger systems represents a considerable challenge. Take the case of Texas, with a peak electricity demand of 85 GW. Last year, Texas operated 23 GW of solar power, with plans to install a further 100 GW over the next decade. There is currently no large power grid in the world with such a dominating level of converter-interfaced power sources. In this tenor, the South Australia power grid experience, relying on fast-acting, energy storage-backed, solar PV power generators and GFM inverters, is an inspiration to follow for larger power grids aiming at their decarbonisation.

Concerning wind power, the most successful form of power generation technology is the so-called variable-speed wind power generation, which makes use of power electronics converters to enable optimum power yields commensurate with the available wind energy resource at any given point in time. Two main flavours of

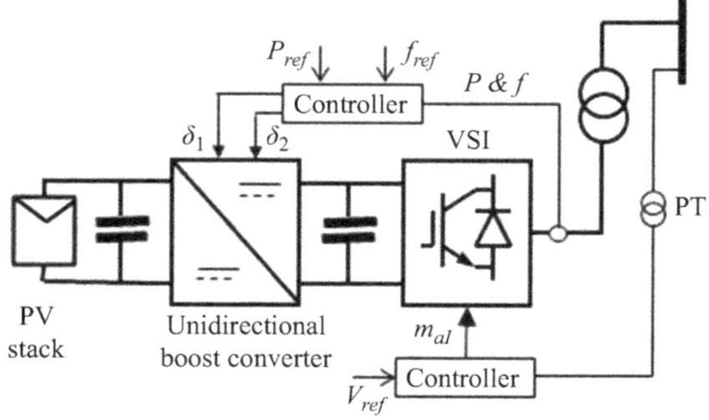

Figure 1.7 Schematic representation of a grid-connected solar PV power generator

variable wind speed generation are used in current wind farm installations: Double Feed Induction Generation (DFIG) and Permanent Magnet Synchronous Generator (PMSG) with Fully Rated Converters (FRC) [30,31]. The schematic diagrams of the DFIG and the PMSG-FRC are shown in Figure 1.8(a) and (b), respectively. The DFIG wind power generator employs a wound-rotor induction machine with slip rings and a back-to-back, VSC-HVDC link. Incidentally, comparing the structure of the UPFC in Figure 1.4(a) and the structure of the back-to-back HVDC link of the DFIG in Figure 1.8(a), they show striking similarities. In the DFIG application, the series converter connects to a three-phase winding (rotor winding) connected in delta as opposed to an AC three-phase busbar, as would be the case in the UPFC application. The series converter injects a given amount of active and reactive power into the rotor winding to make the rotor run at either sub-synchronous or super-synchronous speed according to external operational requirements. The PMSG-FRC operates in a more classical fashion: the back-to-back, VSC-HVDC link acts primarily as a variable frequency converter, changing the frequency of the incoming winds upon the rotor blades and into the synchronous generator's shaft into a "0 Hz frequency" and then into the grid power frequency, i.e., 50 Hz or 60 Hz. Notice that no gearbox sits between the blades and the generator's shaft. Further to this essential frequency modulation process enabled by the back-to-back, VSC-HVDC, there is the all-important voltage and active and reactive power control associated with the smooth operation of the PMSG-FRC wind power installation. Wind power generation is addressed in Chapter 9.

The use of FACTS equipment, controls, and concepts permeated the low-voltage distributed network at an early stage in the development of the technology. The applications were in the areas of power quality enhancement and supply reliability [32]. The three main pieces of equipment are the low-voltage, low-power capacity STATCOM or D-STATCOM, the Dynamic Voltage Restorer (DVR), and the Solid-State Transfer Switch (SSTS). The D-STATCOM provides voltage magnitude and power factor regulation at the high-voltage side of the connecting transformer. The DVR is an SSSC whose main task is to protect sensitive loads from incoming, damaging voltage waveforms containing amplitudes that depart significantly from their nominal values: either voltage swells or voltage dips. In the parlance of power systems engineers, the purpose of the DVR is: "to complete the missing cycles." It should be noted that most of the time, the DVR operates in

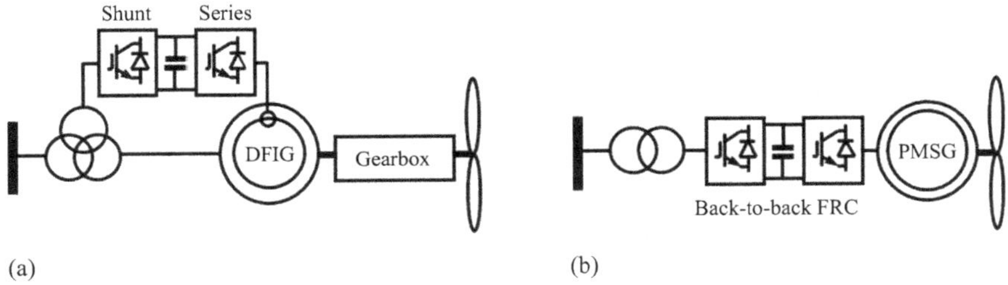

Figure 1.8 Schematic representation of (a) DFIG and (b) PMSG-FRC

passive mode and only becomes active when a menacing voltage waveform approaches the protected load. The SSTS provides redundancy of supply to a sensitive load by incorporating a backup feeder with the capacity to take over the main feeder, with little delay, in cases of faulted operation. Both feeders are linked by a thyristor bridge, which acts as an "instantaneous" switch, alternating between faulty and healthy feeders. Custom Power is addressed in Chapter 6.

Small-scale electrical power generators of various kinds and electrical energy storage systems started becoming economically attractive soon after the consolidation of Custom Power [14]. The amalgamation of these technologies has enabled the power distribution network to supply, with high reliability, one part of its system load using *in situ* renewable power generation, i.e., wind, PV solar, and hydrogen. Hence, the power distribution networks became active instead of passive and over the long-time horizon – an effective net contributor to the decarbonisation of the electricity supply industry. The power electronics valves, converters, and systems used in Custom Power became more sophisticated and expanded their range of applicability in electrical power distribution networks. The so-called switching centres shown schematically in Figure 1.9 embodies an example of what has become possible with the availability of present-day power electronics technology [33]. In essence, these

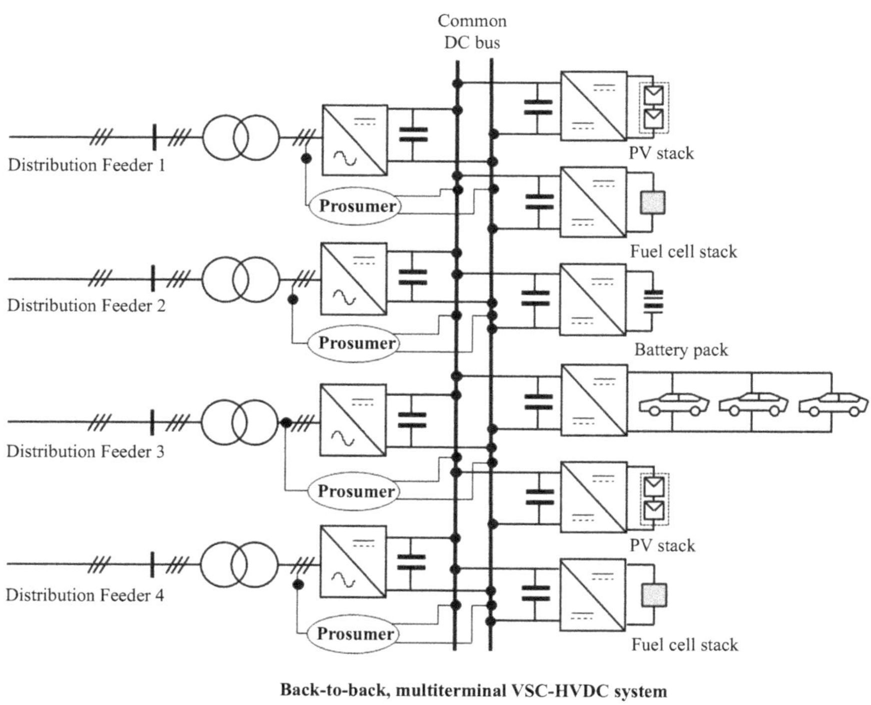

Figure 1.9 Low-voltage electrical power distribution feeders are linked at their far end by a switching centre

switching centres are back-to-back, multiterminal VSC-HVDC systems. This kind of new technology and application has arrived at just the right time to furnish the low-voltage distribution network with the full resilience and operational flexibility required to cope with the high demands imposed by the new generation of electric loads with their high content of electronic circuits, such as the large concentration of electronic lighting and domestic appliances, the advanced motorized processes in industry and the large number of recharging centres that will be required to charge the soon-to-be ubiquitous electric cars. These issues are addressed in Chapter 6 in connection with Active Power Distribution Systems.

An obstacle in developing fully-fledged multiterminal VSC-HVDC systems of arbitrary configuration has been the lack of DC circuit breakers (CB_{DC}). After much R&D efforts in industry and academia, working prototypes are beginning to emerge. A case in point is the CB_{DC} developed by NR in China for use in the first ring-type, four-terminal VSC-HVDC system in the world [25]. Different aspects of multiterminal VSC-HVDC systems are addressed in Chapters 2, 5, and 8.

1.3 Evolution of the electrical power grid

As time and developments go on, an issue of growing concern in nationwide interconnected electrical power systems is the accelerated loss of grid inertia since a large proportion of the conventional, large synchronous generators are coming to the end of their useful life and their power outputs are being replaced with that of MW-level PV solar installations and wind farms. The reason for such a progressive loss of grid inertia is that present-day renewable power generators lack the large rotating masses of the conventional, large synchronous generators directly connected to the AC power grid, i.e., not connected to the AC power through power electronics interfaces. A readily available solution to overcome such a progressive problem is to employ the long-honoured synchronous condenser, which not only contributes inertia to the AC power grid but also provides dynamic reactive power compensation and instantaneous voltage regulation at its point of connection with the power grid [17]. However, this technical solution comes at the expense of additional costs, which must be met, most likely, by the power transmission grid owners as part of their power grid reinforcement and expansion programs.

Meanwhile, the availability of renewable electrical power sources and power electronics technologies have permeated the low-voltage electrical power distribution networks. To a greater or lesser extent, the new generation of electrical power sources is scalable and reliable to be embedded at practically any voltage level of the electrical power network, a feature which is changing for good the face of the electricity distribution industry [34]. These low-voltage distribution networks, which traditionally have been passive power networks, are turning into active power networks – that is, having the capacity to partially generate electrical power to meet its load. Moreover, off-the-shelf power electronics-based controllers are available for the purpose of active and reactive power control, network reconfiguration, the enhancement of power quality, and increased reliability of operation of the low-voltage electrical power network

[33]. In the fullness of time, this may open new business opportunities if this ancillary equipment is owned by an independent company that is contracted to provide such a service, receiving an agreed economic remuneration. There have already been incipient bouts of such services in specific industries, such as paper mills [14,35], but a wide opening of the ancillary services market is not yet within sight.

Throughout this process, the nature of the power system distribution's load has also evolved. The long-lasting industrial, commercial, and domestic loads have modernised to become more versatile and energy efficient by incorporating power electronics circuits and controls. A wide range of specialised electric motors and drive systems have appeared in the market, displacing in many applications the classic three-phase induction motor and the DC motor developed in the Victorian era. The incandescent and fluorescent lighting used at home and offices for over one century has given way to a new family of electronic bulbs, which are reputedly more energy efficient. All modern domestic appliances, such as flat-screen TVs, infrared heaters, kitchen appliances, computers, and printers, employ power electronic circuits. Hence, solid-state electronics are pervasive in today's electrical power distribution systems' load composition, making the power grid more difficult to plan and operate. To exacerbate the technical complexity of the infrastructure of the power grid, electric cars have become technically and economically competitive with petrol cars, and the power grids would need to expand and adapt to provide the great many power grid infeed points that the EVs recharging stations would require at any given country. Once more, planning engineers are looking to power electronics solutions to add further flexibility and resilience to the electrical power distribution grid by investigating the role of the so-called switching centres in the active power grid [33].

It should be remarked that despite the growing technical and structural complexity of the electrical power distribution grid, this is mainly fed from a power grid operating at sub-transmission voltage levels – that is, the electrical power distribution grid may be seen as an extension of the high-voltage electrical power transmission grid, which serves the purpose of feeding into end-users at safe voltage-usage levels. The structure and role of switching centres are at an incipient stage, and yet it is a very promising solution for the flexible incorporation of some kinds of embedded generation, the new kinds of loads, such as EV charging hubs and battery stations. Alternative, fuller solutions would involve the coexistence of aerial or underground AC feeders and DC feeders running in parallel. In such a hybrid arrangement, three-phase and single-phase AC loads would be supplied as normal, and DC loads and Distributed Energy Resources (DERs) would be either directly connected to the DC feeder or through small, dedicated DC/DC converters. Small-scale solar PV power generators, fuel cells, BESSs, and EV battery systems are taken to be DERs. A growing number of industrial and domestic electricity consumers own one or more DERs. As the number and size of DERs grow, the generated power may surpass their own needs and may wish to trade the surplus energy with the electricity supplier company. The concept of a consumer being able to buy and sell electricity has given rise to the term *Prosumer*, which essentially means a customer having the dual role of producer and consumer, following online economic signals. These ideas are elaborated further in Chapter 6.

On the back of a host of disruptive technologies, new concepts and operating philosophies have been developed for the flexible and reliable electricity supply to end-users, which radically depart from the construction and operating practices of classic distribution systems. The microgrid is one such concept, which potentiates the development and technologies of active distribution networks to enable the existence of self-contained and self-reliant power grids, which may operate independently of the high-voltage power grid or be connected to it but work in a predefined, cooperative manner – as opposed to depending on it to be able to operate. As the number and size of renewable energy sources and storage systems grow and become dominant in AC microgrids, one or more of the inverters used to connect such energy resources to the microgrid would need to be provided with grid-forming (GFM) control capabilities, while most of them will be conventional grid-following (GFL) inverters. The microgrid has opened business opportunities for entrepreneurs, private investors, and cooperatives to invest in the electrical energy sector and sell energy to the neighbouring electrical power utility following an open market scheme for electricity [36].

One further step of conceptual development to the microgrid is the so-called Smart Grid [37], where the massive incorporation of voltage, current and temperature sensors, advanced controllers, internet communications, and AI decision-making processes enable the far-sighted concepts of man-less operation and self-healing power grids. The cornerstone enabling technology for all the present and future developments in the electrical power grid is solid-state power electronics. A wide range of power electronics-based equipment is readily available, some of them adopted from former ground-breaking solutions in high-power transmission applications, but many others are indigenous solutions to the technical issues facing low-power distribution systems. Both microgrids and Smart Grids are topics of great research importance in electrical power systems.

On a much wider front than electrical Smart Grids, a global effort from city planners from all corners of the world, supported by key financial global organisations, such as the World Bank and national government, has been honing the concept and practice of the so-called Smart Cities. On this score, demonstrable progress verse on developing smart technological-based approaches to governance and delivery of health, education, and community services from federal and municipal authorities (e-government). Existing urban infrastructure such as roads, water, gas, and electricity networks are being redesigned and upgraded with the use of new materials, equipment, and the latest communications technologies, such as the Internet of Things, blending seamlessly with the local environment and ecosystems to improve on the overall urban quality of life. The Smart Grid for electricity is the key contribution the global electrical power community is making to the Smart Cities drive for the betterment of mankind.

1.4 About the textbook

This book reviews the power electronics theory, operating practices, and equipment that underpin the FACTS technology and its convergence with the long-standing application of HVDC. Note that VSC is the key equipment linking the FACTS and the modern VSC-HVDC technologies. This book offers a timely review of the subject since the presence of power electronics circuits has become ubiquitous in the electrical power network; it encompasses the three key electrical power sectors: generation, transmission, and distribution. The new forms of electrical power production use renewable energy resources (e.g., wind, solar, hydrogen) as precursors of electrical energy production and are supported by large-scale energy storage. To a greater or lesser extent, this process uses power electronics circuits to enable the converted electrical energy to be available in the AC electrical power grid with high availability and controllability. Incorporating power electronics-based equipment in high-voltage AC power transmission has increased power throughputs substantially – in principle at least, up to its thermal limits. Moreover, it has also bestowed great flexibility and controllability upon the power grid.

The new generation of power electronics converters in DC power transmission has enabled the evacuation of large blocks of offshore wind power with utmost supply reliability while keeping power losses at reasonable values. In electrical power distribution systems, power electronics has not only improved power quality at key points of the network but is also proving instrumental for transforming power distribution systems into active power networks, making it possible to incorporate hydrogen-based power generators (fuel cells), solar PV power generators, and battery-based energy storage facilities. In particular, the millions of EVs that will circulate the city streets and motorways of countries in all corners of the world will be serviced by a network of battery recharging points, which are, in essence, arrays of high-efficiency power electronics converters with extended DC buses. The subject is addressed from a modern perspective, including the latest developments in FACTS and HVDC technologies, which enhance the operational control and reliability of the electrical power network supply. These areas include FACTS, HVDC, and storage equipment, as well as examples of installations and analysis of the steady-state and dynamic interactions of the power electronics-based equipment and the AC electrical power grid.

The book has been written to serve the needs of an ample sector of the power engineering community, ranging from utility and equipment manufacturing engineers, researchers, university professors, and PhD and MSc students. We expect the reader to have a sound knowledge of electrical and electronic circuits, algebra and numerical methods, and a working knowledge of electrical power and control engineering – undergraduate students in their final year in an electrical and electronics degree course should benefit from studying the book in its entirety or specific chapters to suit their interests.

Chapter 2 introduces the most popular power electronics-based equipment used in electrical power transmission, distribution, and generation sectors. The chapter presents examples of emblematic FACTS installations and a comprehensive survey of HVDC installations around the world, including classic HVDC, VSC-HVDC, multiterminal HVDC systems, and asynchronous interconnections. The first commercial application of power electronics in electrical power systems was in submarine HVDC transmission, commencing in 1954 in Sweden. Solid-state power electronics had not yet been developed at the time. Instead, mercury-arc bridges were used to pursue rectification and inversion. There are no installations of this kind anymore; either these have been decommissioned or replaced with rectifiers/inverters stations using thyristor valves. A dozen of these installations were built in various locations of the world. As a testimony of this pioneering technology and for historical reasons, these early HVDC installations are listed here. The aim of this chapter is to set the scene for the remainder of the book by describing the main equipment that conforms to the FACTS and HVDC technologies. Their basic principles of operation, power electronics structures, and their role and place in the electrical power network are outlined. The examples of FACTS and HVDC installations aim to give, from the outset, a practical grounding for the ensuing theoretical concepts and computer algorithms that are contained in the book.

Chapter 3 reviews the main structural designs and operational characteristics of the Si-based semiconductor valves employed in MW-level power systems applications, such as the thyristor, the Gate-Turned-On (GTO) thyristor, and the Insulated Gate Bipolar Transistor (IGBT). Until now, Si thyristor is the most popular semiconductor device used in all three sectors of the electricity supply industry: generation, transmission, and distribution. However, forced commutation devices such as Si GTOs and Si IGBTs, which were used relatively sparsely in the 1980s and 1990s, are rapidly increasing in their share of use. Si has played a pivotal role in the expansion and modernisation of the electricity supply industry for more than half a century. Nonetheless, as the global electricity demand continues to increase inexorably in large conurbation areas and the options to generate the required demand locally narrow quite dramatically on environmental grounds, then electricity needs to be transported from remote sites, laying thousands of kilometres away from the large conurbation centres where electricity is required. State-of-the-art, Si-based, HVDC power transmission technologies are being employed to achieve such undertakings. However, the Si power semiconductors employed may be approaching their physical limits. Delving into the future, the chapter draws attention to what may become, in the fullness of time, the natural replacement of Si devices and valves in the handling of larger volumes of electrical power, namely, a range of wide bandgap semiconductor materials, which today are in their experimental and prototype stage.

Chapter 4 carries out a detailed description of the power electronics converters used in MW-level power applications, including multilevel VSC converters. There are two main flavours of HVDC technology: Line-Commutated Converters (LCCs) and Voltage Source Converters (CSC), which are both addressed in this chapter. The former uses thyristor valves and phase-angle firing control, whereas the latter uses IGBT valves and PWM control. The need to rectify the three-phase AC voltage and current waveforms required to feed the DC motors of tramways of the pre-war years provided a strong motivation for developing reliable electrode-graded mercury arc rectifiers, which were also used in long-distance and submarine power transmissions. In the mid-1970s, the availability of semi-controlled, solid-state power electronics valves marked a gradual shift towards using solid-state technology instead of mercury arc bridges. In the early designs, the Graetz bridge and phase-angle firing control developed for mercury-arc bridges continued to be used in thyristor bridges. The intervening years saw the development of higher-rating thyristor valves, better phase-angle firing control systems, and quadripole converter stations. This flavour of the HVDC technology – termed classic HVDC – remains unassailable in long-distance, bulk power transmission. At the turn of the millennium, a new form of HVDC power transmission using VSC bridges aimed at MW-level applications became available. It had rather humble beginnings: low power capacities, low voltages, high power losses, and high harmonic contents, but around 2010, the development of the so-called MMC (Modular Multilevel Converter) bridges propelled the VSC-HVDC technology to a new era of HVDC power systems installations. It is unassailable when tapping large blocks of offshore wind power with the utmost reliability and controllability.

Chapter 5 addresses the main equipment of the FACTS technology, and HVDC presented in Chapters 2 and 4 from the all-important perspective of mathematical modelling aimed at power network-wide analysis with reference to steady-state power flow solutions. The following FACTS equipment has been selected for treatment owing to their more complex modelling: the TCSC and the UPFC. Concerning HVDC, the two options are given full treatment, namely, classic HVDC and VSC-HVDC. The latter is closely linked to the research work carried out by the authors recently and published in top journals. The approach taken to develop the various steady-state models is to extract the dominant constructional and operational electrical characteristics of the relevant FACTS and HVDC equipment detailed in Chapters 2 and 4 and to develop non-linear algebraic equations based on fundamental electric circuit laws (e.g., Ohm and Kirchhoff), which are solved by iteration using the multi-dimensional-space Newton-Raphson method for reliable convergence towards the steady-state solution. The models are flexible and comprehensive. They incorporate all the relevant control mechanisms that this kind of equipment may have to ensure that the power systems solutions are realistic and serve the purpose of carrying out network expansion studies of utility-level power grids. Several examples of power flow solutions are included in the chapter to illustrate what kind of solutions are possible to obtain with the models presented in the chapter. The issue of multiterminal VSC-HVDC systems receives special treatment owing to its current global interest from its algorithmic perspective.

Chapter 6 addresses the issue of electrical power distribution systems, where power electronics equipment, low-capacity renewable power generators, and energy storage have been permeating for the last two decades, transforming the classic power distribution systems from passive power grids into active power grids. The main power electronics equipment is the D-STATCOM, the DVR, and the SSTS, whereas the main electrical power generators are solar PV panels, small-scale wind power generators, and fuel cells. To date, the most popular energy storage system is the Li-ion battery, particularly as the widespread adoption of EVs accelerates on a global scale. All these resources are deployed in the power distribution system in a distributed fashion, and collectively, they have been termed

Distributed Energy Resources (DERs). One or many DERs may be owned and managed by the utility company or by a customer, being a large or small industrial concern, a commercial enterprise, or a domestic customer. This has given rise to the concept of *prosumer*, a new energy actor in the electricity sector that has the dual role of producer and consumer, responding to economic signals. This is opening new business opportunities based on strategic investments and leading-edge technology. To a greater or lesser extent, all these power energy sources and storage systems are connected to the power grid using power electronics converters. The electrical utility company may encourage the prosumer business to flourish by expanding their power grids with DC power infrastructure.

Chapter 7 addresses key aspects of the variable renewable energy's impact on the power grid, emphasising the regulatory aspects that underpin the increased integration of renewable energy using power electronics inverters. It lays out the technical challenges involved in operating the power grid when abrupt meteorological changes occur, subjecting renewable power generators to great stress and compromising the integrity of the power grid. The larger the blocks of variable renewable power are, the more brittle the power grid becomes. This fact was identified at a very early stage in the integration of renewable power into the power grid, and legislation has been developed by regulatory bodies to curtail as much as practicable the adverse impact of renewable power plant outages, leading to large area blackouts. The ensuing legislation is contained in the so-called Grid Codes. Owing to the paramount importance of the topic, international and national Grid Codes have been drafted globally over the past 20 years – this key issue is outlined in the chapter. Energy storage has been identified to be the definite technology to firm the power output of renewable power plants, but further developments are required to make this technology widely accessible at competitive prices. Grid codes also cover the grid connection of energy storage systems. Another negative aspect of integrating large blocks of renewable power into the power grid is the loss of power system inertia, resulting from the withdrawal of the large rotating masses of the conventional synchronous generators in favour of renewable power sources, which have little or no inertia. This is turning into a problem for transmission system operators, and all efforts are being made to resolve these conflicting issues: the topic of virtual inertia is also addressed in the chapter.

Chapter 8 examines the adverse consequences of power imbalances on the operation of the electrical power grid. Power imbalances occur frequently, but most of the time, they are promptly restored either naturally or by the action of the transmission system operators. However, under some unfortunate circumstances, they worsen, leading to catastrophic power system failures. Power system oscillations are most noticeable in the tie lines linking two different power system areas. To put this into context, the chapter presents the basic principles of an application called automatic load frequency control (ALFC), which was developed within the context of inter-area power systems. The daily operation of power systems requires rather taxing power systems simulation studies, which are suitably highlighted in the chapter. Moreover, the limitations that classic power imbalances assessment algorithms exhibit when incorporating DC links to replace AC tie lines are explained. This is followed by developing efficient dynamic AC/DC power algorithms with which to carry out comprehensive power imbalance assessments. These dynamic algorithms use leading-edge models of VSC-HVDC equipment that were recently published by the author in top journals. ALFC formulations of point-to-point VS-HVDC links are explained first, and then the discussion moves onto the multiterminal VDC-HVDC systems. Several numerical simulations are presented to underpin the theoretical discussion.

Chapter 9 discusses the importance of power generation using the wind as a primary energy resource, which is increasing worldwide. Indeed, the UK, which experiences one of the best global wind regimes in land and surrounding seas, has plans to count on more than 50 MW of offshore wind power by 2030, making wind power the backbone of its power grid and *en route* for its decarbonisation. To a similar extent, other European countries with access to North Sea waters hold similar plans. The evacuation of wind power from MW-level wind farms requires the use of state-of-the-art power transmission technology, more often than not, resorting to point-to-point VSC-HVDC links and FACTS equipment to reinforce wind power collecting substations, which tend to be weak points in the AC power grid, and yet need to comply with very stringent Grid Code legislation. The all-important role of Grid Codes in connection with wind power generation is discussed in this chapter. The two main areas of the wind power industry, the onshore and offshore wind power sectors, are covered in this chapter, together with the main variable speed wind generator technologies. In this category, the main characteristics of the so-called DFIG and PMSG-FRC are discussed, placing emphasis on the power electronics structures that support their variable speed attributes. Looking to the near and medium-term futures, the area of multiterminal VSC-HVDC systems is discussed as the ultimate power transmission system better poised to evacuate the extra-large amounts of wind power that will be generated in offshore wind power installations. In this context, GFM and GFL inverters are addressed in their respective roles of enabling power grids dominated by converter-interfaced power generators.

Chapter 10 – The widespread incorporation of power electronics equipment in the electrical power network at the generation, transmission, and distribution systems levels has a marked impact on the theory and practice of power systems protection and relaying equipment. The incrustation of distributed power generation sources at the power network's low and very low voltage points has called for a revision of traditional power system protection practices because of the very high risks of reverse power flows arising in the power distribution system, but so is the impending, future coexistence of AC and DC power systems and equipment at the generation, transmission, and distribution levels. It is explained and justified why the long-enduring protection equipment and methods that have enabled AC electrical power networks to expand and operate so well for more than a century do not apply to protect DC power networks. The main culprits are the lack of zero crossing, phase, sequence components, and nominal frequency in the DC current. Hence, fault current interruption is a serious inconvenience, which has prompted the need to use new techniques in DC protection. The chapter outlines the current research efforts in this all-important area of electrical power systems. Travelling wave principles, together with pattern recognition techniques, are emerging as strong candidates for developing reliable algorithms to detect and locate DC faults. The chapter includes a very useful summary – in the form of an appendix – of relay methods and principles available for AC power circuits.

Chapter 11 highlights the salient points of the theory, concepts, and case studies addressed in the individual chapters and draws the key conclusions within the context of the entire book. The closure offers a view of the expanding role that FACTS equipment and HVDC systems will play in enhancing the secure operation of the electrical power network.

References

[1] Fairly P. The Transformer. *IEEE Spectrum*, vol. 60, no. 12, 2023, pp. 31–33.

[2] Charette RN. Can Flow Batteries Finally Beat Lithium? *IEEE Spectrum*, vol. 61, no 2, 2024, pp. 22–27.

[3] Schaffarczyk A (ed.). *Understanding Wind Power Technology: Theory, Deployment and Optimization*. New York: Wiley, 2014.

[4] Kang C, Kirschen D, and Green TC. The Evolution of Smart Grids. *Proceedings of the IEEE*, vol. 111, no. 7, 2023, pp. 691–693.

[5] Acha E, Roncero-Sanchez P, de la Villa-Jaén A, Castro LM, and Kazemtabrizi B. *VSC-FACTS-HVDC: Analysis, Modelling and Simulation in Power Grids*. New York: Wiley, 2019.

[6] Fairley P. High-Voltage DC Power Roars Ashore in Europe. *IEEE Spectrum*, vol. 61, no. 1, 2024, pp. 28–31.

[7] Kroposki B, and Hoke A. A Path to 100 Percent Renewable Energy. *IEEE Spectrum*, 2024, pp. 50–57.

[8] Arrillaga J. *High Voltage Direct Current Transmission*. Stevenage: IET, 1998.

[9] Hingorani NG. High-Voltage DC Transmission: A Power Electronics Workhorse. *IEEE Spectrum*, vol. 33, no. 4, 1996, pp. 63–72.

[10] Acha E, Agelidis VG, Anaya-Lara O, and Miller TJE. *Power Electronic Control in Electrical Systems*. Amsterdam: Elsevier, 2002.

[11] Hingorani NG, and Gyugyi L. *Understanding FACTS: Concepts and Technology of Flexible AC Transmission Systems*. Piscataway, NJ: IEEE Press, 1999.

[12] Hingorani NG. Flexible AC Transmission. *IEEE Spectrum*, vol. 30, no. 4, 1993, pp. 40–45.

[13] Acha E, Fuerte-Esquivel CR, Ambriz-Pérez H, and Ángeles-Camacho C. *FACTS: Modelling and Simulation in Power Networks*. New York: Wiley, 2004.

[14] Hingorani NG. Introducing Custom Power. *IEEE Spectrum*, vol. 32, no. 6, 1995, pp. 41–48.

[15] Maza-Ortega JM, Acha E, García S, and Gómez-Expósito A. Overview of Power Electronics Technology and Applications in Power Generation,

Transmission and Distribution. *Journal of Modern Power Systems and Clean Energy*, vol. 5, no. 4, 2017, pp. 499–514.

[16] Christl N, Hedin R, Sadek K, *et al.* Advanced Series Compensation (ASC) with Thyristor Controlled Impedance. *International Conference on Large High Voltage Electric Systems (CIGRÉ)*, paper 14/37/38-05, Paris, Sep. 1992.

[17] Miller TJE (ed.). *Reactive Power Control in Electric Systems*. New York: Wiley Interscience, 1982.

[18] Gyugyi L. Power Electronics in Electric Utilities: Static Var Compensators. *Proceedings of the IEEE*, vol. 76, no. 4, 1988, pp. 483–494.

[19] Helbing SG, and Karady GG. Investigations of an Advanced Form of Series Compensation. *IEEE Transactions on Power Delivery*, vol. 9, no. 2, 1994, pp. 939–947.

[20] Gyugyi L. Dynamic Compensation of AC Transmission Lines by Solid State Synchronous Voltage Sources. *IEEE Transactions on Power Delivery*, vol. 9, no. 2, 1994, pp. 904–911.

[21] Schauder CD, Stacey E, Lund M, *et al.* AEP UPFC Project: Installation, Commissioning and Operation of the 160 MVA STATCOM (Phase 1). *IEEE Transactions on Power Delivery*, vol. 3, no. 4, 1998, pp. 1530–1535.

[22] Renz BA, Keri A, Mehraban AS, *et al.* AEP Unified Power Flow Controller Performance. *IEEE Transactions on Power Delivery*, vol. 14 no. 4, 1999, pp. 1374–1381.

[23] Sumi Y, Harumoto Y, Hasegawa T, Yano M, Ikeda K, and Matsuura T. New Static Var Control Using Force-Commutated Inverters. *IEEE Transactions on Power Apparatus and Systems*, vol. 100, no. 9, pp. 4216–4224, 1981.

[24] Asplund G. Application of HVDC Light to Power System Enhancement. 2000 IEEE PES Winter Meeting Conf. Proc., 23–27 Jan. 2000.

[25] NR Electric Co., Ltd. NR's VSC-HVDC Solution ±500 kV Zhangbei DC Grid for Boosting Large Scale Hybrid Renewables Penetration. Available: https://www.nrec.com/en/web/upload/2019/05/14/15578229444108hgcuh.pdf.

[26] Castro LM, and Acha E. On the Dynamic Modelling of Marine VSC-HVDC Power Grids Including Offshore Wind Farms. *IEEE Transactions on Sustainable Energy*, vol. 11, no. 4, 2020, pp. 2889–2900.

[27] Servotte J, Acha E, and Castro LM. Smart Frequency Control in Power Transmission Systems Using a BESS. 2015 IEEE Innovative Smart Grid Technologies – Asia (ISGT ASIA), Bangkok, Thailand, 2015, pp. 1–7.

[28] Moriconi F, Koshnick N, De la Rosa F, and Singh A. Modeling and Test Validation of a 15 kV 24 MVA Superconducting Fault Current Limiter. *IEEE PES T&D 2010*, New Orleans, LA, USA, 19–22 Apr. 2010.

[29] Messo T, Puukko J, and Suntio T. Effect of MPP-Tracking DC/DC Converter on VSI-Based Photovoltaic Inverter Dynamics. *IET Conference on Renewable Power Generation (RPG 2011)*, Edinburgh, Scotland, UK, 6–8 Sep. 2011.

[30] Hughes FM, Anaya-Lara O, Jenkins N, and Strbac G. Control of DFIG-Based Wind Generation for Power Network Support. *IEEE Transactions on Power Systems*, vol. 20, no. 4, 2005, pp. 1958–1966.

[31] Qu R, Liu Y, and Wang J. Review of Superconducting Generator Topologies for Direct-Drive Wind Turbines. *IEEE Transactions on Applied Superconductivity*, vol. 23, no. 4, 2013, p. 5201108.

[32] Anaya-Lara O, and Acha E. Modeling and Analysis of Custom Power Systems by PSCAD/EMTDC. *IEEE Transactions of Power Delivery*, vol. 17, no. 1, 2002, pp. 266–272.

[33] Mata-Diaz FJ, Mushtaq U, Gross AM, *et al.* DC-Bus Energy Management of a Converter-Interfaced Renewable Energy Source Comprising an Energy Storage System. *2021 IEEE Madrid PowerTech*, 28 Jun–2 Jul 2021, pp. 1–6.

[34] Singh M, Khadkikar V, Chandra A, and Varma RK. Grid Interconnection of Renewable Energy Sources at the Distribution Level with Power Quality Improvement Features. *IEEE Transactions of Power Delivery*, vol. 26, no. 1, 2011.

[35] Burden, A. Caledonian Paper DVR – The Utility Perspective. *IEE Half Day Colloquium on Dynamic Voltage Restorers – Replacing Those Missing Cycles*, Glasgow, UK, 1998, Sec. 2, pp. 1–2.

[36] Ahmad S, Shafiullah M, Ahmed CB, and Alowaifeer M. A Review of Microgrid Energy Management and Control Strategies. *IEEE Access*, vol. 11, 2023.

[37] Aho J, Amin M, Annaswamy A, *et al. IEEE Vision for Smart Grid Controls: 2030 and Beyond Roadmap*. IEEE Standard, 2013.

Chapter 2
Flexible transmission systems equipment and installations: FACTS and HVDC

Enrique Acha[1], Claudio R. Fuerte-Esquivel[2] and Francisco De La Rosa[3]

Solid-state power electronics has permeated all three sectors of the electrical power supply industry, namely, generation, transmission, and distribution. The ascendency of power electronics circuits and equipment over the long-enduring electromechanical equipment and servomechanisms in the electrical power grid continues to grow relentlessly. The main drivers behind this trend are the following: (i) the high level of development reached by solid-state power electronics at the MW-level; (ii) the increasing operational complexity of the electrical power grid and the need for more effective control to reign on the otherwise more brittle and unstable power grid; (iii) the opportunity to integrate remote electricity markets to increase the renewable power mix and to take advantage of differential tariffs; (iv) the direct electricity supply of large conurbation and industrial areas with vast energy resources that lie thousands of kilometres away; (v) the linking of AC power grids with different operating frequencies, and AC power grids of different strengths; (vi) to reign on the vast amounts of money and resources wasted due to poor power quality in today's large and complex industrial process. This chapter emphasises power transmission, including power electronics applications and installations in AC and DC power transmission.

2.1 Introduction

The global power industry continues to experience unprecedented growth and technological innovation. Some of the components driving these developments are the unquenching global demand for electricity, decarbonising action of the electricity sector, increasing liberalisation of energy markets, and "green jobs" opportunities. Integration to the grid of large-scale photovoltaic and wind power installations brings about new challenges in the operation of the otherwise

[1]Electrical Energy Consultant, Power Systems Laboratory, Bristol, United Kingdom
[2]Faculty of Electrical Engineering, University of Michoacan, México
[3]Senior Electric Power Systems Consultant, San Antonio, TX, USA

conventional power grid. For one, the variable nature of the solar and wind primary energy resources is responsible for the intermittent operation of PV plants and wind farms. Fully-fledged, MW-level battery storage is available from several vendors to ameliorate the intermittence of the power output and voltage of these kinds of renewable power plants. Still, economics are not yet in favour of the widespread use of such a solution. This issue is addressed in more detail in Chapter 7.

Likewise, an issue of growing concern is the accelerated loss of grid inertia since the conventional, large synchronous generators are coming to the end of their useful life, and their power output is being replaced with that of MW-level PV solar installations and wind farms. The main technical reason behind this concern is that present-day renewable power generators lack the large rotating masses of the conventional synchronous generators directly connected to the AC power grid, i.e., not connected to the AC power through power electronics interphases. A readily available solution to overcome such a progressive problem is to employ the long-honoured synchronous condenser, which not only restores inertia to the AC power grid but also provides dynamic reactive power compensation and instantaneous voltage regulation at its point of connection. Once again, this comprehensive technical solution comes at the expense of additional costs, which must be met, most likely by the power transmission grid owners as part of their power grid reinforcement and expansion. This issue is discussed in Chapter 7.

A structural characteristic of present-day, grid-connected renewable power generators is the use of power electronics-based interfaces of various kinds. These interfaces are essential elements for the grid connection of these kinds of new generators and the correct functioning of the current and future power grid. However, the faultless operation of the power grid requires well-designed and maintained power converters, the lack of power grid resonances' build-up, and, in no case, adverse interactions of the generators' power converters and the power grids' protection systems. Although such issues may seem remote, they are already causing concern among power grid operators and planners, particularly in power grids containing large concentrations of PV solar and wind power installations in offshore sites or weak locations of the power grid. Chapter 7 outlines these issues.

2.2 FACTS equipment and installations

Ever since the outcome of the SCR (Silicon-Controlled Rectifier) valve in the early 1950s at the Bell Telephone Laboratories in NYC and at GE in Schenectady, NY, USA, there has been a drive to employ solid-state semiconductor valves for power rectification [1] and reactive power control [2]. More recently, with the development of more advanced semiconductor valves, the range of power systems applications has widened to encompass the instantaneous regulation of power phase angle, series impedance, power factor, active and reactive power control, and more advanced forms of voltage magnitude control and power rectification [3].

The various kinds of power electronics equipment may be classified in several forms: (a) according to the type of semiconductor valves, converter, and control employed; (b) according to the main power grid application intended; (c) according to the voltage and power levels used; (d) according to their connection to the power grid: shunt, series, shunt-series, and cascade connections [4]. The last classification is used below to provide an entry-level description of all the FACTS and HVDC equipment known until now, regardless of whether or not the actual installations exist at this point in time – some of them may exist today only at the prototype stage. Still, a technical and economic justification may exist for their future fully-fledged development.

2.2.1 Shunt-connected equipment

Two distinct types of FACTS equipment are connected in shunt with the power grid and serve the primary purpose of voltage magnitude regulation at its point of connection with the power grid, namely, the Static Var Compensator (SVC) and various flavours of the Static Compensator (STATCOM) [5]. Both kinds of shunt-connected FACTS equipment have shown quite effective in damping voltage and frequency oscillations, hence, improving overall power systems stability. Nonetheless, the STATCOM is a more recent technological development than the long-enduring SVC and, as one would expect, it shows superior performance in several counts, such as a smoother voltage control over the full range of operating conditions and a faster speed of response to damp power systems oscillations. Contrary to the SVC, it yields maximum reactive current throughout the full range of voltage values, an operational characteristic that renders the STATCOM a beneficial device to discourage the inception of wide-area blackouts at the root [6].

It should be noted that some forms of SVC appeared in the power systems scene at least two decades before the FACTS initiative was put forward in the early 1990s [7]. In some respects, the power electronics-based SVC may be considered the forebear of some of the equipment developed under the auspices of the FACTS initiative [8]. Also, an early utility-level, 20 MVA STATCOM using GTO valves was developed and commissioned in Japan in 1981 [9]. It was installed in the Kansai electric power grid serving the Kansai region in Southern Central Japan's main island, Honshu. It was built at the Mitsubishi's industrial premises.

Static Var Compensator (SVC)

In the 1960s, Dr. Friedlander introduced the first SVC based on saturated reactors to control voltage magnitudes at weak network points [10]. The advent of power electronics and their use to electronically control the level of shunt reactive compensation overcame the technical limitations of saturated reactors, giving place to a drastic increase in SVC installations in many power systems worldwide. One of the earliest thyristor-controlled static compensators was installed in 1978 at the Rimouski substation in the Gaspe region, in the 230 kV system of Hydro-Québec [2]. This was deployed within the following context: Hydro-Québec – the fourth largest hydropower producer in the world – contains several long-distance, high-voltage transmission lines requiring a high degree of dynamic, shunt VAR compensation. At the time, the James Bay project was drawn to tap into the vast hydro resource in the north of the Québec province, involving the power transport of 11,000 MW at 735 kV over 1,000 km. Such technical requirements called for dynamic shunt compensation at five locations along the length of the line. It should be noted that before 1978, this function was carried out solely with rotating synchronous condensers. However, in-depth planning studies in the James Bay project indicated that the SVC technology was mature enough to be employed in such challenging operating scenarios. Nonetheless, for performance evaluation and in preparation for the 735 kV James Bay project, the decision was taken to deploy two SVCs of equal rating in the 230 kV system of Hydro-Québec – in the Gaspe region, as opposed to going directly for implementation in the more challenging James Bay system.

As shown schematically in the one-line diagram of Figure 2.1, an SVC is connected in shunt with the AC system through a step-up transformer. The primary function of the SVC is to supply/absorb reactive power to support a specified voltage magnitude at the high-voltage side of its connecting transformer. It comprises a bank of Thyristor Controlled Reactors (TCR) in parallel with a bank of Thyristor Switched Capacitors (TSC). The TCR consumes variable reactive power up to its design limit. However, the TCR generates harmonic currents since it uses phase control to finely regulate the amplitude of the current waveform passing through the linear inductors of the three-phase TCR. Suitable transformer connections and passive filtering are required to prevent these harmonic currents from escaping towards the power grid – more detail of this is provided in Chapter 4.

The SVC structure shown in Figure 2.1 typifies well all the SVCs used in high-voltage installations worldwide since the late 1970s. Moreover, most of its design features are reproduced in lower voltage, lower power applications, such as flicker

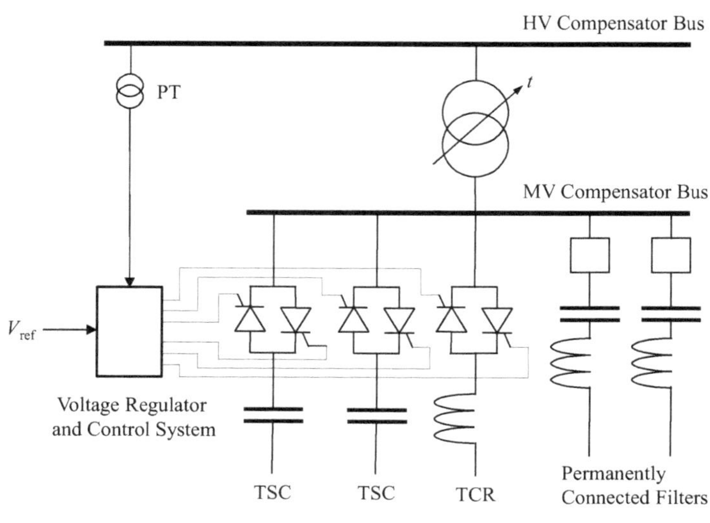

Figure 2.1 SVC schematic representation, adapted from [2]

mitigation in electric arc furnaces, and power factor control and load balancing in high-speed electric traction installations. In this respect, it would be fair to say that the SVC has become ubiquitous equipment in high-voltage power transmission and high-power industry-level installations around the globe and that an exhaustive list of existing SVC equipment in service today may be a near-impossible task to fulfil.

Below are examples of SVC installations in which SVCs have provided solutions to challenging operating scenarios that would have been technically difficult or economically taxing to solve using conventional practices in vogue at the time of the SVC installation.

The Rimouski SVC in Hydro-Quebec [2]

The Rimouski SVC was commissioned in 1978 in the Hydro-Québec 230 kV system in the Gaspé region, operating at 60 Hz. It was connected through a 100 MVA, 230/24 kV step-down power transformer with tap-changing facilities. The SVC comprises a three-phase, delta-connected 93.6 MVAR TCR in parallel with a three-phase, star-connected 93.6 MVAR fixed capacitor bank. Figure 2.2 shows this arrangement in simplified form.

Figure 2.3 shows the SVC voltage-current characteristic, as seen from the 230 kV bus. The slope of the control range is 3%.

As seen from the V–I characteristic shown in Figure 2.3, operation in the continuous steady-state range, a voltage change of −3% yields the rated capacitive reactive power of 84 MVAR. Conversely, a voltage change of +3% yields the rated inductive reactive power of 88 MVAR. Comprehensive system studies indicated

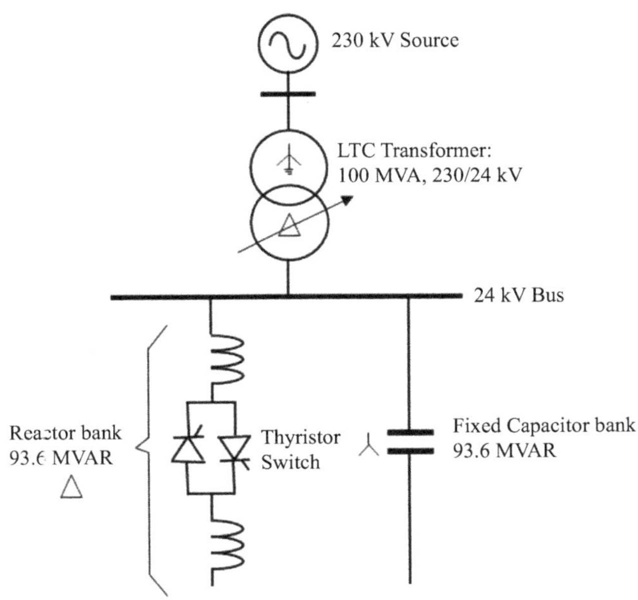

Figure 2.2 Simplified diagram of a TCR in parallel with a fixed capacitor, adapted from [2]

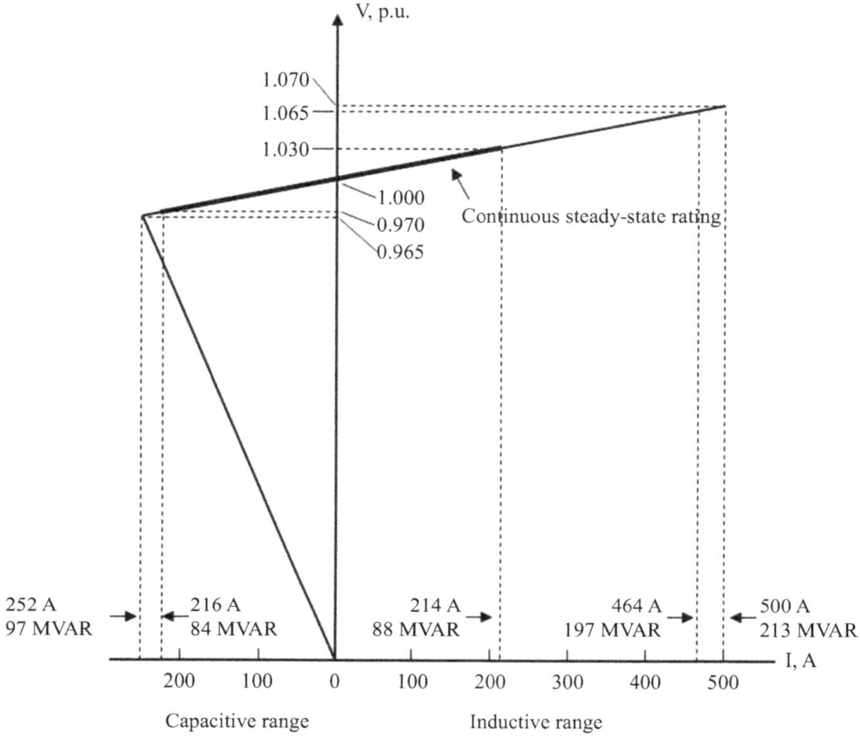

Figure 2.3 Static voltage-current characteristic, from [2]

that a value of SVC slope of 3% was suitable for this application, assuming a typical loading cycle including load rejection and fault recovery studies. In the inductive range, temporary voltage excursion of up to 5 min would be withstood in the voltage range of 1.030 p.u. to 1.065 p.u. and transient voltages of 1 s in the voltage range of 1.066 p.u. to 1.070 p.u. In the capacitive range, temporary voltages will be tolerated for 1 s in the voltage range of 0.969 p.u. to 0.965 p.u. These operating values are typical design values employed in SVC installations.

In this installation, the interfacing 100 MVA power transformer incorporates tap-changing facilities with ±10% adjustment in 33 steps with a response time of 3 s per step. Its impedance is rated at 11%. The 93.6 MVAR capacitor bank comprises three three-phase groups, each 31.2 MVAR, with independent switching and connected at 24 kV in star (floating). The three-phase reactor consists of six air-core inductors, each of 6.6 mH, connected, as shown in Figure 2.4. The three phases are connected in a delta to contain the triplet harmonic currents.

Note that the thyristor controllers and the linear inductors have the same current-carrying ratings in each phase. A reactor has an inductance of 6.6 mH, which at 60 Hz operation yields a reactance of 2.49 Ω and is rated at 1,300 A rms continuously, 2,790 A rms for 5 min, and 3,800 A rms for 1 s. These values make

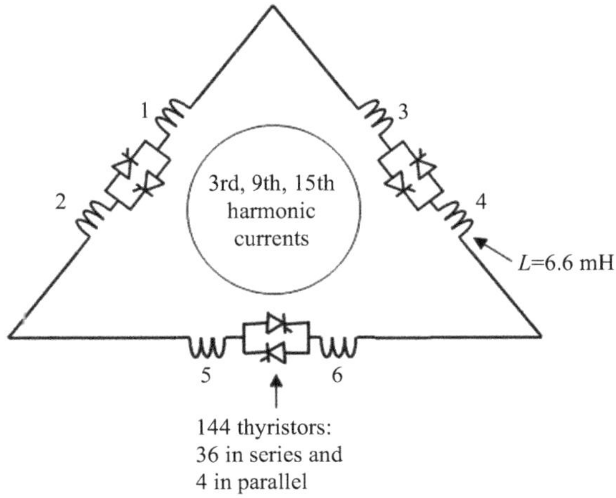

Figure 2.4 Three-phase TCR arrangement, adapted from [2]

provisions for the following harmonic currents components: 3rd, 5th, 7th, 9th, 11th, 13th, and 15th (note that beyond these harmonic orders, the TCR harmonic currents become negligibly small). This calls for four strings of 36 thyristors connected in parallel.

The thyristor valves are placed on insulating boards with snubber circuits, heat sinks, and their respective gate drive circuits. This assembly, together with the control system, the protection system, the station service supply, and the thyristor cooling system, are all housed in the same thermally insulated, metal-sheathed chamber. The thyristor valves are cooled down with recirculating forced air.

The NGC Saint John's Wood Substation in Central London [11]

In the 1970s, like all major cities worldwide, London started experiencing an unabated growing demand for electricity. Deregulation of the electrical industry in the UK in the early 1980s had not really improved much its distribution system infrastructure in the country, and by the middle of that decade, NGC planner engineers were seriously concerned about the possibility of large area blackouts.

By early 1990, the NGC Saint John's Wood substation in Central London had been identified as the critical node in the entire NGC's power grid that required dynamic reactive power support. The provision of dynamic VARs using the SVC topology shown schematically in Figure 2.1 was adopted since it represented the newest and most advanced technology at the time. In this installation, the TCR was parallelled with a Thyristor Switched Capacitor (TSC), in contrast to the fixed bank of capacitors used in the Rimouski SVC of Hydro-Québec.

In this SVC installation within the Saint John's Wood substation, there is a dedicated building to house the TCR thyristor valves, the TSC thyristor valves, the SCADA and protection and control systems, the air-core inductors, and the banks of capacitors.

The St. John's Wood substation was fitted with two identical 256 MVAR SVCs to control voltage for the City of London. Cables rated at 400 and 275 kV were fed into the substation, with the significant 132 kV out-feed to London Electricity's adjacent substation. Figure 2.5 shows the electrical system diagram of the substation.

The 275 kV cable feeds into an SF_6 bus duct and the high-voltage side of the 192 MVA, 275/16 kV three-phase power transformer, protected by a 275 kV surge arrester and its associated relay system. The 275 kV and 16 kV buses are protected by voltage differential relays with over-current backup. The capacitor banks are fitted with unbalanced current detection to prevent unit over-voltage. Inverse time/overcurrent relays are in place for fault protection. Ground faults in the 16 kV bus are detected in the neutral of the zig-zag low-voltage earthing transformer – illustrated in the schematic diagram of the SVC installation shown in Figure 2.6. If a ground fault is detected, the voltage differential relay or the ground over-current relay will operate.

Inverse over-current relays clear fault currents in the reactors or thyristor bridges, and failure to do so causes the 230 kV circuit breaker to trip off the whole installation. These issues will be covered in-depth in Chapter 10.

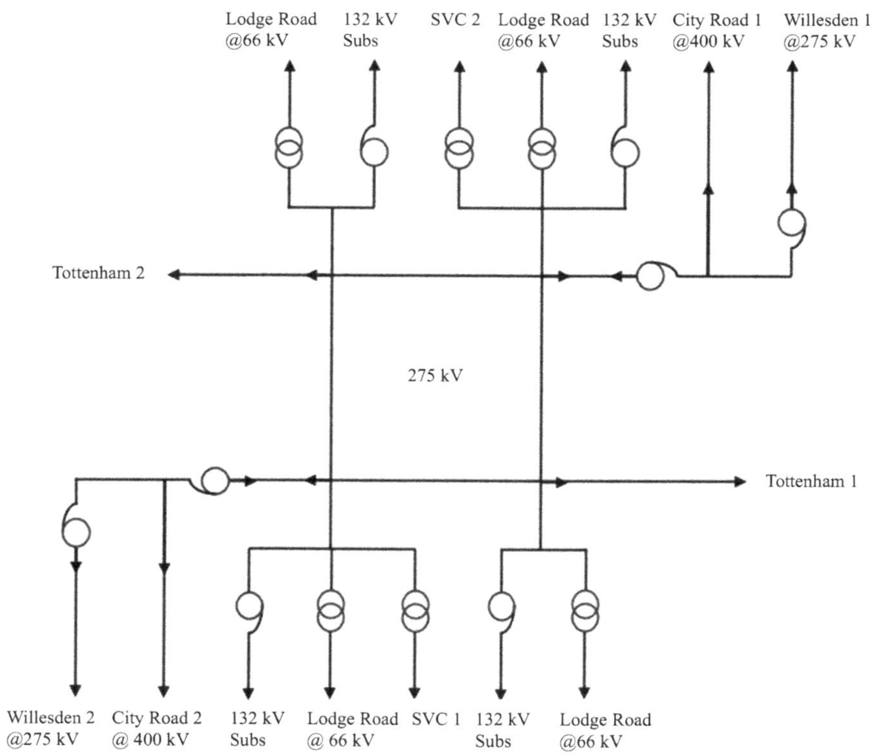

Figure 2.5 Electrical system diagram of the St. John's Wood substation with two identical 256 MVAR SVCs

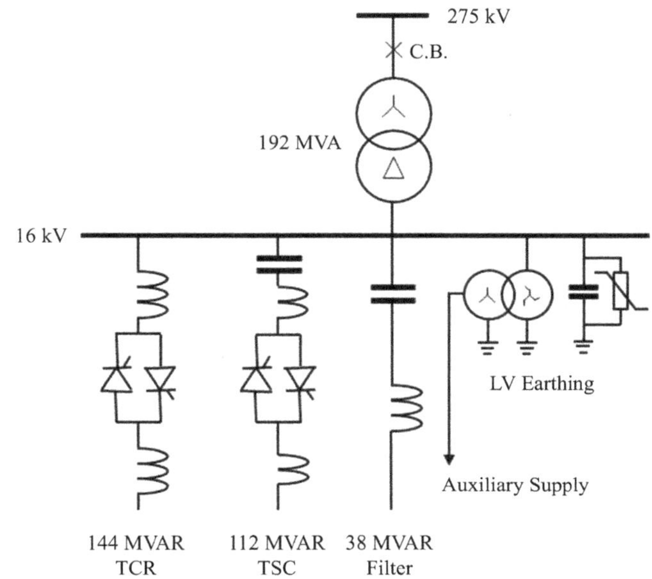

Figure 2.6 Simplified schematic diagram of one of the two identical 256 MVAR SVCs in the St. John's Wood substation

Ever since the installation of the SVC in London Electricity's Saint John's Wood Substation, the electricity demand in the London area has continued to undergo very substantial growth despite the cyclical slowdowns in the global economy in general and the UK in particular. The Saint John's Wood substation remains a critical node. Together with the City Road substation, it continues to supply the significant demand spots in central London. In 2004, it underwent a major upgrade in preparation to be fed from a 400 kV underground XLPE cable coming from the National Grid's Elstree substation via a purpose-built, 20 km long, deep cable tunnel [12]. Comprehensive power systems studies indicated that no new static var capacity was required even though the substation was due to receive larger power imports through the 400 kV XLPE cable. Nonetheless, the transformer had to be changed to accommodate the higher incoming voltage.

The Mexican power grid case [13,14]

Since the early 1980s, planning engineers of the national Mexican power utility (CFE) conducted technical studies to assess the impact of SVCs on the steady-state and dynamic performance of the Mexican power grid [13]. These studies demonstrated a great many benefits of using dynamic shunt compensation equipment in a system characterized by its longitudinal structure where long transmission lines feed into large load centres and voltage collapse phenomena are a concern.

Additional power plants were built to keep pace with the growing demand and alleviate voltage problems in crucial control areas. Nonetheless, the transmission network in the northern part of the country tended to experience high voltages during periods of minimum demand. On the other hand, the voltage magnitudes at the main tie-lines fell below acceptable operating limits during periods of peak demand. In this context, the installed SVCs maintained voltage magnitudes at acceptable ranges for safe power system operation.

Following the initial success of the first SVC installations in the Mexican power grid and over the years, many more SVCs have been installed in this power grid. Figure 2.7 shows the locations of the main SVCs installed in the 115 kV, 230 kV, and 400 kV transmission system levels. The names and main characteristics of the SVCs are listed in Table 2.1 [14]. These SVCs are set to control voltage magnitude at the high voltage node of the SVC's transformer in coordination with other fixed reactive power compensation equipment based on their characteristic sensitivity (kV/MVAR). The control strategy ensures that a specified margin of reactive power reserve is available in cases of disturbances or during scheduled maintenance tests of the equipment.

Figure 2.7 shows SVCs 14 and 15 are in proximity (i.e., 20 km). These devices interact with each other in a coordinated fashion through a master-slave control scheme using a dedicated communication channel. Furthermore, SVC 25, located at the bottom tip of the Baja California peninsula, and SVC 15, located in the northwest control area, are equipped with a Power Oscillation Damping (POD) controller, which utilises the power flow and frequency as control input signals, to strengthen the isolated power grid in Baja California, which is essentially a large microgrid. Figure 2.8 shows the main functions of an SVC equipped with a POD controller.

SVC 7 located within the substation Cerro de Oro at the Eastern control area has some design characteristics that merit additional comments. It was designed to have a dynamic compensation range of $\pm$300 MVAR based on the parallel connection of

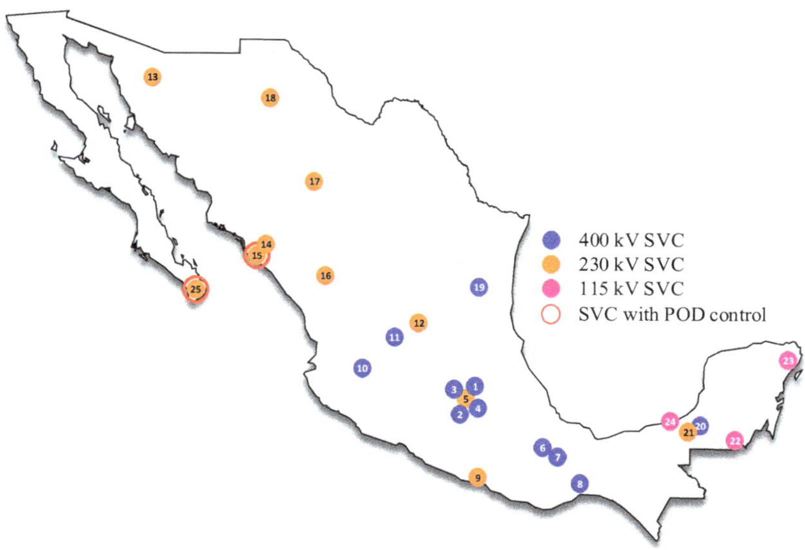

Figure 2.7 Location of main SVCs in the Mexican power grid, from [14]

Table 2.1 Capacity of main SVCs in the Mexican power grid

	Region	SVC name	MVAR	kV
1	CENTRAL	TEXCOCO	−90/300	400
2		TOPILEJO	−90/300	400
3		NOPALA	−90/300	400
4		LA PAZ	−300/300	400
5		CERRO GORDO	−90/300	230
6	EASTERN	TEMASCAL TRES	−300/300	400
7		CERRO DE ORO	−300/300	400
8		IXTEPEC POTENCIA	−300/300	400
9		PIE DE LA CUESTA	−50/200	230
10	WESTERN	ACATLÁN	−90/300	400
11		CAÑADA	−200	400
12		LA PILA	−50/200	230
13	NORTHWESTERN	SANTA ANA	−50	230
14		CULIACAN TRES	−100/100	230
15		HIGUERA	−300/300	230
16	NORTHERN	J. ORTIS MARTINEZ	−50/150	230
17		CAMARGO DOS	−50/200	230
18		MOCTEZUMA	−90/300	230
19	NORTHEASTERN	GUEMEZ	−90/300	400
20	PENINSULAR	ESCARCEGA	−300/300	400
21		ESCARCEGA	−50/150	230
22		XUL-HA	−10/40	115
23		NIZUC	−30/100	115
24		CIUDAD DEL CARMEN	−15/50	115
25	BAJA CALIFORNIA	EL PALMAR	−50/150	230

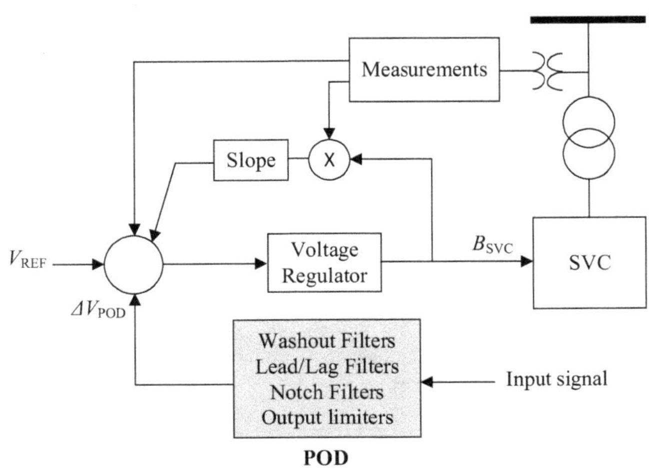

Figure 2.8 Main functions of SVC equipped with POD controller, adapted from [14]

three TCRs rated at 120 MVAR each, three TSCs rated at 80 MVAR each, and two harmonic filters tuned to filter out the 5th and 7th harmonic currents, respectively. The SVC bus operates at 16.6 kV and it is connected to the 400 kV bus using three single-phase transformer units connected in a star-delta configuration. The SVC control system involves three different modes of operation: (i) automatic voltage control, which includes a POD regulator and the reactor switching control; (ii) manual control; (iii) coordinated control with SVC 6 in TEMASCAL located 12 km apart.

According to the Program for the Development of the National Electrical System [15], the expansion projects for the period 2024–2030 comprise 11,475 MVAR of dynamic reactive power compensation (SVC and STATCOM), fixed shunt capacitors and reactors, and series capacitors. The expansion projects will increase the reliability of electrical supply in the metropolitan area of Mexico City and enhance the stability of the Mexican power grid, minimising the risk of voltage collapse under conditions of maximum demand in the central region. Moreover, power imports from the Pacific region into the central region will be maintained at suitable levels to ensure reliability and quality of supply.

2.2.1.2 STATCOM

This is the modern counterpart of the SVC. Like the SVC, its main function is to supply/absorb reactive power to support a specified voltage magnitude at the high-voltage side of its connecting transformer. Figure 2.9 shows the one-line schematic circuit of the STATCOM, made up of a converter bridge termed Voltage Source Converter (VSC), the smoothing inductor, the interfacing transformer, and the Pulse Width Modulation (PWM) control system. The VSC and the PWM control technique are covered in detail in Chapter 4.

There is widespread agreement that the operational behaviour of the STATCOM is superior to that of the SVC. The VSC and the STATCOM perform

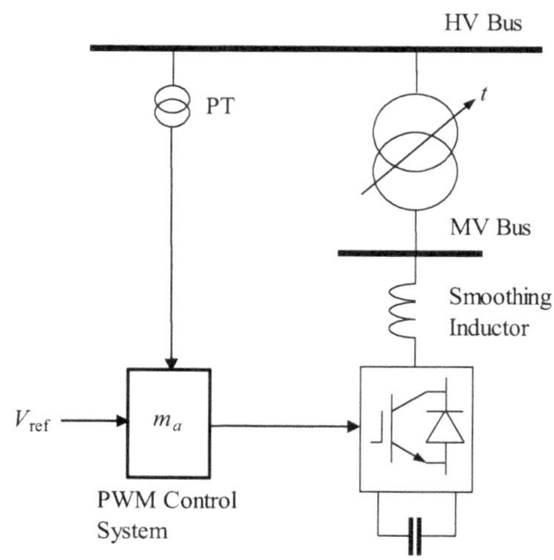

Figure 2.9 STATCOM schematic representation

like variable voltage sources [16]. Basically, they behave like a fast synchronous condenser with no moving parts. The STATCOM does not use bulky inductors and banks of capacitors to absorb and generate reactive power, respectively. The generation/absorption process is carried out electronically by the action of the VSC voltage control to yield either leading or lagging VAR operation to satisfy operational requirements [17]. Note that in an ideal, lossless system, the current injected by the STATCOM is orthogonal to the bus voltage because no active power exchange occurs due to the isolated DC bus of the VSC. The smoothing inductor is used between the VSC and the step-up transformer to eliminate the high-order harmonics. The DC capacitor is used to support and stabilise the DC voltage.

The first high-power STATCOM installation

It is reported that the first high-power STATCOM in the world was installed in 1995 at the Sullivan substation of the Tennessee Valley Authority (TVA). Its design, construction, and commissioning represented a major milestone in a far-sighted collaboration established by EPRI, Westinghouse, and TVA [18,19].

The Sullivan substation serves the needs of the NE Tennessee area. It is supplied from the 500 kV bulk power grid. As shown in Figure 2.10, the 500 kV bus is tied to a 161 kV bus through a 1,200 MVA, step-down, tap-changing transformer connected in a star-star configuration.

Several 161 kV transmission lines span from the substation, which, at the time of the commissioning of the TVA STATCOM, met a winter peak load of 900 MW. An 84 MVAR switchable bank of capacitors is connected to the 161 kV bus. This was a critical node susceptible to undergoing voltage collapse and in dire need of dynamic reactive power support – the slow response of the tap-changing transformer and the switchable capacitor bank was not adequate to deter instances of severe

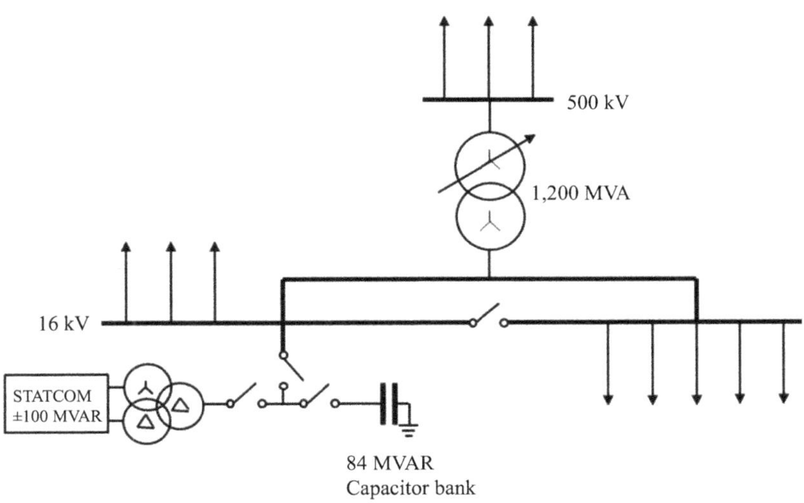

Figure 2.10 One-line diagram of the ±100 MVAR STATCOM installed at the TVA's Sullivan substation, adapted from [18]

voltage disturbances; hence, the ±100 MVAR STATCOM was selected to provide a state-of-the-art solution to such pressing issue. The STATCOM increased the reactive power controllable range of the Sullivan substation from 100 MVAR inductive to 184 MVAR capacitive, given that the STATCOM and the switchable bank of capacitors were operated in a coordinated fashion. The STATCOM was charged with the voltage regulation of the 24-h load cycle so that there was little need for the tap-changer servomechanism of the 1,200 MVA transformer to move from its nominal set operating point.

The ±100 MVAR STATCOM installed at the Sullivan substation represented state-of-the-art power electronics technology in the 1990s: it used forced-commutated valves, e.g., GTO, and a multi-pole converter design. As shown schematically in Figure 2.11, the design used eight inverter poles, each comprising 16 GTO valves – note that this illustration is a per-phase representation.

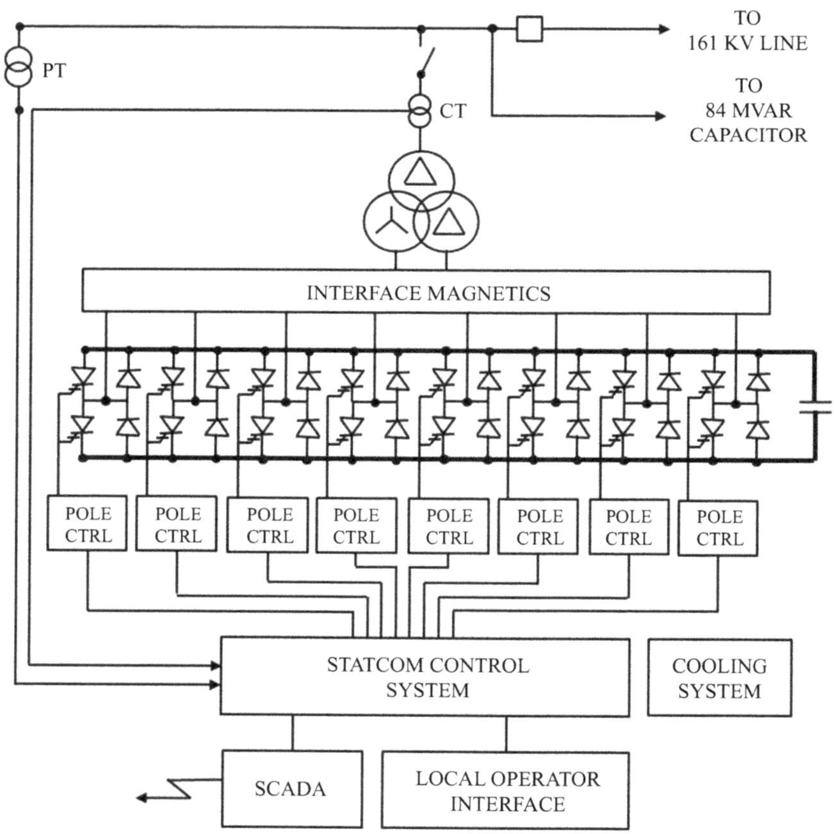

Figure 2.11 One-line diagram of the ±100 MVAR STATCOM installed at the TVA's Sullivan substation, showing the high-pulse converter design, adapted from [18]

Each inverter pole is, in essence, a two-level converter, and as such, it produces a square-voltage waveform, but in this design, each square-wave output is progressively phase-shifted from one pole to the next by a pre-selected phase angle. Magnetic summing circuits combine the eight phase-shifted square-wave pole voltages into two separate voltage waveforms, arranged to exhibit a 30° phase shift between them. One of the two voltage waveforms feeds on the star-connected winding, whereas the other feeds on the delta-connected winding. Owing to the natural 30° in the star-delta connection, a quasi-sinusoidal output voltage waveform comprising 48-pulse is formed in the 161 kV delta-connected primary winding.

A sample of STATCOM installations and applications

The past two decades have witnessed an accelerated increase in the number of STATCOM installations worldwide to provide a cumulus of application solutions ranging from improving power quality and grid code compliance in the industry sector to providing dynamic reactive power support and improving the power system's transient response at the transmission operating level. Table 2.2 exemplifies some of these installations and their application purposes.

The STATCOM technology has reached a mature status, with all major electrical equipment manufacturers offering their proprietary designs. To a greater or lesser extent, these are scalable designs that find applicability in a wide range of applications and ratings that range from several hundreds of MVARs to under 1 MVAR. Below are examples of typical STATCOM applications.

High-voltage, high-power transmission application

A STATCOM was installed in 2017 at the Rourkela substation of India's 400 kV power transmission grid, with a dynamic range of 600 MVAR plus 250 MVAR mechanically switched components. This is one of the largest STATCOMs in the world. Its function is to enable an almost instantaneous control of voltage and frequency at the substation, leading to increased availability of uninterrupted power.

Nearshore windfarm support

The Westermost Rough wind farm, located 8 km off the Yorkshire Coast near Hull, UK, is a nearshore wind farm that connects to the mainland using submarine AC power cables. The park comprises 35 turbines rated at 6 MW. Two 25 MVAR STATCOM units were installed at the onshore grid connection point to provide dynamic reactive power compensation. This ensures full wind farm export capability while complying with grid code requirements.

Onshore windfarm support

The Harestanes wind farm near Moffat in Scotland is an onshore wind farm. It is rated at 136 MW with 68 turbines. To ensure grid code compliance, increased network stability, and full export capacity, two 34 MVAR STATCOM units were installed in the wind farm substation. A key function of the STATCOM is to dampen possible system resonance build-up between the wind farm and the connecting AC power system.

Table 2.2 STATCOM installations

Name	Manufacturer	Nominal rating	Main operational purpose	Installation year and key characteristics
East Claydon Substation, Buckingham, UK [20]	Alstom T&D Ltd.	±75 MVAR 400 kV	Steady-state and dynamic reactive power support.	Commissioned in 2000. First commercial application of a relocatable STATCOM using a novel topology employing GTO-based single-phase convertors connected in series, with independent control, to make up the individual branches of modular, multilevel, three-phase converters. This arrangement may be considered the forebear of modern multilevel converters [5].
Essex Substation, Burlington, VT, USA [21]	Mitsubishi Electric Power Products, Inc.	−41/+133 MVAR 115 kV	Fast dynamic reactive power compensation during critical contingencies.	Commissioned in 2001. Coordinated control between the STATCOM and seven capacitor banks for voltage profile adjustment during the daily load cycle.
Dongguan Substation, China Southern Grid [22]	Rongxin Power Electronic Co.	2 × ±100 MVAR 500 kV	Steady-state constant voltage control. Dynamic reactive power support.	Commissioned in 2011. A unified controller operates the two parallel STATCOMs, or independent operation.
Collington Substation, North Carolina, USA [23,24]	Mitsubishi Electric Power Product, Inc.	±125 MVAR 115 kV	Dynamic voltage recovery at the transmission bus to 70% of nominal in 2.5 seconds following a N-1 contingency event.	Commissioned in 2017. The STATCOM has supplementary controls that allow voltage balancing, susceptance adjustment, reduction of losses when operating near zero output power, and cyclical gain adjustment.
Alabama Power Co., Industrial bus, Alabama, USA [25]	ABB	0/+200 MVAR 34.5 kV	Amelioration of power quality problems generated by a steel plant: flicker, voltage unbalance, low-power factor, and harmonics.	Commissioned in 2017. The STATCOM control responds in less than 100 μs, enabling an almost instantaneous speed of response to any power quality disturbance.
Llano Sánchez Substation, Panama [26]	Hyosung Heavy Industries Corp.	2 × ±60 MVAR 230 kV	Dynamic voltage regulation in a radial system.	Commissioned in 2022. Manual or automatic voltage control and adaptive droop control scheme. Following a line trip, the voltage recovers to its reference value in less than 250 ms.

An oil pipeline installation support

The capacity expansion of an oil pipeline in British Columbia, Canada, included three additional pumping stations with high-capacity electric motors. To compensate for the effects of heavy load changes, three ±12 MVAR STATCOM units with 280% overload and active filtering capabilities were installed at the substation. The STATCOM equipment enables swift reactive power compensation and increased transfer power capacity of the long, radial transmission line feeding the electric motors.

A mine installation support

A gold mine in Matachewan, Ontario, Canada, started operations in 2012. The mine demanded uninterrupted, large blocks of electricity to power the mine's working machinery, fed over very long cable lengths, with total power quality while keeping energy costs at reasonable values. The installation incorporates six 4.5 MVAR STATCOM units to provide reactive power to the grid in either capacitive or inductive modes to keep voltage levels at healthy values.

2.2.2 Series-connected equipment

An established approach to enhance long-distance transmission line operational capability is to install a series variable compensator, thus affecting the net transmission line's series impedance. A series impedance with a dominant inductive reactance will enlarge the effective distance of the compensated transmission line, whereas a series impedance with a dominant capacitive reactance will shorten it. In most practical situations, the aim is to reduce the electrical length of the series-compensated transmission line. In this respect, two different equipment options seem practical at present: The Thyristor-Controlled Series Capacitor (TCSC) and the Thyristor-Switched Series Compensator (TSSC). Both equipment flavours will enable the change of the series impedance of the compensated transmission line up to the rated value of the series compensation equipment on an almost instantaneous basis. In principle, the TCSC will exert impedance regulation in both the inductive and capacitive ranges in a continuous manner. In contrast, the TSSC will only operate in the capacitive range and in a discrete manner. Both equipment options are elaborated in Sections 2.2.2.1 and 2.2.2.2.

2.2.2.1 TCSC

The main operational function of the TCSC is to vary the net series impedance of the compensated transmission line up to the maximum rating of the TCSC, with very little delay. As a result of this, the active power flowing along the transmission line can be maintained at a specified value under a range of different operating conditions. Just like the classic series bank of capacitors, the TCSC increases the stability margin of the system but in an adjustable fashion. Moreover, the TCSC has been shown to damp Synchronous Series Resonance (SSR) and power systems oscillations [27].

An actual TCSC installation consists of one or more blocks connected in tandem, with the basic TCSC kernel comprising a capacitor, a TCR, and a varistor connected in parallel, as shown schematically in Figure 2.12. The varistor is used for overvoltage protection purposes. It plays no role during steady-state operation.

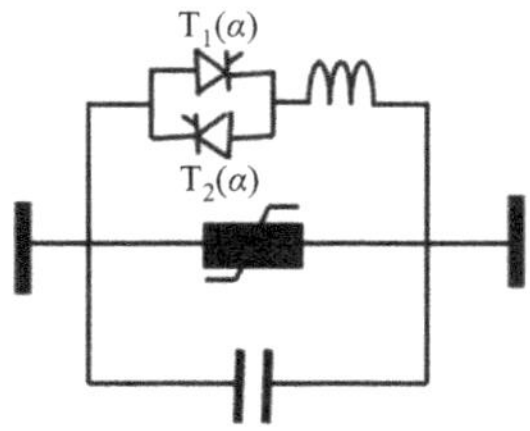

Figure 2.12 Basic TCSC kernel

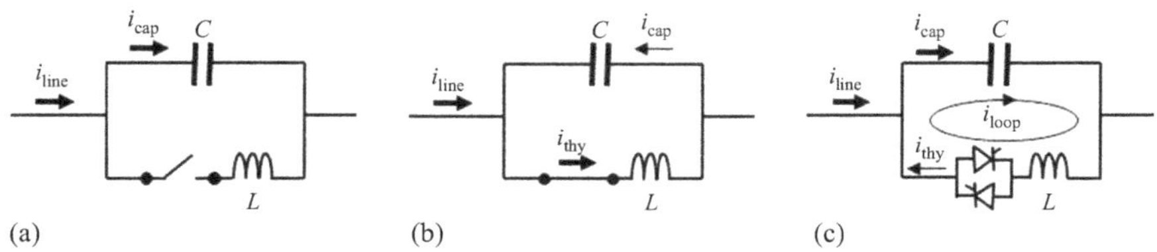

Figure 2.13 Operating modes of the TCSC: (a) Thyristor-blocked; (b) thyristor bypassed; and (c) thyristor vernier mode

Like the TCR used in the SVC, the TCSC's controlling element is the thyristor valve, shown in Figure 2.12 as the bidirectional thyristor pair T_1 and T_2. In a practical TCSC installation, many thyristors (typically 10 to 40) are connected in series to meet the required blocking voltage levels [28]. Moreover, the thyristor grid must be formed to withstand the continuous flow of line current.

A TCSC has three basic modes of operation: (i) thyristor blocked, (ii) thyristor bypassed, and (iii) thyristor vernier mode. As illustrated in Figure 2.13, in thyristor-blocked mode, the TCSC's impedance is X_C; in thyristor-bypassed mode, the TCSC's impedance is the parallel combination of X_L and X_C, with most of the transmission line current flowing through the thyristors. In the vernier mode, the thyristor valve is operated with a firing signal α in the range $90^\circ < \alpha < 180^\circ$. Depending on the design values of X_L and X_C, and the chosen value of α, the TCSC's impedance may be either capacitive or inductive. The relationship between X_L, X_C, and α is a highly non-linear one [29]. In the vernier mode, a circulating current takes place within the *L*-*C*-Th loop. The direction of the circulating current in Figure 2.13(c) corresponds to the capacitive mode. If the TCSC is acting in the inductive mode, then the direction of the circulating current reverses.

Basically – in the vernier mode – the TCSC would exhibit the behaviour of a variable series reactance governed by the TCR's firing angle α. Depending on the firing angle, the TCSC may operate in capacitive or inductive modes. However, there is little incentive for steady-state, inductive operation since this would increase the electrical length of the transmission line, with adverse consequences on stability margins, extra power losses, and excessive harmonic voltage distortion [30]. Note that in this case, the TCR harmonic currents do not tend to escape towards the network; instead, the harmonics are trapped inside the TCSC because of the low impedance of the capacitor compared to the network's equivalent impedance. Notwithstanding this, excessive harmonic generation for prolonged periods will pose great stress on the *L*-*C*-Th elements of the TCSC, and operation in the inductive range ought to be kept to a minimum. Perhaps a case of great practical interest would be the TCSC operation in a bang-bang control mode (feedback controller that switches abruptly between two states), to prevent any possible resonance build-up at the sub-synchronous and super-synchronous ranges. The two abruptly switched states would be between a capacitive operating point and an inductive operating point.

The thyristor vernier mode of the TCSC enables the adjustment of the reactance of the TCSC to perform an effective regulation of the power flow through the series compensated transmission line. In this respect, the TCSC may have the representation of a variable series reactance, which is a function of the TCR's firing angle, as illustrated in Figure 2.14.

The most complex TCSC installation until now is a TCSC installed in the Slatt substation of the Bonneville Power Administration (BPA) in the Northwestern part of the USA. The Slatt TCSC is a multimodule thyristor-controlled device that comprises six identical modules connected in tandem, as shown in Figure 2.15. Each module includes a capacitor, a varistor, a reactor, and a bidirectional thyristor valve.

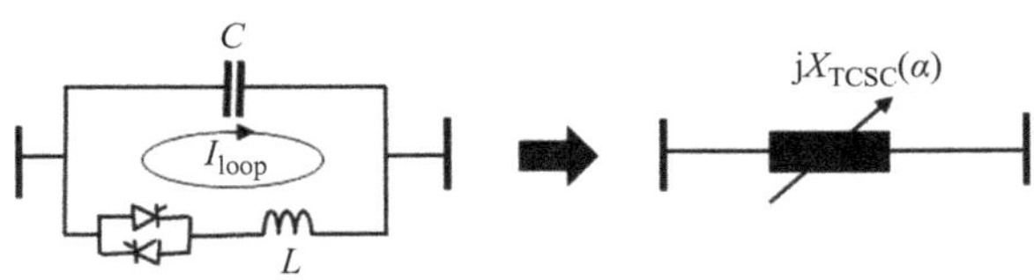

Figure 2.14 Series variable reactance interpretation of the TCSC operation in the thyristor's vernier mode

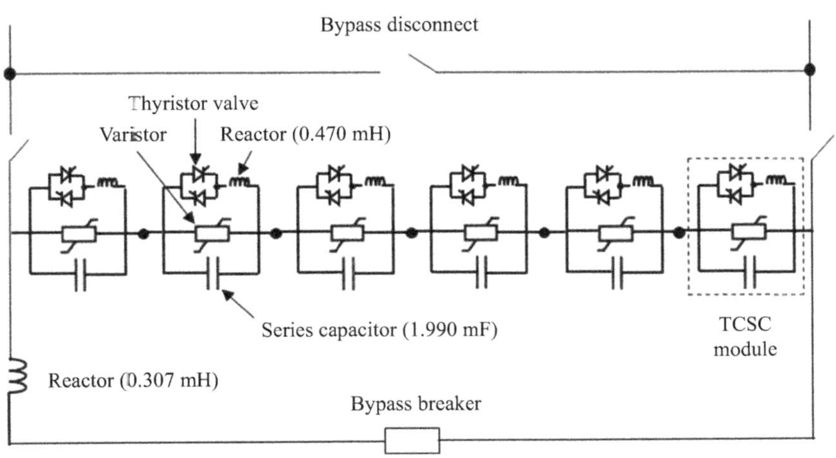

Figure 2.15 Thyristor-Controlled Series Compensator (TCSC) – layout of one phase of the TCSC installed in the Slatt substation, adapted from [29]

This installation belongs to the program of FACTS initiatives developed by EPRI in cooperation with key electrical power utilities in the USA and large electric equipment manufacturers, in the late 1980s. This TCSC was designed and installed by General Electric (GE) to series compensate the Slatt-Buckley 500 kV transmission line of BPA. As the host utility of this TCSC installation, BPA is responsible for its operation, monitoring, and maintenance. The main objectives of this TCSC are to damp sub-synchronous resonance build-up, power flow balancing to reduce power system losses, and to increase power system stability margins in a fully dynamic fashion [31].

In addition to the Slatt TCSC, several other TCSCs have been installed in countries that have long transmission distances, such as the USA, Brazil, Sweden, China, and India. These TCSCs exhibit a hybrid structure comprising fixed series banks of capacitors and one or two TCSC kernels, as illustrated in Figure 2.16 for the case of a symmetrically placed one kernel. The former provides the bulk of the capacitive series compensation requirements of the compensated transmission line and the latter is used to provide an almost instantaneous series impedance controllability.

Below, is a succinct description of the key characteristics of selected TCSC installations.

The TCSC in Kayenta, USA [32]

In 1992, the first TCSC was commissioned in the Kayenta substation, located in north-eastern Arizona, in the middle of a 320 km transmission line rated at 230 kV. Three capacitor sections were used: (i) one conventional 55 Ω, 165 MVAR bank; (ii) one conventional 40 Ω, 120 MVAR bank; (iii) one controllable section of 15 Ω, 45 MVAR. The thyristors are 100 mm in diameter and rated at 3.5 kA continuous current and 5.5 kV blocking voltage. Once it had been commissioned, no major faults were reported in the installation. However, the TCSC system has been decommissioned but the fixed bank of capacitors is still used.

The TCSCs in Serra Meza-Imperatriz, Brazil [33]

A TCSC system was built in early 1999 to compensate for a 1,020 km long-distance transmission line linking the Brazilian South-Southeast and North-Northeast power networks. The installed generation capacities of these systems are 48 GW and

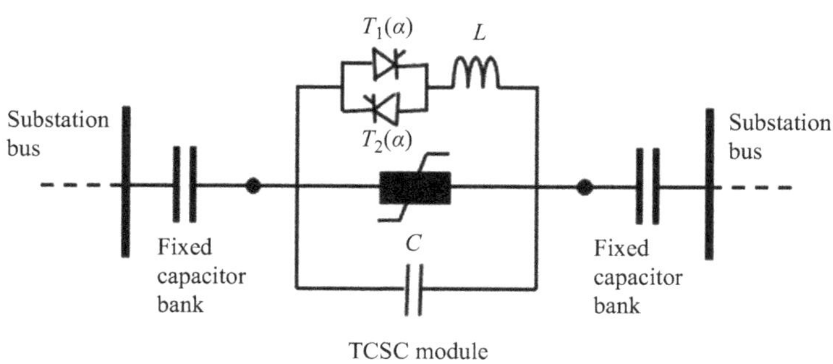

Figure 2.16 Hybrid TCSC structure

14 GW, respectively. The two power systems are made up of mainly hydroelectric power plants, which concentrate more than 95% of the total Brazilian production and consumption. The TCSC system upgrade enables the smooth transfer of power up to 1,300 MW in either direction.

The backbone of the interconnection is a single 500 kV transmission line from Imperatriz in the North and Serra da Meza in the South, which has a transmission capacity of 1,300 MW. As illustrated in the schematic one-line diagram in Figure 2.17, the installation uses six conventional, large capacitor banks and two small TCSCs. The six fixed banks of capacitors account for 54% of the total transmission line reactance. Each of the two TCSC modules accounts for 6% of the total line reactance in steady-state conditions. The two TCSC modules were located at the Imperatriz and Serra da Mesa substations, respectively. The shunt compensation is 95%, being a combination of line and bus reactors.

The two TCSCs' main role is to dampen 0.2 Hz inter-area oscillations, enabling the reliable operation of the whole Brazilian power system, ranging from no load to maximum power flow in either direction [34]. The TCR thyristors employed are 100 mm in diameter and rated at 3.5 kA continuous current and 5.5 kV blocking voltage.

The TCSC in South China, China [32]

The South China power grid comprises several long-distance HVDC and HVAC transmission lines. The increasing demand for electricity in the South required the construction of a TCSC system to enhance the operation of the AC transmission system. It uses one fixed segment of 29.2 Ω, 2,000 A, 350 MVAR, and one controlled segment of 4.15 Ω to 12.5 Ω, 2,000 A, to provide 35% compensation and 5% compensation, respectively. The thyristor valves are rated at 2 kA continuous current and 10 kV blocking voltage.

The TCSCs in Gorakhpur and Purnia, India [35]

The double-circuit, 400 kV transmission circuit Gorakhpur-Muzaffarpur-Purnia, was upgraded in 2006 with series capacitive compensation at the Gorakhpur and Purnia ends. A fraction of the 1,020 MW produced at the Tala hydropower station in Bhutan is transmitted to the Gorakhpur substation, 475 km away from Purnia in the eastern part of the country. The one-line diagram of the transmission system is shown in Figure 2.18.

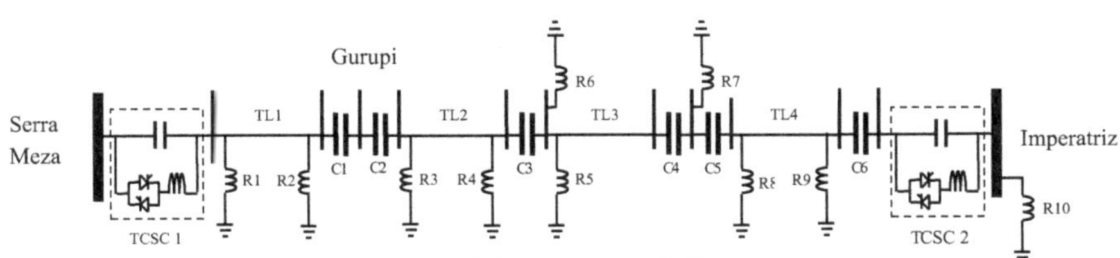

Figure 2.17 Layout of the TCSC-upgraded transmission system Serra Meza-Imperatriz, adapted from [33]

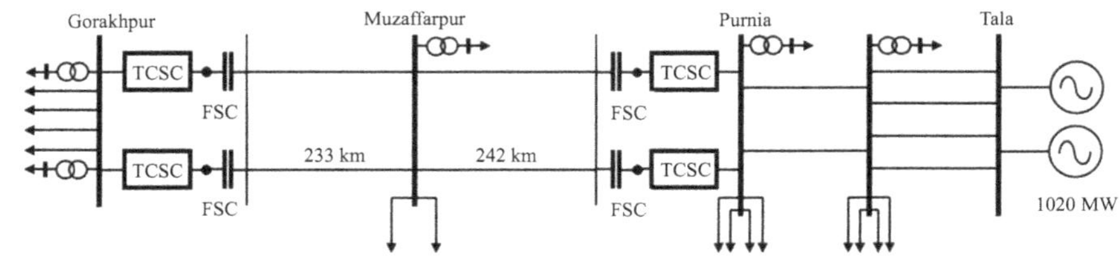

Figure 2.18 Layout of the TCSC-upgraded transmission system Gorakhpur-Muzaffarpur-Purnia, adapted from [35]

The hybrid TCSC at the Gorakhpur substation uses large, fixed capacitor blocks of 23.32 Ω, 716 MVAR, which compensates for 40% of the Gorakhpur-Muzaffarpur transmission line's series impedance. Dynamic analysis indicated that the series, fixed capacitor blocks were successful in ensuring system transient stability. However, the system experienced sustained inter-area, low-frequency (about 0.3 Hz) oscillations. To damp out these inter-area oscillations, TCSC modules were installed at the Gorakhpur substation to provide 5% to 15% variable series compensation during steady-state operation. At the Purnia end, similar operating requirements existed, and a second hybrid TCSC was installed at the Purnia substation. This comprises fixed capacitor blocks of 24.2 Ω, 743 MVAR, to warrant stable steady-state and transient operation. TCSC modules like those installed at the Gorakhpur substation were installed at Purnia. The thyristors are light-triggered with a diameter of 12.5 cm and rated at 11.2 kV.

A survey of TCSC installations carried out by CIGRE WG B4.49 in 2013 identified the installations listed in Table 2.3.

Potential TCSC installations

There are many other power grids in the world that, owing to their elongated power transmission structures, would benefit from the technical advantages that the TCSC technology would bring into their daily operations. Still, the economics may not yet be in favour of their implementation. Preliminary technical studies that elucidate the trade-offs of their implementation have been carried out for a number of power grids around the globe. Cases in point are the South Korean power grid [36,37], the UK electrical power grid [38], the Finnish power grid [39], and the Mexican and Argentinian transmission systems where long-distance transmission lines exist with only fixed series banks of capacitors. In Argentina, the Chocon dam in the Patagonia region in Argentina exports 600 MW to the central region using an 1,100 km, 500 kV, double circuit transmission line with two sets of classic series compensation units symmetrically placed along the length of the line [40]. In the Mexican case, there are several series-compensated transmission lines in the 400 kV transmission corridor that span from the large hydroelectric resource in the deep southeast of the country to the largest load area located in the centre of the country [41]. These series-compensated transmission lines use fixed banks of capacitors. This transmission corridor would be a good candidate for incorporating one or more TCSCs at key corridor locations.

Table 2.3 *TCSC installations [32]*

Year of Installation	Location	Configuration	Rating	Purpose
1992	Kayenta, WAPA, USA.	Two fixed capacitors plus one TCSC.	230 kV. Each FC: 165 MVAR. TCSC: 45 MVAR, 1 kA.	Increase power transfer. SSR damping.
1993	Slatt system, BPA, USA.	Six TCSC modules in tandem.	500 kV. TCSC system: 208 MVAR, 2.9 kA.	Power control. SSR damping. Stability control. Fault performance investigation.
1997	Stöde, Svenska, Kraftnat, Sweden.	Fixed capacitors plus one TCSC.	400 kV. FCs: 493 MVAR. TCSC: 148 MVAR, 1.5 kA.	SSR control.
1999	Imperatriz, Electronorte, Brazil.	One TCSC.	500 kV. TCSC: 108 MVAR, 1.5 kA.	System oscillations damping.
1999	Serra da Meza, Furnas, Centrais Eléctricas, Brazil.	One TCSC.	500 kV. TCSC: 107 MVAR, 1.5 kA.	System oscillations damping.
2002	Pingguo, State Power South Co., Guangzhou, China.	Two fixed capacitors plus two TCSCs.	500 kV. FCs: 350 MVAR. TCSCs: 55 MVAR, 2 kA.	System oscillations damping.

(Continues)

Table 2.3 (*Continued*)

Year of Installation	Location	Configuration	Rating	Purpose
2004	Chengxian, Gansu Electric Power Co., China.	One TCSC.	220 kV. 86.7 MVAR, 1.1 kA.	System oscillations damping. Improving power transmission capability.
2004	Raipur, Power Grid Corporation of India, Ltd.	Two fixed capacitors plus two TCSCs.	400 kV. FCs: 394 MVAR. TCSCs: 71 MVAR, 1.55 kA.	System damping oscillations.
2006	Gorakhpur, Power Grid Corporation of India, Ltd.	Two fixed capacitors plus two TCSCs.	420 kV. FCs: 716 MVAR. TCSCs: 112 MVAR. 3.929 kA.	System damping oscillations.
2006	Purnia, Power Grid Corporation of India, Ltd.	Two fixed capacitors plus two TCSCs.	420 kV. FCs: 743 MVAR. TCSCs: 112 MVAR. 3.931 kA	System damping oscillations.
2009	Fengtun, Northeast China Grid Co., China.	One TCSC.	500 kV. TCSC: 326 MVAR, 2.33 kA.	System damping oscillations.

2.2.2.2 TSSC

The TSSC comprises a series of paralleled capacitor-thyristor modules connected in tandem, which may be used to modify the effective electrical length of the compensated transmission line in a discrete-wise manner. Figure 2.19 exemplifies an *n*-module structure for one phase of the TSSC placed at the midpoint of a transmission line of reactance value X_{TL}.

A block capacitor becomes active by turning off its associated thyristor valve and inactive otherwise, i.e., when the thyristor valve is on, the capacitor is bypassed. Note that no harmonic distortion is produced by such a switching action. This arrangement yields controllable series compensation up to the rated capacity value in a discrete, stepwise manner. The larger the number of active modules, the higher the degree of series capacitive compensation and, accordingly, the shorter the electrical length of the compensated transmission line.

Although the TSSC is an interesting device due to its simple design and harmonic-free operation, this concept has not been adopted for use at the utility level so far.

2.2.3 Shunt-Series-connected equipment

Several power electronics equipment that use more than one converter or transformer, arranged in shunt and series connections, are used to make up equipment designed to perform more than one operational control task. The Unified Power Flow Controller (UPFC) is an excellent example of this, where two VSC converters are used, forming a shunt-series assembly capable of exerting simultaneous control of voltage magnitude, and active and reactive power control. The electronic phase shifter (EPS) is another case in point, where the suitable combination of a shunt thyristor bridge and a series-connected transformer enables active power regulation by exerting phase angle control. Both the UPFC and the EPS are addressed below.

2.2.3.1 UPFC

The unified power flow controller (UPFC) comprises two voltage source converters connected back-to-back, sharing a common DC link, as shown in Figure 2.20. This power electronics structure may be seen as the combination of a STATCOM and a static synchronous series compensator (SSSC) working in a coordinated fashion to control the active power flowing through the parallel DC and AC links. Each converter independently exchanges reactive power with the electrical power system. The series-connected converter produces an AC voltage $\overline{V}_{se}$ of variable

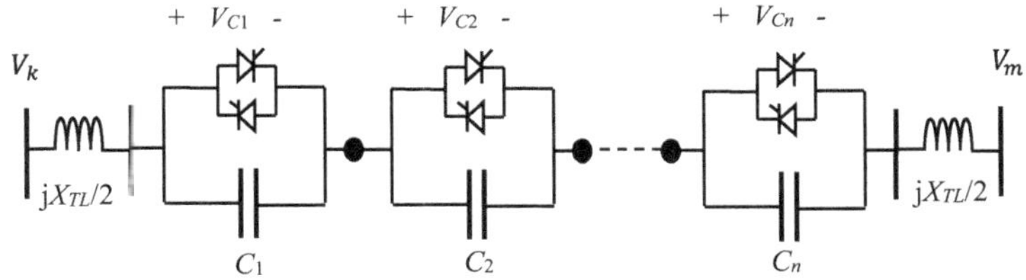

Figure 2.19 Layout of one phase of a mid-point TSSC-upgraded transmission line

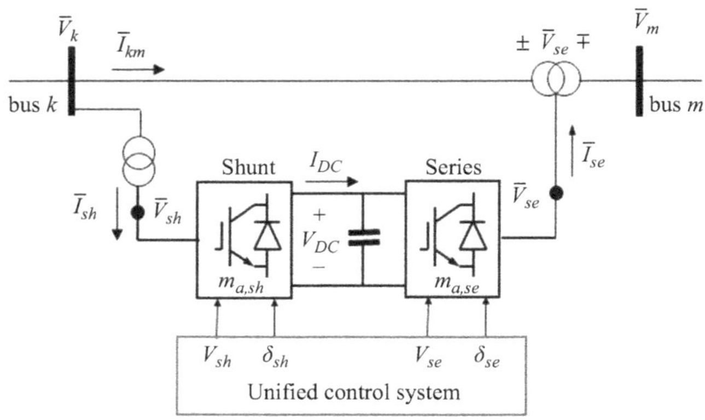

Figure 2.20 UPFC: two back-to-back VSCs with one VSC connected to the AC network using a shunt transformer and the second VSC connected to the AC network using a series transformer

magnitude and phase angle, which is injected in series to the nodal voltage, $\overline{V}_m$. Note that the controlled voltage does not depend on the line current $\overline{I}_{km}$ and transmission phase angle $\theta_k - \theta_m$, thus enabling the independent control of active and reactive power flows injected into node m. Concerning the shunt converter, it allows for the control of the voltage magnitude at node k by locally generating or absorbing reactive power. This converter acts as a shunt-controlled voltage source, denoted by $\overline{V}_{sh}$. A key role of the shunt converter is to support the active power drawn by the series converter through the back-to-back DC link.

Series converter control. It has several control modes of operation exerted directly at node m:

(i) Voltage regulation is exerted by injecting a voltage $\overline{V}_{se}^{(1)}$ in-phase or anti-phase with the nodal voltage $\overline{V}_m$, as illustrated in Figure 2.21(a).
(ii) Injecting a voltage $\overline{V}_{se}^{(2)}$ in quadrature with the line current angle, as shown in Figure 2.21(b), to mimic the control action of a variable series compensator (SSSC).
(iii) Regulating only the nodal voltage's phase angle θ_m by injecting a series voltage $1\angle \pm \theta_{se}$, as depicted in Figure 2.21(c). Note that if the value of θ_{se} is orthogonal to θ_k then the UPFC performs the role of a phase-shifting transformer.
(iv) Exerting simultaneously the previous three control actions: $V_{se} = V_{se}^{(1)} + V_{se}^{(2)} + V_{se}^{(3)}$, as exemplified in Figure 2.21(d).

Shunt converter control. It has two control modes of operation exerted directly at node k:

(i) Constant reactive power.
(ii) Constant voltage magnitude.

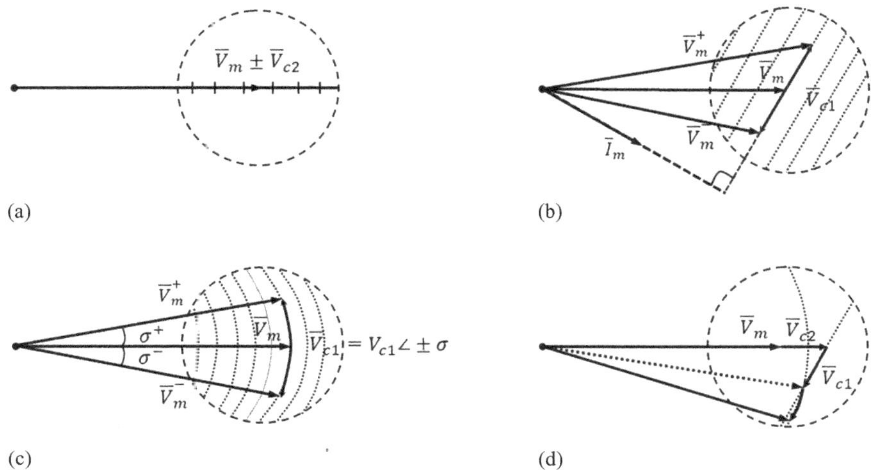

Figure 2.21 Phasor diagrams illustrating the general concept of: (a) Magnitude voltage control; (b) impedance line compensation; (c) phase angle regulation; and (d) simultaneous control. Note that for the purpose of drawing these phasor diagrams, the phase angle θ_m is taken to have a value of 0°, from [42]

The former involves injecting/absorbing a predetermined amount of reactive power irrespective of the system's operating conditions, and the latter injects/absorbs reactive power dynamically to maintain the voltage magnitude at the system node at a specified value.

UPFC installations and applications

The unique operational characteristic of the UPFC to simultaneously provide shunt and series compensation, enabling dynamic voltage and power flow control, has raised great interest among power utility planning engineers worldwide, but despite all its great technical merits, this has so far resulted in a limited number of utility-level installations. Three early UPFC installations employed GTO-based voltage source converters [43]. These are described in Table 2.4. Two more recent applications have used IGBT valves; these are described separately.

A 500 kV, IGT-based UPFC in the Suzhou Southern Power Grid [44]

It is reported that in 2015, the electrical energy demand in the Suzhou and Wujiang area was approximately 10,270 MW, with a predicted growth of 3,000 MW by 2020, approaching the maximum demand allowed by the existing electrical infrastructure, which stood at 15,000 MW. The electrical energy is supplied by the Suzghou Southern Power Grid (SSPG) company using three 500 kV AC transmission lines and the Jinsu DC transmission line, which carries power from the Jinping Hydropower Station. The DC link operates at full capacity during the summer and at substantially reduced power

Table 2.4 GTO-based UPFC installations

Installation	Project description	Main purpose	Characteristic
St. Inez 138 kV substation, Kentucky, USA. Commissioned in 1999 [43,74].	The first stage came into operation in July 1997 as a STATCOM rated at ±160 MVAR. The second started operations in June 1998 with a series converter of ±160 MVAR. The 765 kV network was upgraded, with the installation of a 765kV SVC.	Dynamic reactive power support and power flow control.	World's first UPFC installation. Four different control modes: ±160 MVAR STATCOM ±320 MVAR STATCOM ±160 MVAR SSSC ±320 MVAR UPFC
Kang-jin 154 kV substation, Jeonnam, South Korea. Commissioned in 2003 [74,75].	It was put into operation in 2003 and taken out of service in 2010. Maintaining the GTO firing control boards was difficult.	Automatic control of voltage and power flows. Under critical contingency scenarios, the series converter control mode was switched to constant voltage injection.	The 80 MVA UPFC comprised two ±20 MVA shunt converter modules and two ±20 MVA series inverter modules.
Marcy 345 kV substation, NY, USA. Commissioned in 2004 [74,76]	In February 2001, two 100 MVA STATCOMs were installed. The associated series converters started operation in July 2004.	Used primarily as STATCOM to provide steady-state and dynamic voltage control.	First convertible static compensator. It has four control configurations: STATCOM, SSSC, UPFC, and interline power flow controller (IPFC).

flows of 10 to 20 percent of the nominal power during the winter period due to water shortages. This causes significant operational problems in the face of N-1 contingencies and rapidly changing power flows due to the randomness of the renewable energy sources embedded in the system and a deficit of reactive power support for voltage regulation. The area served by SSPG is illustrated in Figure 2.22.

Two traditional solutions were considered to alleviate the operational problems brought about by the increasing demand. These options were: (i) increasing the conductors' capacity in the transmission lines connecting the Meili and Mudu substations; (ii) building a new 60 km transmission corridor to link the two substations, as illustrated in Figure 2.22. The former solution was not feasible since it would be insufficient to transfer the required amount of electrical power demanded at the Mudu substation. Likewise, the latter solution was deemed unsuitable due to the marked increase of short circuit current at the Shipai substation. The planning engineers turned their attention to the use of FACTS solutions. The decision was made to install a UPFC on the existing Meili-Mudu transmission lines, with the topological structure shown in Figure 2.23.

The UPFC entered operation in December 2017. Each converter is rated at 250 MVA and the transformers at 300 MVA. The DC bus operating voltage is ±90kV. The technical-economic advantages brought about by the UPFC installation are the following:

1. The dynamic reactive power compensation capacity was increased substantially, reducing the possibility of commutation failures in the LCC-HVDC inverter at the Tongli converter station.

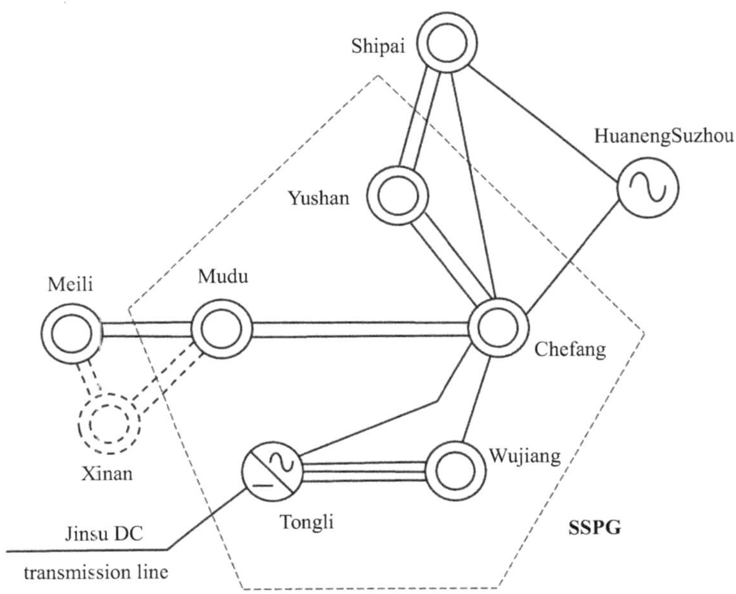

Figure 2.22 Network structure of the 500 kV SSPG in 2015 considering a new transmission corridor in broken lines, adapted from [44]

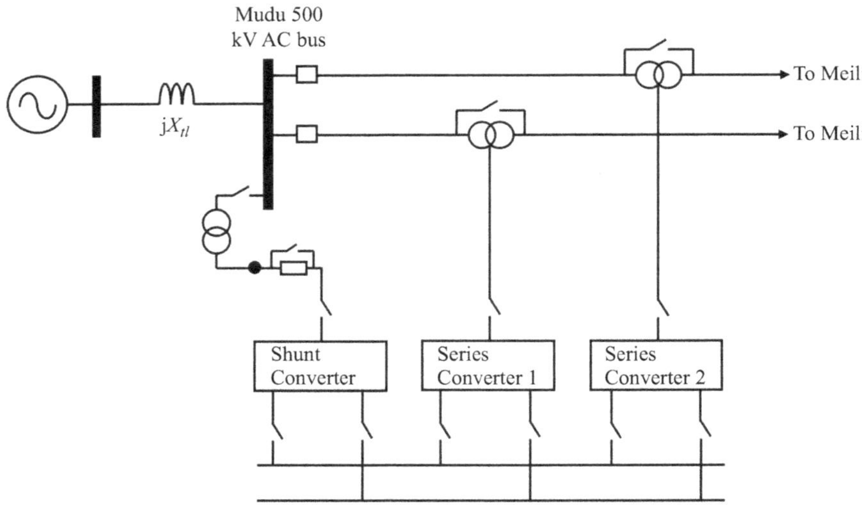

Figure 2.23 Power system configuration of the Suzhou 500 kV UPFC project, adapted from [44]

2. An increase in power transmission capacity and power flow control at the Meili-Mudu 500 kV AC transmission lines alleviates transmission line overloading and reduces load shedding in the event of commutation failures in the LCC-HVDC converters or an N-1 operating condition.
3. The short-circuit current at the Shipai substation did not increase.
4. The overall cost of the project was 37% lower than the cost of adding new transmission lines, providing a cost-effective solution to the operational problems outlined above.

A 200 kV, IGBT-based UPFC in the Shanghai Power Grid [45]

It is reported that in 2019, the growing demand for electric power in the Shanghai region was primarily met by importing electric power generated by large hydropower plants in Southwest China, using four long-distance, high-power LCC-HVDC links. Note that the four LCC inverters associated with these links are sinks of reactive power. The power imports accounted for a high percentage of the peak load. Hence, the local generators had limited control over reactive power and voltage magnitude in the wider Shanghai region. This issue was exacerbated by significant demand fluctuations and extensive use of high-power AC cables, leading to voltage stability problems in the Shanghai metropolitan power grid. An additional problem was that unequal power transfers occurred in three parallel AC transmission lines owing to their different impedances, causing these transmission lines to overload. To address such operational challenges and to optimise the operation of the Shanghai network, it was decided to install a UPFC at the 220 kV Yunzaobang substation, in a corridor of three parallel AC transmission lines, as shown in the circuit diagram shown in Figure 2.24.

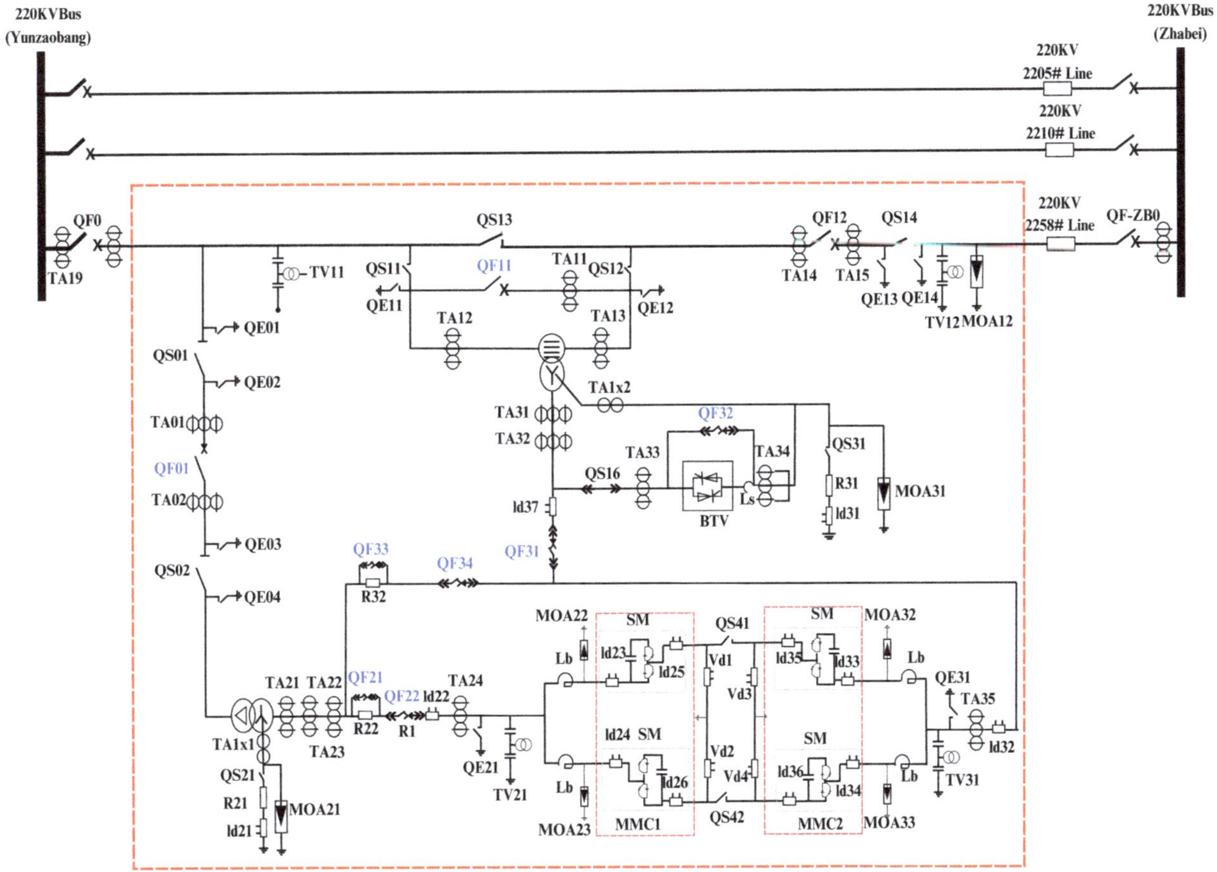

Figure 2.24 Circuit diagram of 220 kV UPFC Shanghai grid project, from [45]

The UPFC began operation in 2017. The UPFC converters are rated at 50 MVA, and the DC bus is ±20.8 kV. The series and shunt transformers are rated at 50 MVA and 100 MVA, respectively. The series converter's circuit breakers system allows its disconnection, enabling a STATCOM operating mode for the case when it is only desired to control the nodal voltage magnitude at the AC node or provide dynamic shunt compensation control at the Shanghai metropolitan power grid. The UPFC's series converter provides independent active and reactive power flow control, enabling the regulation of these power flows to target values or achieving a balanced distribution of power transfer in parallel lines under different operating conditions.

2.2.3.2 EPS [4]

This EPS regulates the voltage phase angle of the output voltage ($\overline{V}_{OUT}$) in Figure 2.25(a) using a transformer whose primary winding is connected in series with the power grid (i.e., series transformer), an auxiliary shunt excitation transformer, a thyristor bridge, and a controller. The controllable incremental voltage drawn from the excitation transformer is added in quadrature to the incoming nodal

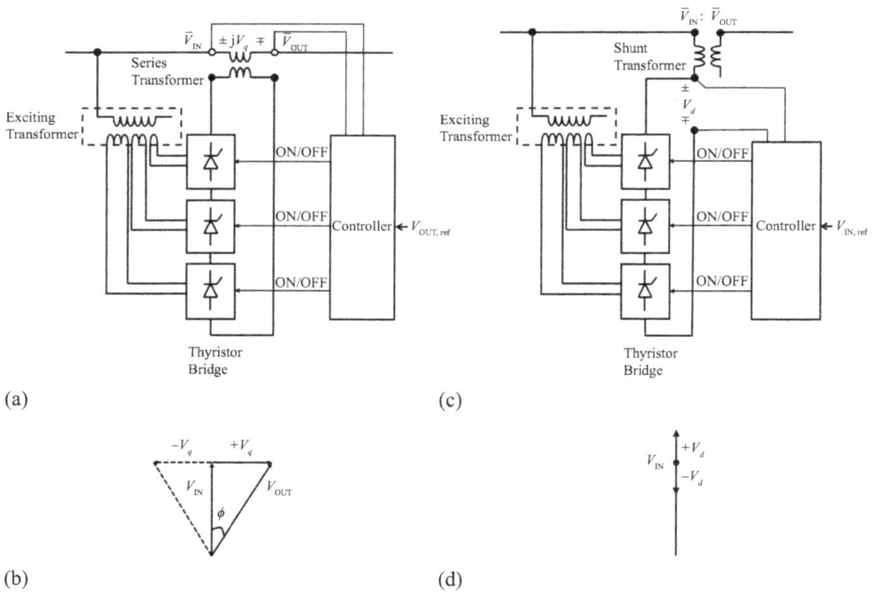

Figure 2.25 Power electronics controllers using an ancillary, shunt, and excitation transformer to modify the operating states of the main transformers in an almost instantaneous fashion, using an array of thyristor switches. (a) EPS structure; (b) associated phasor diagram; (c) ETC structure; and (d) associated phasor diagram.

voltage to modify the phase angle of the incoming nodal voltage. This action is appreciated from the phasor diagram in Figure 2.25(b).

Note that the output voltage of the excitation transformer is discrete but variable in both magnitude and polarity. By controlling the voltage phase angle, the phase shifter controls the active power flow through the series transformer.

On the one hand, employing a similar power electronic setup to the EPS but applied directly to the primary winding of the main transformer – as shown in the schematic diagram in Figure 2.25(c), an electronic tap changer (ETC) is realised. The output voltage of the excitation transformer is added in phase to the primary winding of the main transformer ($\overline{V}_{IN}$), as illustrated in the phasor diagram of Figure 2.25(d). Also, in this case, the output voltage of the excitation transformer is discrete but variable in both magnitude and polarity. In both power electronic controllers, the thyristor switches are operated using an ON/OFF control, and therefore, no harmonic distortion is introduced.

Neither the EPS nor the ETC seem to have been used so far in any practical installation, but only conventional versions based on mechanically driven servomotors.

2.3 HVDC equipment and installations

Nowadays, there are two practical ways of transporting large blocks of electrical power in the electricity supply industry: the three-phase AC and DC forms. By and large, the former is predominant, but the latter is unassailable in specific power system applications, such as in submarine power transmission, the evacuation of offshore wind power, long-distance transmission involving distances above 700 km, the interconnection of power systems exhibiting different operating frequencies, e.g., 50/60 Hz, and cases where high power controllability is a top requirement, such as the interconnection of a strong power grid and a weak power grid. These issues will be detailed in several points of this section.

Present-day DC transmission systems use rectification/inversion processes of AC waveforms that employ power electronics equipment, methods, and control techniques. This contrasts with the low-voltage DC power systems that existed in the early days of the electrical power industry – the last decade of the 19th century – which were radial and covered rather short power distribution distances. Power electronics did not exist at the time in any form; instead, these power systems employed dedicated DC dynamos to generate electricity at the right voltage level to supply specific types of DC loads: streetlights, trams, electric motors, and lights at homes and public buildings. The DC dynamos were placed as near as practicable to the load points in order to keep power losses at reasonable values. The irruption of the fully-fledged power transformer circa 1900 gave AC power distribution systems an unassailable advantage over those early DC power distribution systems [46].

However, changing socio-economic conditions in the post-war years and the advent of non-solid-state power electronics technology led to an incipient revival in the use of DC power technology. The need was to feed the entire Island of Gotland from mainland Sweden, in 1953. This was the first commercial HVDC link in the world, starting operations in 1954, rated at 100 kV DC, 20 MW, employing electrode-graded arc-valves. In 1970, the rated voltage of this link was increased to 150 kV DC, 30 MW, when solid-state valves (thyristors) became available. The valves in the upgraded system were hybrid, comprising the series combination of the original "vapour" valves and the new kind of solid-state valves. In 1983, this submarine power transmission corridor was repowered with the addition of a 130 MW monopole operating at 150 kV. In 1987, an additional monopolar link of equal ratings was added to this undersea transmission corridor [1].

Two quite distinct flavours of the HVDC technology are available at present: the HVDC transmission based on the use of thyristor valves and phase control and a more recently emerged form of HVDC transmission that uses forced-commutated valves (e.g., IGBT, GTO) driven by PWM control [5]. The former is known as LCC-HVDC or CCS-HVDC or Classic HVDC, whereas the latter is known as VSC-HVDC. The acronyms LCC stand for line current commutated, and CSC stands for current source converters, whereas VSC stands for voltage source converter. At present, LCC-HVDC is unassailable in the transport of extra-large blocks of electrical power, owing to the very robust nature of the high-power thyristor valves and quadrupole bridges employed – based on the so-called Graetz bridge. Concerning the VSC-HVDC technology, its latest generation of equipment uses IGBT valves and modular, multilevel converter (MMC) topologies. The MMC design employs low-frequency valve-switching, which enables a marked reduction in switching power losses compared to the initial two and three-level converters used in VSC-HVDC designs [47]. One further additional advantage of modern MMC designs is to reduce harmonic distortion waveform to a trickle, sparing the need for bulky filtering equipment [48]. The full impact of the irruption of the VSC-HVDC option in the electricity supply industry is just beginning to be dimensioned. For one, a long-pursued objective of power systems engineers working in HVDC transmission has been to be able to make up DC power grids of arbitrary configurations with the same simplicity that AC power grids are formed. Until quite recently, with LCC-HVDC, this was possible in only a rather contrived manner, that is, forming radial DC power systems, but not meshed DC power systems. Nowadays, VSC converters enable such an option, albeit a few teething problems are yet to be solved, such as the availability of reliable and cost-effective DC circuit breakers [1].

Based on the capacities of past and present HVDC installations, Figure 2.26 offers a comparative view of the relative capacities of the three types of valves that have been used in HVDC installations, namely, graded-electrode, mercury-arc, classic thyristor, and IGBT. This is with respect to the year in which they were integrated into this area of power systems application. It should be noted that electrode-graded, mercury-arc valves are no longer used in existing HVDC schemes intended for MW-level transport installations [1].

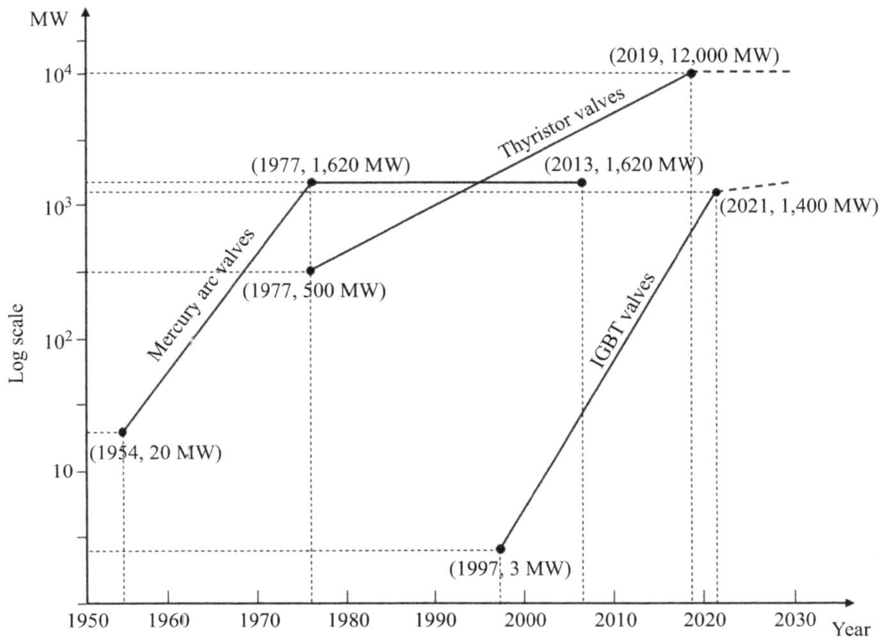

Figure 2.26 Capacities of power electronics valves vs. time of their application in HVDC installations – note that log scale is in the ordinate axes

It is illustrated in Figure 2.26 that electrode-graded, mercury-arc valves started being used in 1954 and ceased to be used in 2013. They were employed for the first time in the submarine Gotland HVDC cable rated at 20 MW. The HVDC link with the highest capacity with this kind of valve was the Nelson River Bipole 1, rated at 1,620 MW, commissioned in 1977. Under the continuous advancement of solid-state power electronics, the electrode-graded, mercury-arc HVDC installations were either decommissioned or reconverted to use thyristor valves. The last HVDC installation to have used electrode-graded, mercury-arc valves, the Inter-Island New Zealand HVDC link, was totally refitted with thyristor valves in 2013.

Thyristor valves were used for the first time in the 500 MW Square Bute HVDC link in 1977. From 1977 until 1997, all new HVDC installations employed thyristor valves. That year, a new form of HVDC power transmission using IGBT valves was prototyped in central Sweden using a relatively small capacity installation rated at only 3 MW [47]. Ever since, new HVDC installations have used either thyristor valves or IGBT valves, depending on the intended application. As will be shown below in connection with the great many HVDC installations scattered around the world, the former is used for extra-long distance, bulk-power transmission applications, whereas the latter is preferred in applications where near-instantaneous controllability is required, such as the evacuation of offshore wind power. The highest capacity HVDC installation in the world, which is an

LCC-HVDC, is rated at a staggering amount of 12 GW. It corresponds to the Changji–Guquan HVDC link, which spans 3,300 km across China. It was commissioned in 2019 [49]. By comparison, the ratings and lengths of VSC-HVDC systems are more modest but progressing rapidly on both counts. These installations are particularly suited to submarine HVDC transmission. The North Sea link between Norway and the UK and the Viking link between Denmark and the UK are good cases in point. The former employs a submarine cable of 730 km, and the installation is rated at 1,400 kW, ±515 kV. This link was commissioned in 2021 [50]. The latter employs a 740 km submarine cable, and the installation is rated at 1,400 kW, ±525 kV. This link was commissioned in 2023 and is currently the world's largest submarine power cable.

2.3.1 LCC-HVDC links

The most basic HVDC transmission structure employing thyristor valves is the so-called point-to-point, monopolar DC link with ground or seabed return. The rectifier and inverter station contain six thyristor valves arranged in the classic Graetz bridge. This DC link is illustrated in Figure 4.18 of Chapter 4. Although the monopolar scheme was quite popular in the late 1950s and early 1960s, the structures of modern HVDC links are far more complex. Standard point-to-point HVDC schemes are bipolar, as opposed to monopolar, and the converter stations are quadrupoles; this is in the interest of higher power transfer yields and increased reliability [1]. This is exemplified in Figure 2.27 in a simplified manner.

Some HVDC-specific applications call for using more than two HVDC converter stations to form multiterminal HVDC systems. The LCC-HVDC technology, intrinsically based on current balances, limits the multiterminal HVDC structures to be connected in series (tandem). This restriction does not apply to the VSC-HVDC technology, which enables meshed DC power grids. This all-important fundamental difference between the two flavours of the HVDC technology is detailed in Section 2.3.3.

In existing HVDC systems, the cable linking the rectifier and inverter stations may be longer than 3,000 km (as in China) or only a few meters long. In the latter

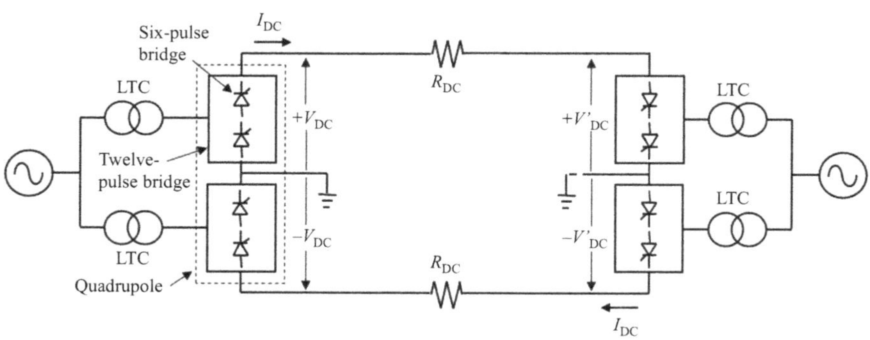

Figure 2.27 A bipolar, quadrupole, LCC-HVDC transmission system

case, the two converters are housed in the same building substation and the DC link is termed a back-to-back scheme – also known as a zero-distance scheme. One possible application of a back-to-back link is to act as a frequency coupler to link two AC power systems with different power frequencies, say, 50 Hz and 60 Hz (as in Japan). Another application of the back-to-back link is the connection of two AC power systems exhibiting quite different short-circuit ratios, i.e., a strong and a weak system [51].

2.3.2 VSC-HVDC links

In 1997, a new form of carrying DC power at the MW level was trialled. The rectification/inversion processes employed two-level voltage source converters using IGBT valves driven by sinusoidal-PWM control. It was a rather small installation rated at only 3 MW, 180 kV. A 10 km overhead transmission line linked the two VSCs, running between Hellsjön and Grängesberg in central Sweden [47].

This experimental installation showed great potential, and a second experimental VSC-HVDC installation was deployed in a small, onshore wind park at Tjæreborg Enge, Denmark, in 2000. This was rated at 7 MW, 9 kV [52]. This HVDC link operates alongside an existing AC transmission line that connects the wind park to the Danish power grid. Either the DC or the AC lines can be used to evacuate the wind park's power output, but the former yields better regulation of power peaks. The fast power regulation afforded by the VSC technology, demonstrated by this experimental installation, would show the way forward for the connection of the large offshore wind farms that would be installed during the following decades in the North Sea. This is a niche area where the VSC-HVDC technology is unassailable at present [53].

The bipolar scheme in Figure 2.28 is a simplified schematic representation of the point-to-point VSC-HVDC link. It comprises a twelve-pulse, IGBT-based rectifier station, an IGBT-based inverter station, a DC cable, and interfacing transformers at both ends of the link. Contrary to LCC-HVDC, the transformers

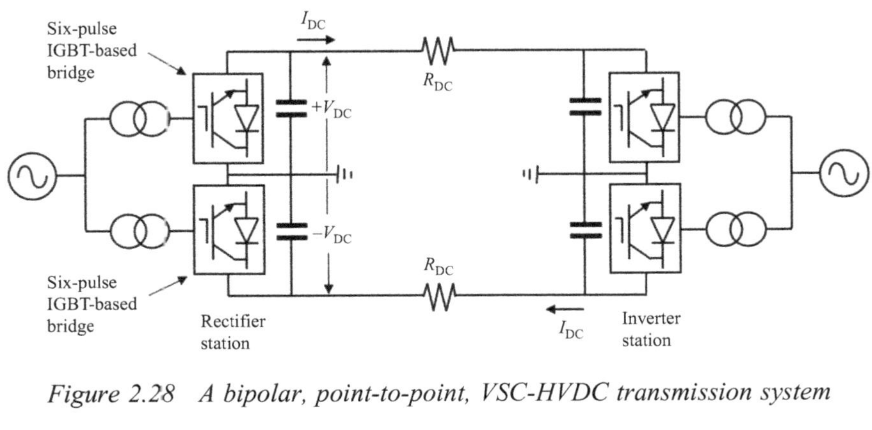

Figure 2.28 A bipolar, point-to-point, VSC-HVDC transmission system

employed here are not tap-changing transformers. The VSC modules making up the rectifier and inverter stations are addressed in depth in Chapter 4.

The early VSC converters used in this application are of the two-level and three-level kinds. The switching frequencies employed are in the range of 1 to 2 kHz. Despite all their worth, major handicaps limited the immediate, widespread use of this technology over the Classic HVDC technology: (i) they were more expensive than installations of comparable ratings; (ii) the output voltage waveforms of these converters are far away from being sinusoidal waveforms – in fact, they are a rectangular train of pulses. Refer to Chapter 4 for a full assessment of this issue; (iii) the switching power losses per converter station are 2% or higher compared to 0.7% incurred by thyristors.

Nonetheless, the small footprint and high controllability afforded by the VSC-HVDC technology recommended itself for exploring its use in key applications despite their relatively high-power loss and, in some cases, large filtering system requirements – the back-to-back VSC-HVDC in Eagle Pass, Texas commissioned in 2000 is a good example of this [54]. Owing to their small footprint, they were ideally suited to feed oil rigs in the North Sea from mainland Norway. This is exemplified by the 70 km HVDC submarine link to feed the offshore platform Troll A off the coast of Norway [55]. Exploiting the high controllability exhibited by the VSC-HVDC technology, a 105 km submarine link was commissioned in the Baltic Sea in 2006 to export power between Finland and Estonia. The link was rated at 350 MW, 150 kV. The VSC-HVDC link enabled the secure export of electrical power between power grids of markedly different strengths [56]. The southern part of the Finnish power grid is highly meshed and at the time, the northern part of the Estonian power grid was weakly meshed.

In total, ten VSC-HVDC installations were built in the period 1997–2009. The converters employed were either 2-level or 3-level VSCs. All of them are installed by the major equipment manufacturer ABB. Up to this point, the highest rating VSC-HVDC link was the BorWin 1 rated at 400 MW, ±150 kV, linking the offshore windfarm BorWin 1 in the North Sea to the power grid in mainland Germany, spanning 200 km. However, in 2010 a new breed of VSC design was incorporated for the first time in a MW-level VSC-HVDC installation, which ameliorated very significantly the two major handicaps exhibited by the two-level and three-level VSC designs: high switching power losses and poor output voltage waveforms. This installation, carried out by Siemens using its brand new MMC VSC proprietary design, was the Trans Bay Cable in Northern California, USA: an 85 km underwater cable between San Francisco and Pittsburgh, California, rated at 400 MW and 200 kV [57]. The MMC VSC design provided a springboard for a renewed era in HVDC power transmission. In the last 14 years alone, more than 60 new VSC-HVDC links have been built in many parts of the world, most of them in Europe. At present, there are four manufacturers of MW-level power electronics converters, namely, Hitachi-ABB, Alstom, Siemens, and RXHK (China), each having their proprietary brands of MMC VSC equipment.

The AC voltage output of an MMC SVC is not a rectangular train of pulses as is the case in a two-level VSC, but rather an AC voltage waveform, which

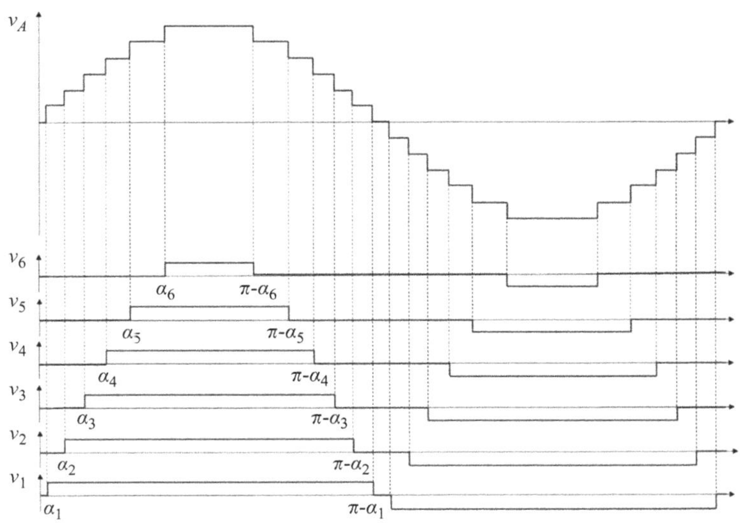

Figure 2.29 Synthesized voltage waveform in the phase A output of an MMC VSC

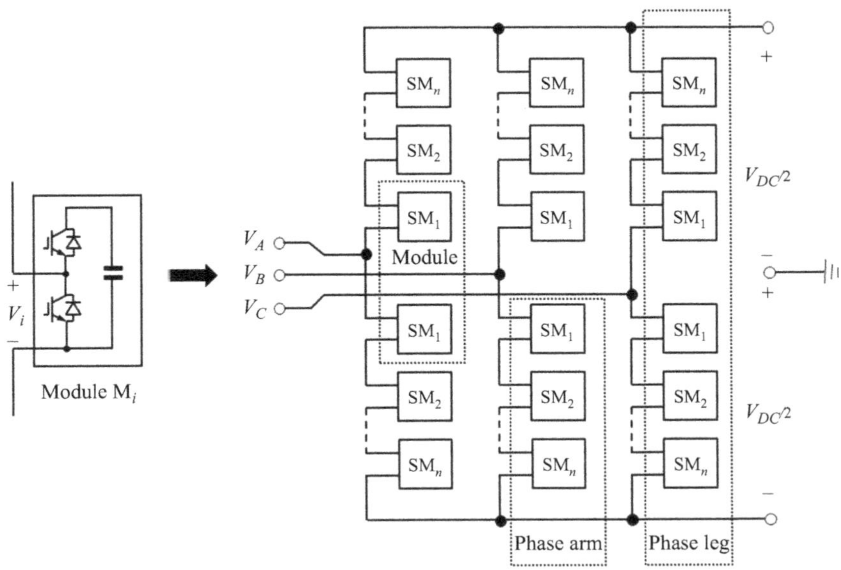

Figure 2.30 Schematic diagram of a three-phase MMC VSC

resembles as much as practicable a sinusoidal waveform, as illustrated in Figure 2.29 for phase A of a 6-level MMC scheme in Figure 2.30. The stepwise voltage waveform is made up of the voltage summation of individual voltage steps with preprogrammed, progressive firing delays with switching frequencies in the range of 150–200 Hz. In this waveform example, the firing angles are $\alpha_1 \ldots \alpha_6$. It should be noted that in utility-level applications, the number of levels is much higher to achieve a quasi-sinusoidal waveform, avoiding the need for tuned filters.

The low switching frequencies employed in MMC VSCs contrast with the switching frequencies employed in 2-level and 3-level VSCs, which are in the range of 1 to 2 kHz. One significant outcome of this is that the power losses were reduced to 1% as opposed to the 2% incurred by the 2-level and 3-level VSCs. Note that the 1% switching power loss is just above the 0.7% switching power loss incurred by thyristor-based power converters of comparable ratings [57].

The new VSC design was a modular, multilevel converter (MMC), based on a ground-breaking VSC design put forward in 2003 [58]. The basic structure is shown in the schematic diagram of Figure 2.30. This MMC VSC topology, together with alternative multilevel topologies, is discussed in Chapter 4.

2.3.3 Multiterminal HVDC systems

There are many practical applications where a DC power grid may have advantages over an AC power grid, such as submarine power grids. However, the formation of meshed, submarine DC power grids was not possible with the HVDC equipment and methods available before the turn of the millennium. Before then, the only options available for submarine DC power transmission were radial, muti-terminal HVDC systems [1]. An illustrative example of a three-terminal HVDC system connected in series is shown in simplified diagrammatic form in Figure 2.31.

The early HVDC systems used graded-electrode, mercury-arc valves, and converters that acted as current sources. Hence, the HVDC systems of this kind worked on current balances, limiting the structures to be radial, like the one shown in the schematic diagram in Figure 2.31 for the case of three terminals. During the 1970s, thyristor valves came to replace electrode-graded, mercury-arc valves in HVDC transmission applications. It should be noted that, in principle, at least, additional converters in series could be added to the schematic diagram of Figure 2.31. However, in practice, no current multiterminal HVDC installation

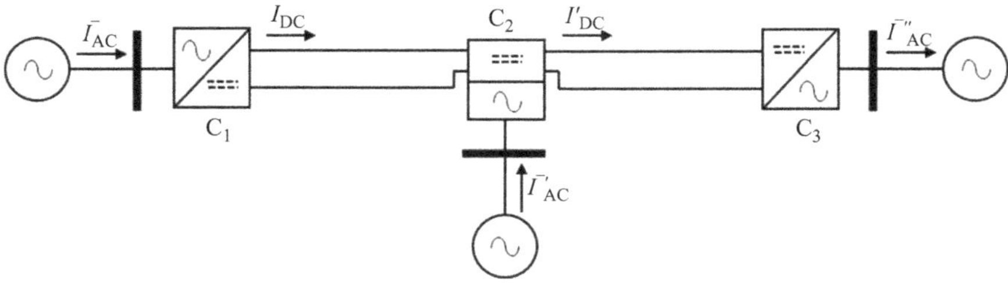

Figure 2.31 Radial, three-terminal HVDC system with three converter stations connected in series on their DC sides

exists with more than three converter stations, except the ring-connected *Kangbao-Zhangbei-Fengning-Changping* DC system in China that uses four VSC converters, commissioned in 2021 [59]. This is a ground-breaking development in HVDC technology at least on two counts: (i) the superior versatility of VSC converters over LCC converters has been demonstrated in practice since the formation of these kinds of meshed installations is not possible with LCC converters; (ii) working prototypes of DC circuit breakers have been demonstrated in this installation. Key points of the multiterminal HVDC systems that are either in operation or have been in operation, are penned below, following a chronological order.

(i) *The Kingsnorth–Beddington–Willesden DC system* [60]
The three-terminal Kingsnorth–Beddington–Willesden DC system in the Kent-London areas in the UK was commissioned in 1974. This HVDC system was rated at 640 MW and ±266 kV. It used electrode-graded, mercury-arc converters. The link was built to increase power injections into the Beddington and Willesden substations in London from Kingsnorth, Kent, without increasing short-circuit levels, which were already high in the London area. The converter station at Willesden was upgraded to use thyristor valves in 1981, but the three-terminal HVDC link was dismantled in 1987 following substantial reinforcements of the AC power grid in the London area.

(ii) *The Sardinia-Corsica-Italy (SACOI 2) DC system* [61]
In contrast to the multiterminal DC link in *(i)*, in 1992, a 300 MW, 200 kV, thyristor-based, three-terminal HVDC system was commissioned between Sardinia, Corsica, and mainland Italy in the Tyrrhenian Sea. It was termed SACOI 2 and was built on an existing point-to-point, monopole, HVDC link, with electrode-graded, mercury-arc converters, termed SACOI 1, commissioned in 1967; it connected the Sardinia AC power grid to the power grid in mainland Italy, passing through Corsican territory. SACOI 1's purpose was to provide frequency support to the Sardinian AC power grid by drawing on the power/frequency control capabilities of the DC link. The link was rated at 200 MW, 200 kV. SACOI 2 incorporates a 50 MW terminal at Corsica to supply the local electricity requirements. The total length of the HVDC system is 385 km, comprising overhead transmission lines and submarine cables. The submarine cable between Bastia, Corsica, and Salivoli, Tuscany, is 120 km long. SACOI 2 has been operating satisfactorily for more than 30 years, but new requirements have called for a major upgrade and modernisation of this three-terminal HVDC link, mainly to facilitate the use of renewable power sources and make the participating power grids more efficient, reliable, and sustainable. In September 2023, the Italian Ministry of the Environment and Energy Security gave the go-ahead to the SACOI 3 project. This is the name of the upgraded multiterminal HVDC system, which is set to increase its rating to 400 MW by erecting two new power converter stations next to the existing ones at both ends of the HVDC system, namely, Codrongianos, Sardinia, and in Suvereto, Tuscany.

Other than that, the existing infrastructure of SACOI 2 will be used to minimise the impact on the environment and ecosystems.

(iii) *The Québec-New England DC system* [62]
In 1992, a three-terminal, thyristor-based HVDC system with 2 GW capacity was commissioned in North America. It is a 1,500 km link between Québec, Canada, and New England, USA, which provides electrical energy to densely populated areas in Montreal and Boston with clean hydropower. The three HVDC terminals are in the large hydro-resource at Radisson, in Montreal, and in Boston.

(iv) *The Nanao Island DC system* [63]
The three radial, three-terminal HVDC terminal systems described in (i)–(iii) were the only multiterminal HVDC systems built before 2013. That year, a new breed of multiterminal HVDC systems came into being, using VSC converters and IGBT valves. The Nanao HVDC link in China was the first of such systems. It holds a radial, three-terminal HVDC structure like the one illustrated in Figure 2.31, but with IGBT-based converters instead of thyristor-based converters. Two VSC terminals are in the island of Nanao, namely, Quingao and Jinning, and the third terminal is in Guangdong (mainland). The installation includes 32 km of underground and sea cables. The main purpose of this radial, three-terminal VDC-HVDC link was to harvest wind power generated in the island and to transport it to the mainland. The converter's ratings are 50 MW, 100 MW, and 200 MW, respectively. The operating voltage of the bipolar DC link was $\pm$160 kV.

(v) *The Kangbao-Zhangbei-Fengning-Changping DC system* [59]
An obstacle in developing multiterminal HVDC systems of arbitrary configuration has been the lack of DC circuit breakers (CB_{DC}). After much R&D efforts in industry and academia, working prototypes are beginning to emerge. A case in point is the CB_{DC} developed by NR in China for use in the first ring-type, four-terminal VSC-HVDC system in the world. The layout of this VSC-HVDC system is shown in Figure 2.32.

The VSC terminals are in Kangbao and Zhangbei in the wind and solar power regions, in Fengning – the site of a sizeable pumped-storage hydro station, and in Changping which provides smooth electricity for Beijing's load centre. This is the first DC transmission system to use DC circuit breakers. It uses 16 DC circuit breakers, four per bipole transmission line. Note that in the illustrative Figure 2.32, the DC lines are overhead bipoles operating at $\pm$500 kV. The four SVC-type converters are rated at 3,000 MW, 3,000 MW, 1,500 MW, and 1,500 MW, respectively.

The DC CBs are rated at 500 kV and 3 kA. The breaking current is 25 kA with a breaking time of less than 3 ms. Fast overcurrent protection is essential in the protection of VSC-HVDC converters due to the rather low overcurrent withstanding capabilities of the IGBT valves. Hence, the sampling frequency of the DC measurement devices used in this installation is 100 μs [59]. These are early days in the development of DC CBs, and the

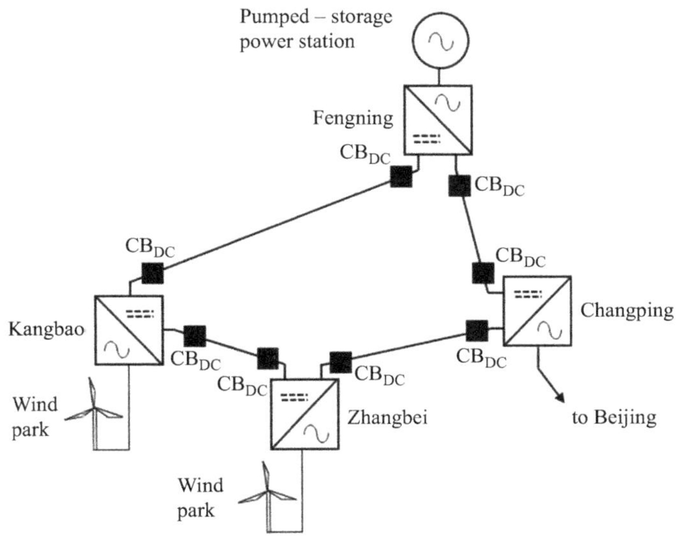

Figure 2.32 Four-terminal, VSC-HVDC system in a ring configuration, adapted from [59]

HVDC community will eagerly be waiting for performance reports and statistics of this multiterminal HVDC system and its protection system – CD CBs included – to gain confidence in the new ground-breaking technology before its widespread adoption in many other parts of the world [64].

All the major manufacturers of HVDC equipment are working on DC circuit breakers. A case in point is Hitachi Energy's proprietary units: one that has been verified at 350 kV and a unit at 525 kV lined up for verification procedures. A large German manufacturer is pursuing an alternative approach to DC protection, which is to embed the interrupting capabilities in the converters themselves. This involves developing MMCs with full-bridge modules capable of opposing and stopping GWs of DC power. To begin with, the price to be paid is the use of twice as many IGBTs per module, with their high price tag and high power loss [64].

It should be remarked that this kind of meshed HVDC system is only possible with VSC-type converters and is ruled out using LCC-HVDC converters – the ones that use thyristors and electrode-graded, mercury-arc valves.

(vi) *The Wudongde Multiterminal DC system* [65]

This is the first UHV (ulta-high voltage), multiterminal HVDC system, operating at ±800 kV. It was built to tap into the large hydropower resource in Yunnan in the West region and send it East to the Guanxi and Guangdong regions. It is a radial, three-terminal, bipolar HVDC system using one thyristor-based station and two VSC stations. The thyristor-based Kunming station – rated at 8 GW – sends DC power to the VSC stations at Liubei and

Longnen, rated at 3 GW and 5 GW, respectively, over 1,452 km of OH transmission lines.

(vii) *The Shetland Interconnector* [64]

This submarine VSC-HVDC link rated at 600 MW was built to evacuate the output power of the 443 MW Viking wind farm in Shetlands. The wind farm comprises 103 DFIG turbines. The link is a bipole operating at ±320 kV. At full wind farm capacity output, 90% of it is carried to the Scottish mainland using a 345 km long cable, of which 320 km runs undersea. The island's local demand is 44 MW.

The link between the VSC station in the Upper Kergord Valley, Shetland, and Staxigoe, Caithness at the most northern top of the Scottish mainland, was commissioned in 2024. It has created a three-terminal VSC-HVDC system by adding onto an existing VSC-HVDC cable further south, commissioned in 2018. This VSC-HVDC link between Caithness/Spittal and Moray/Blackhillock is 160 km long, rated at 1,200 MW, ±320 kV. This forms a Y-connected DC point, with the central node being a super DC bus in Caithness/Spittal with potential capabilities for connecting future DC links to evacuate offshore wind power from the North Sea and hydropower from Norway.

2.3.4 HVDC links around the world

Since 2010, slightly more than 100 HVDC installations, including LCC-HVDC and VSC-HVDC links, have been commissioned worldwide. It is significant that following the irruption of the MMC VSC technology in the MW-level power electronics scene, the market share of the VSC-HVDC technology has detonated. Out of the 100 HVDC links installed since 2010, there has been an almost 50% split between the use of these two flavours of the HVDC technology. However, a clear trend is emerging: (i) in transmission distances longer than 1,000 km and power ratings above 1,500 MW, LCC-HVDC links are unassailable; (ii) submarine links above 70 km long tapping onto the output of offshore windfarms are clearly within the realm of VS-HVDC transmission, due to the very fast controllability afforded by the VSC technology; (iii) in the area of power grids interconnections using asynchronous back-to-back links, the VSC-HVDC technology is starting to dislocate the LCC-HVDC technology; (iv) the VSC-HVDC is beginning to be used to form meshed DC power grids, something that is ruled out with the use of the LCC-HVDC technology; (v) in bilateral energy transactions between countries, VSC-HVDC is being favoured over AC tie-lines and LCC-HVDC links, as illustrated by the various VSC-HVDC links that have been built recently or are in the planning stage, in Europe. These points are illustrated below with reference to the list of HVDC installations.

Reputedly, the first power electronics valves suitable for high-voltage, high-power applications were developed rapidly following the end of WW2 in the form of electrode-graded, mercury-arc valves. These early valves were employed to build the first commercial HVDC transmission system in the world, linking mainland Sweden and the Isle of Gotland, in 1954. The success of this pioneering

Table 2.5 HVDC links using electrode-graded, mercury-arc valve power converters

Link (Country)	Year	Rating		Length (km)	Main characteristics
		Power (MW)	Voltage (kV)		
Sweden-Gotland (Sweden)	1954	20	100	96	Monopole submarine cable with seabed return.
English Channel (UK – France)	1961	160	±100	64	Bipole submarine cable laid out across a busy shipping corridor.
Volgograd-Donbass (USSR)	1965	720	±400	470	Overhead transmission line using a bipole configuration.
Inter-Island Link New Zealand	1965	600	±250	610	Bipole with 40 km of submarine cable and 570 km of overhead transmission line.
Konti-Skan (Sweden – Denmark)	1965	250	250	87	Monopole with earth and sea return, partly overhead and partly submarine.
Sakuma Interconnect (Japan)	1965	300	±125	0	Back-to-back bipole to interconnect two AC power grids operating at 50 Hz and 60 Hz, respectively.
SACOI 1 (Italy)	1967	200	200	121	Point-to-point, monopole DC link from Sardinia to Italy, passing through Corsica.
Vancouver Island (Canada)	1968	312	132	74	Monopole DC link running in parallel with two 132 kV AC cables. In 1979 a second pole of equal power was added and the link reconverted to use thyristor bridges.
Pacific Intertie (USA)	1970	1,440	±400	1,372	Long-distance, bipole DC link running in parallel with two AC 500 kV transmission lines. This was replaced in 1986 by a 1,920 MW link using thyristor valves.
Kingsnorth scheme (UK)	1974	640	±266	59 and 82	Bipole, 3-terminal DC link between Kingsnorth–Beddington–Willesden in the UK to reinforce the existing AC supply to areas of high load density in London, with no increase of short-circuit level.
Nelson River Bipole 1 (Canada)	1977	1,620	±450	895	Long-distance, bipole DC link. It used the largest electrode-graded, mercury-arc valves ever built, rated at 150 kV and 1.8 kA. This link was subsequently upgraded with thyristor valves to have a capacity of 6,500 MW.

installation encouraged the construction of similar installations worldwide during the following two decades, albeit with progressively higher ratings and longer transmission distances. Eleven HVDC links were built using electrode-graded, mercury-arc valves from 1954 to 1977 – Table 2.5 lists these installations.

These 11 HVDC installations have all been decommissioned or upgraded to use thyristor valves instead of the original electrode-graded, mercury-arc valves. They hold significant historical value in HVDC development.

The advent of reliable solid-state valves suitable for high-voltage, high-power applications – acting as a disruptive technology in power electronics – replaced the use of electrode-graded, mercury-arc valves in HVDC transmission links during the 1970s. Throughout that decade, both flavours of valves were employed in newly built HVDC schemes, but towards the end of the decade, the solid-state option prevailed on economic and reliability grounds [1]. The Nelson River Bipole 1 in Manitoba, Canada, was the last HVDC system employing electrode-graded, mercury-arc valves. Since then, a gradual but relentless move toward the use of thyristor valves occurred. This point is summed up by the inter-island New Zealand HVDC transmission system, the longest-serving electrode-graded, mercury-arc HVDC bipole. It was commissioned in 1965, and fully replaced by a thyristor-based HVDC bipole in 2013, increasing its original power-carrying capacity to 1,240 MW from its original 600 MW [66]. Perhaps this was the last high-voltage, electrode-graded, mercury-arc HVDC link to have existed in the world.

Compared to thyristor valves, the main drawbacks of electrode-graded, mercury-arc valves were their propensity to arc backs, scalability and modularity issues, longer maintenance times, and costlier manufacturing processes [1]. In both cases, the valves' firings employed a simple phase angle control, which was highly susceptible to commutation failures. The current version of the phase angle control method uses Phase Lock Oscillator (PLO) control to ensure reliable commutations [67]. Current thyristor capsules comprise wafer diameters in the range of 100 to 150 mm diameters, having repetitive peak off-state voltages of 7.5 and 8 kV and average on-state currents of 2,100 and 3,500 A, respectively. Both electrically triggered (EET) and light-triggered (LTT) thyristors are currently available. The latter option is designed with a light-sensitive gate that turns the device on via a light source. The ratings of individual thyristors would continue to increase. Still, progress in valve architecture, where individual devices are connected in series and in multibridge configurations, enabling the transmission of practically unlimited power, has somehow diminished the urgency for developing even higher-capacity thyristors. This case is epitomised by the largest ever-built power transmission system: the 12 GW Changji–Guquan HVDC link spanning 3,300 km across China and operating at $\pm$1,100 kV [68].

2.3.4.1 Long-distance, bulk-power transmission

Engineering praxis indicates that on-land power transmission involving distances longer than 700 km ought to be carried out using DC power transmission rather than AC power transmission based on system stability issues, reduced power losses, and overall economic grounds [1]. However, this threshold distance is substantially reduced if the HVDC link is wholly undersea or involves one or more submarine sections. The Inter-Island HVDC link in New Zealand is an excellent example of a hybrid land-submarine HVDC link [66].

In the early days of bulk-power transmission, the term applied to power quantities in the range of 500 MW. However, contemporary bulk-power transmission installations are in the range of 4,000 MW to 12,000 MW, involving transmission distances well beyond 700 km. For all their worth, VSC-HVDC links cannot compete at present with LCC-HVDC transmission in extra-long distance, ultra-high voltage, GW-level power transmission applications. This is epitomised by the Changji-Guquan HVDC link, rated at 12 GW and 1,100 kV, spanning 3,293 km across China, the largest and most advanced transmission system in the world. It uses leading-edge high-voltage, high-current, low-loss power converters, substations, and transmission line equipment, to achieve such record-breaking power throughputs. The link was built to export electrical power from the abundant energy resource region of Xinjiang, Changji in Northwest China to Xuancheng, Anhui in East China [68]. Table 2.6 lists the 10 largest HVDC installations in the world.

Approximately 50 LCC-HVDC links have been identified to exist, built in the period 1977–2024. It should be noted that AC power transmission would not have been competitive against the DC option for such long transmission distances, nor the new form of HVDC transmission based on the use of forced-commutated valves (e.g., IGBT), which are not yet developed to the point of being used in bulk-power, long-distance applications. At present, the largest VSC-HVDC transmission system is the Viking link between the UK and the Danish power grids using a submarine cable spanning 740 km.

2.3.4.2 Submarine HVDC power transmission systems

When the requirement is to transmit electrical power across seas, HVDC has been the preferred option, mainly if the undersea distances are more than 70 km. These include HVDC installations, which are partly undersea and partly over land. The undersea transmission distances are not yet extra-long distances since cable technology is still developing. The longest undersea transmission cable is just over 700 km long [69]. A preliminary search indicates that there are 79 DC links that can be classified under this category: 37 use thyristors, and 42 use IGBTs. It is significant that most LCC-HVDC links were built before 2010 (27 out of 37). On the other hand, most VSC-HVDC links have been built after 2010 (38 out of 42), with most of these using the MMC VSC technology. A niche application of VSC-HVDC is the evacuation of offshore wind power. Table 2.7 lists the VSC-HVDC links used in offshore wind power applications.

2.3.4.3 Asynchronous and frequency converters HVDC links

A third identified area of application of the HVDC technology is the interconnection of geographically close but electrically independent power grids. These power grids may have the same or different operating frequencies and may be strong or weak. It is on record that at least 52 such installations are in existence, scattered around 17 countries [70]. Table 2.8 gives a list of key back-to-back HVDC installations.

Table 2.6 Ten largest HVDC links in the world

Link (Country)	Year	Rating		Length (km)	Main characteristics
		Power (MW)	Voltage (kV)		
Changji – Guquan (China)	2019	12,000	±1,100	3,300	Today, this is the longest, largest, and most advanced HVDC link in the world.
Xingu – Rio (Brazil)	2019	4,000	±800	2,550	It starts at the Belo Monte hydropower resource and arrives at the Estreito substation in the Rio de Janeiro State.
Rio Madeira (Brazil)	2013	7,100	±600	2,375	It was built to export power from the Madeira River dam in the Amazon Basin to a major load centre in Porto Belo in eastern Brazil.
Hami – Zhengzhou (China)	2014	8,000	±800	2,210	It runs from Dananhu power station in Xinjiang in the Northwestern part of the country, to Henan in central China.
Jinping – Sunan (China)	2013	7,200	±800	2,090	It was built to transport electricity from the Guandi and Jinping hydroelectric power plants in the Sichuan province in Western China to the densely populated Jiangsu province on the East Coast of China.
Xingu – Estreito (Brazil)	2017	4,000	±800	2,077	It runs between the Belo Monte dam in the Para State and the Estreito substation in the city of Ibiraci in the Minas Gerais State.
Xiangjiaba – Shanghai (China)	2010	6,400	±800	1,980	It runs between the Xiangjiaba dam in the Sichuan province, to Shanghai City – a major load centre in the country.
NW Yunnan – Guangdong (China)	2018	5,000	±800	1,953	It starts in the Yunnan province and runs for almost 2,000 km to feed electricity into the Guangdong region – a major load centre in the southeast of the country.
Dianxibei – Guangdong (China)	2017	5,000	±800	1,950	It carries hydropower from the Yunnan province in Southwestern China to the Pearl River Delta region in Guangdong, one of the most densely urbanized regions in the world and a major economic and manufacturing hub.
Raigarh – Pugalur (India)	2019	6,000	±800	1,830	It carries power from the Jindal Tanmar coal-based power plant near Raigarh in Chhattisgarh, to the densely populated region of Pugalur in Tamil Nadu, India.

Table 2.7 HVDC links for the evacuation of offshore wind power

Link (Country)	Year	Rating		Length (km)	Converter topology
		Power (MW)	Voltage (kV)		
BorWin 1 (Germany)	2009	400	±150	200	Two-level from ABB.
BorWin2 (Germany)	2015	800	±300	200	MMC from Siemens.
DolWin 1 (Germany)	2015	800	±320	165	MMC from ABB.
HelWin 1 (German)	2015	576	±250	130	MMC from Siemens.
HelWin 2 (Germany)	2015	600	±320	130	MMC from Siemens.
SylWin 1 (Germany)	2015	864	±320	205	MMC from Siemens.
DolWin 2 (Germany)	2017	900	±320	135	MMC from ABB.
DolWin 3 (Germany)	2017	900	±320	160	MMC from Alstom.
Caithness-Moray link (UK)	2018	1,200	±320	160	MMC from ABB
BorWin 3 (Germany)	2019	900	±320	200	MMC from Siemens.
Rudong Connector (China)	2022	1,100	±400	100	Built by RXHK (China).
Sheyang Connector (China)	2022	1,100	±400	83	Built by RXHK (China).
DolWin 6 (Germany)	2023	900	±320	90	MMC from Siemens.
DolWin 5 (Germany)	2024	900	±320	130	MMC from Hitachi Energy-ABB.
Shetland Interconnector (UK)	2024	600	±320	260	MMC from Hitachi Energy-ABB.
Dogger Bank A (UK)	2024	1,200	±320	207	MMC from Hitachi Energy-ABB.
Dogger Bank B (UK)	2024	1,200	±320	207	MMC from Hitachi Energy-ABB.

Table 2.8 A sample of key back-to-back HVDC links

Link (Country)	Year	Rating		Valve type	Main characteristics
		Power (MW)	Voltage (kV)		
Sakuma (Japan)	1965 1993	300	125	Electrode-graded, mercury-arc Thyristor	First back-to-back HVDC link. It joined Japan's Western and Eastern power grids. The original link employed mercury-arc valves, replaced in 1993 by

(Continues)

Table 2.8 (Continued)

Link (Country)	Year	Rating		Valve type	Main characteristics
		Power (MW)	Voltage (kV)		
					thyristor valves. Three other back-to-back LCC-HVDC links exist in Japan: Shin-Shinano, Minami-Fukumitsu, Higashi-Shimuzu.
Eel River (Canada)	1972	320	230	Thyristor	This was the first all-thyristor, back-to-back HVDC link. It enabled the export of hydropower from Hydro-Québec to the Eastern North American power grid. There are four other back-to-back LCC-HVDC links in Canada: Chateauguay, Madawasca, McNeill, and Outaouais.
Mackinac (USA)	2014	200	70	IGBT	First back-to-back, VSC-HVDC of large size. Preferred over the LCC-HVDC technology to provide voltage and reactive power control with the future connection of wind power. Also, the stabilisation of a very weak network and a technical feature called black-start. There are 15 other back-to-back links in the USA, 14 of which are LCC-HVDC.
Vindhyachai (India)	1989	500	176	Thyristor	This back-to-back LCC-HVDC links India's Northern and Western 400 kV AC power grids. There are four other back-to-back LCC-HVDC links in India: Chandrapur, Vizag 1, Sasaram, Vizag 2.
Yu (China)	2017	4×1,250	±420	IGBT	Four back-to-back, VSC-HVDC links located in Chongqing, Southwestern China. In addition, five back-to-back LCC-HVDC links exist in China: Lingbao, Heihe, Gaoling, Shandong, and North-Central.

Most of these back-to-back installations use thyristor valves, except seven, which use IGBT valves. However, it is significant that the last five installations have used IGBT valves instead of thyristor valves.

2.3.4.4 DC links for bilateral energy transactions

In the last decade, several VSC-HVDC links have commenced operation or are in the construction and planning stage in Europe. These are underground or submarine DC cables linking the power grids of EC countries, asynchronously. They are used to conduct bilateral green energy transactions and to warrant security supply between EC nation-states at all times. Perhaps the clearest examples of this policy are the Celtic Interconnector between France and Ireland, following Brexit, and the Harmony Link between Poland and Lithuania, to further integrate the infrastructure of the Baltic States into Europe (Table 2.9).

Table 2.9 A sample of key VSC-HVDC links for bilateral energy transactions and security of supply in Europe

Link (Country)	Year	Rating		Length (km)	Main characteristics
		Power (MW)	Voltage (kV)		
INELFE (France-Spain)	2015	2,000	±320	64	INELFE was Europe's first onshore VSC-HVDC link, running from Santa Llogaia (Spain) and Baixas (France), designed as two identical parallel links each rated at 1,000 MW. It is used to trade electricity in a bilateral trading fashion.
Fréjus Road Tunnel (France-Italy)	2019	1,200	±320	190	This HVDC link was designed to interconnect the French and Italian power grids in an asynchronous manner. It enables the bidirectional power flow between the two grids.
Nemo Link (Belgium-England)	2019	1,000	400	140	This is the first link between Belgium and the UK. It has an annual transmission capacity of 8.76 TWh.
ALEGrO (Belgium-Germany)	2020	1,000	320	90	The DC link uses an underground cable between Lixhe, Belgium, and Oberzier, Germany. The link is used for the European electricity market (day-ahead), intraday capacities, and the allocation of long-term capacities.
IFA-2 (France-England)	2021	1,000	±320	204	Operated by NGC in England and RTE in France. IFA-2 enables the exchange of renewable energy between the two countries and advances the integration of the European power grid.

(Continues)

Table 2.9 (Continued)

Link (Country)	Year	Rating		Length (km)	Main characteristics
		Power (MW)	Voltage (kV)		
North Sea Link (Norway-England)	2021	1,400	±515	730	It is a commercial DC cable jointly owned and operated by Statnett and National Grid. It is a bidirectional DC link but is mostly used to import hydropower into the UK.
NORD.LINK (Norway-Germany)	2021	1,400	±525	623	This is a commercial DC cable built to transport the full production of the Mågeli hydropower plant in Norway to the north of Germany. The sole client is the German state-owned railways company, which has plans to become carbon-neutral by 2040.
Viking Link (Denmark-England)	2023	1,400	±525	740	This is currently the longest submarine power cable in the world. It serves the purpose of exchanging wind power between Denmark and England.
Green Link (Irland-Wales)	2024	500	±320	200	The DC cable running along the Irish Sea was built to export wind power from Ireland into Wales.
Celtic Interconnector (Ireland-France)	2026	700	±320	575	The link is built to enhance the security of supply for Irish customers. It will provide Ireland's only direct energy connection to an EU Member. It will have the ability to import and export 700 MW of electricity.
NeuConnect (Germany-England)	2028	1,400	±525	725	The project is financed by British and German private investors. It will be used to exchange excess wind power in either direction. Both countries have been expanding their wind turbine capacities and production has become more volatile. Electricity prices have been consistently lower in Germany.
Harmony Link (Lithuania-Poland)	2028	330	±320	700	The link is built to enhance the security of supply for Lithuanian customers. It will provide a direct energy connection to an EU Member.
FAB link (France-England)	2030	1,400	±320	210	This will be the third DC link between France and England. The other is IFA built in 1986 and IFA-2 built in 2021. This link will take the total import/export capacity between France and England to 4,400 MW.

2.4 Conclusions

The chapter has presented the main applications of MW-level power electronics converters embedded in electrical power grids, emphasising practical installations relating to power transmission. The AC and the DC options have been considered. These applications fall under the umbrella titles of FACTS and HVDC, respectively. AC voltage and active and reactive power control are the main concerns of FACTS. In contrast, HVDC focuses on transmitting power over very long distances, submarine power transmission, and asynchronous power grid interconnections. The main semiconductor valves employed in both FACTS and HVDC applications are Si thyristors and Si IGBTs. AC nodal voltage control is carried out with the long-enduring SVC and the newer, more advanced option termed STATCOM. Owing to the very large number of installations, it would be quite a difficult task to carry out a comprehensive survey of SVCs and STATCOMs installed around the world; instead, this chapter has described a series of crucial installations to exemplify the great progress achieved with the seamlessly merging of leading-edge power electronics technology and the conventional electrical power grid, as well as the wide range of applicability of this equipment. In the same tenor, more complex equipment such as the TCSC and the UPFC have been addressed. In both cases, comprehensive surveys of installations have been conducted, detailing the main characteristics of the equipment employed in each installation and the specific operational issues that they have resolved.

The first power electronics installations in the electrical power grid were HVDC. FACTS installations followed some thirty years later. This started in the middle of the last century with the availability of mercury-arc rectifiers and the need to carry electrical power under sea waters for several dozen kilometres – in retrospect; we may conclude that this was the first application wave of power electronics in electrical power systems. It was well understood at the beginning of last century, in academic circles, the theoretical and practical limitations that AC power transmission has for transporting electrical energy beyond 1/8 of the 50 Hz or 60 Hz wavelengths owing to the so-called Ferranti effect. It was also understood that transporting electrical power at 0 Hertz would be carried out without the Ferranti effect being present, but no practical means of rectification of the AC waveforms were available at the time. Once the principle of rectification/inversion of the waveform using mercury arc rectifiers was demonstrated in the submarine installation of Sweden-Gotland, it paved the way for power transmission beyond 1/8 of the waveform without resorting to series capacitive compensation. This groundbreaking technology was also used to interconnect AC power grids having different operating frequencies in Japan, e.g., 50 Hz and 60 Hz. Despite the great success achieved by the early HVDC installations and the need to implement similar installations in other parts of the world, mercury-arc rectifiers were expensive to manufacture and prone to commutation failures while in operation. However, the opportunity for global expansion of these installations presented itself with the arrival of a second groundbreaking development in power electronics: the entrance of solid-state power electronics at the MW level in the mid-1970s. This came in the form of the so-called SCR or thyristor. This is a development area that has not ceased to grow in terms of more powerful and versatile power electronics valves and applications in the greener and more reliable electrical power grid. Examples are the transport of GWs of hydro-electrical power across thousands of kilometres to reach some of the largest conurbations and industrial hubs, e.g., China, Brazil, and India. The third wave of applications arrived with the use of forced-commutation semiconductor valves, e.g., GTO and IGBT. This has resulted in even more advanced power electronics-based equipment and DC systems capable of tapping GWs of offshore wind power with utmost reliability, such as in the North Sea. This is carried out using point-to-point, MMC-VSC-HVDC systems. Perhaps a fact of greater significance is the intrinsic flexibility that VSC converters exhibit to make up multiterminal, meshed DC systems and to combine seamlessly with LCC converters to make up hybrid multiterminal DC systems. The applicability of these concepts has been demonstrated by industry-level prototypes built during the last decade in China. This is paving the way for a global electrical power grid, a fact that is beginning to happen with some of the mega interconnection projects currently on the drafting table: the DC super-grid in the North Sea covering an area of 570,000 km^2 and harbouring the largest repository of wind resource in the world, the Xlinks Morocco-UK Power Project spanning 3,800 km [71], and the Australian-ASEAN Power Link spanning 5,000 km [72].

References

[1] Arrillaga J. *High Voltage Direct Current Transmission*. Stevenage: Institution of Engineering and Technology, 1998.

[2] Miller TJE. *Reactive Power Control in Electric Systems*. New York: Wiley Interscience, 1982.

[3] Hingorani NG, and Gyugyi L. *Understanding FACTS: Concepts and Technology of Flexible AC Transmission Systems*. Piscataway, NJ: IEEE Press, 1999.

[4] Maza-Ortega JM, Acha E, García S, and Gómez-Expósito A. Overview of Power Electronics Technology and Applications in Power Generation, Transmission and Distribution. *Journal of Modern Power Systems and Clean Energy*, vol. 5, no. 4, 2017, pp. 499–514.

[5] Acha E, Roncero-Sanchez P, de la Villa-Jaén A, Castro LM, and Kazemtabrizi B. *VSC-FACTS-HVDC: Analysis, Modelling and Simulation in Power Grids*. New York: Wiley, 2019.

[6] Hingorani NG. Flexible AC Transmission. *IEEE Spectrum*, vol. 30, no. 4, 1993, pp. 40–45.

[7] Friedlander E. Static Network Stabilization, Recent Progress in Reactive Power Control. *GEC Journal of Science & Technology*, vol. 33, no. 2, 1966, pp. 58–65.

[8] Gyugyi L. Power Electronics in Electric Utilities: Static VAR Compensators. *Proceedings of the IEEE*, vol. 76, no. 4, 1988, pp. 483–494.

[9] Sumi Y, Harumoto Y, Hasegawa T, Yano M, Ikeda K, and Matsuura T. New Static Var Control Using Force-Commutated Inverters. *IEEE Transactions on Power Apparatus and Systems*, vol. 100, no. 9, 1981, pp. 4216–4224.

[10] Reichert K. Controllable Reactive Compensation. *International Journal of Electrical Power and Energy systems*, vol. 4, no. 1, 1982, pp. 51–61.

[11] Modern Power Systems, Hull, England, June 1994, pp. 45–49.

[12] Modern Power Systems. London Extension Uses Biggest Ever XLPE Cable. *Modern Power Systems*. Hull, England, April 2001.

[13] Aboytes F, Arroyo G, and Villa G. Application of Static Var Compensators in Longitudinal Power Systems. *IEEE Trans. on Power Apparatus and Systems*, vol. PAS-102, no. 10, 1983, pp. 3460–3466.

[14] National Center for Energy Control. Program for the Expansion and Modernization of the National Transmission Network and elements of the General Distribution Networks that correspond to the Wholesale Electricity Market. PRODESEN 2023–2037, Mexico.

[15] La Secretaría de Energía publica el Programa de Desarrollo del Sistema Eléctrico Nacional 2024–2038 | Secretaría de Energía | Gobierno | gob.mx (www.gob.mx)

[16] Gyugyi L. Dynamic Compensation of AC Transmission Lines by Solid State Synchronous Voltage Sources. *IEEE Transactions on Power Delivery*, vol. 9, no. 2, 1994, pp. 904–911.

[17] Xu L, Anaya-Lara O, Agelidis VG, and Acha E. Development of Prototype Custom Power Device for Power Quality Enhancement. *9th International Conference on Harmonics and Quality of Power*, 1–4 Oct. 2000, Orlando, Fl, USA.

[18] Schauder C, Gerhardt M, Stacey E, *et al.* Development of a ±100 Static Condenser for Voltage Control of Transmission Systems. *IEEE Transactions on Power Delivery*, vol. 10, no. 3, 1995, pp. 1486–1496.

[19] Schauder C, Gerhardt M, Stacey E, *et al.* Operation of ±100 MVAR Statcon, *IEEE Transactions on Power Delivery*, vol. 12, no. 4, 1997, pp. 1805–1811.

[20] Hanson DJ, Horwill C, Loughran J, and Monkhouse DR. The Application of a Relocatable STATCOM-based SVC on the UK National Grid System. In *Proceedings of the IEEE/PES Transmission & Distribution Conference and Exhibition*, Yokohama, Japan, Feb. 2003, pp. 1202–1207.

[21] Reed G, Paserba J, Croasdaile, *et al.* STATCOM Application at VELCO Essex Substation. *Proceedings of the IEEE/PES Transmission & Distribution Conference & Exhibition*, Atlanta, GA, USA, Aug. 2002, pp. 1133–1138.

[22] Rao H, Xu S, Zhao Y, *et al.* Research and Application of Multiple STATCOMs to Improve the Stability of AC/DC Power Systems in China Southern Grid. *IET Generation, Transmission & Distribution*, vol. 10, no. 3, 2016, pp. 3111–3118.

[23] Bailey A, Faulkner B, Chou H, *et al.* Installation and Commissioning of Mitsubishi Electric's MMC STATCOM (SVC-Diamond®) at Dominion Energy's Clover & Rawlings Substations. *Proceedings of the 2022 Grid of the Future*, CIGRE US National Committee, Chicago, IL, USA, Nov 2022, pp. 1–5.

[24] Sullivan D, Buterbaugh B, Paramalingam J, *et al.* Application of FACTS Devices for a Dynamic Power System within the USA. *Proceedings of the 2018 International Power Electronics Conference*, Niigata, Japan, May 2018, pp. 2329–2334.

[25] Hasler JP, Sneed T, Holmberg M, Lund J, and Näslund M. Power Quality Analysis and IEC Standard Evaluation Using Measurements and Simulations

in a STATCOM Application. *Proc. of the 47 CIGRE Session*, Paris, France, Aug. 2018, Paper C4-114.

[26] Halpen L. STATCOM in Panama-First Application of this Technology in Central America. *Proceedings of IEEE 40th Central America and Panama Convention (CONCAPAN)*, Panama, Panama, Nov. 2022, pp. 1–6.

[27] Hingorani NG. A New Scheme for Subsynchronous Resonance Damping of Torsional Oscillations and Transient Torque—Part I. *IEEE Transactions on Power Apparatus and Systems*, vol. PAS-100, no. 4, 1981, pp. 1852–1855.

[28] Larsen EV, Clark K, Miske SA, and Urbanek J. Characteristics and Rating Considerations of Thyristor Controlled Series Compensation. *IEEE Transactions on Power Delivery*, vol. 9, no. 2, 1994, pp. 992–1000.

[29] Christl N, Hedin R, Sadek K, *et al.* Advanced Series Compensation (ASC) with Thyristor Controlled Impedance. *International Conference on Large High Voltage Electric Systems (CIGRÉ)*, paper 14/37/38-05, Paris, Sep. 1992.

[30] Helbing SG, and Karady GG. Investigations of an Advanced Form of Series Compensation. *IEEE Transactions on Power Delivery*, vol. 9, no. 2, 1994, pp. 939–947.

[31] Kinney SJ, Mittelstadt WA, and Suhrbier RW. Test Results and Initial Experience for the BPA 500 kV Thyristor Controlled Series Capacitor Unit at Slatt Substation, Part I – Design, Operation and Fault Test Results. *Flexible AC Transmission Systems (FACTS 3): The Future in High Voltage Transmission Conference*, Baltimore, MD: EPRI, Oct. 1994.

[32] WG B4.49. Performance Evaluation and Applications Review of Existing Thyristor Control Series Capacitor Devices – TCSC. CIGRE Report 554, Oct. 2013.

[33] Gama C. Brazilian North-South Interconnection Control-Application and Operating Experience with a TCSC. *1999 IEEE Power Engineering Society Summer Meeting. Conference Proceedings (Cat. No. 99CH36364)*, Edmonton, Alberta, Canada, vol. 2, 18–22 July 1999, pp. 1103–1108.

[34] Gama C, Angquist L, Ingestrom G, and Noroozian M. Commissioning and Operative Experience of TCSC for Damping Power Oscillation in the Brazilian North-South Interconnection. *Paper 14-104, CIGRE Session 2000*, Paris.

[35] ABB TCSC Thyristor Controlled Series Compensation – Keeping grids together. A02-0158.pdf (abb.com).

[36] ABB to Improve Reliability of Power Grid in South Korea. Trade Press Release, Zurich, Switzerland, 26 Feb. 2018. ABB to improve reliability of power grid in South Korea.

[37] Minh-Quan T, Minh-Chau D, Seok-Ju L, *et al.* Analysis and Mitigation of Sub-synchronous Resonance in a Korean Power Network with the First TCSC Installation. In *Special Issue on Grid Services in Electric Power Systems published in Energies*. 2020.

[38] Hor C, Finn J, Thumm G, and Mortimer S. Introducing Series Compensation in the UK Transmission Network. *9th IET International Conference on AC and DC Power Transmission (ACDC 2010)*, London, 2010, pp. 1–5.

[39] Datta U. Enhanced Operational Performance of SIIRTOVERKKOMALLI Using Static Compensators and BESS Equipment. M.Sc. Thesis, Tampere University of Technology, Tampere, Finland, 8 June 2016.
[40] Jones KM. AC System Network Control Methods Compensation. *Symposium on Power System Dynamics, Control and Compensation*. New Castle Upon Tyne: IEE Northeastern Centre, 1974.
[41] Aboytes F, and Arroyo G. Security Assessment in the Operation of Longitudinal Power Systems. *IEEE Transactions on Power Systems*, vol. 1, no. 2, 1986, pp. 225–232.
[42] Angeles-Camacho C. Phase Domain Modelling and Simulation of Large-Scale Power Systems with VSC-based FACTS Equipment. *PhD Thesis*, March 2005. The University of Glasgow, Glasgow, Scotland, UK.
[43] Schauder C, Stacey E, Lund M, *et al.* AEP UPFC Project: Installation, Commissioning and Operation of the ±160 MVA STATCOM (Phase I). *IEEE Transactions on Power Delivery*, vol. 13 no. 4, 1998, pp. 1530–1535.
[44] Wang X, Wang H, Yang J, *et al.* Application of 500 kV UPFC in Suzhou Southern Power Grid. *Journal of Engineering*, vol. 2019, no. 16, 2019, pp. 2580–2584.
[45] Cui Y, Yu Y, Bao W, *et al.* Analysis of Application Effect of 220 kV UPFC Demonstration Project in Shanghai Grid. *Journal of Engineering*, vol. 2019, no. 16, 2019, pp. 758–762.
[46] Edison Tech Center. History of Transformers. History of Transformers (edisontechcenter.org)
[47] Asplund G, Eriksson K, Svensson K. DC Transmission Based on Voltage Source Converters. *CIGRE SC14 Colloquium*, Johannesburg, South Africa, 1997.
[48] Yajun MA, Hua L, Zhe W, and Zuyao Z. Stability Analysis of Modular Multilevel Converter Based on Harmonic State-Space Theory. *IET Power Electronics*, vol. 12, no. 2019, pp. 3859–4087.
[49] Changji-Guquan UHVDC Transmission Project – NS Energy (nsenergybusiness.com)
[50] Cable to the UK – Projects – Statnett (archive.org)
[51] ABB Mackinac – ABB HVDC Reference Projects in North America (HVDC References) (archive.org)
[52] Tjaereborg | Hitachi Energy
[53] TenneT TSO GmbH – BorWin2 (archive.org)
[54] Paulsson L, Ekehov B, Halén S, *et al.* High-frequency Impacts in a Converter-based Back-to-back Tie: The Eagle Pass Installation. *IEEE Transactions on Power Delivery*, vol. 18, no. 4, 2003, pp. 1410–1415.
[55] ABB commissions power link to Valhall offshore platform (archive.org)
[56] Abrahamsson B. Estlink HVDC Light in 19 months. 2007 *IET Power Convention*, 2007, pp. 1–34.
[57] Westerweller T, Friedrich K, Armonies U, Orini A, Parquet D, and Wehn S. Trans Bay Cable – World's First HVDC System Using Multilevel Voltage-sourced Converter, *CIGRÉ Session, Paris*, 2010, paper reference B4-101.

[58] Lesnicar A and Marquardt R. An Innovative Modular Multilevel Converter Topology Suitable for a Wide Power Range. 2003 *IEEE Bologna Power Tech Conference Proc.*, Bologna, Italy, 2003, vol. 3, pp. 6.
[59] NR's VSC-HVDC Solution ±500 kV Zhangbei DC Grid for boosting large scale hybrid renewables penetration 1557822944108hgcuh.pdf (nrec.com)
[60] Calverley TE, Gavrilovic A, Last FH, and Mott CW. The Kingsnorth–Beddington–Willesden DC Link, *CIGRÉ Session, Paris*, 1968, paper 43–04.
[61] Arcidiacono V, Corsi S, Pincella C, *et al.* System Commissioning Tests for the SACOI-2 HVDC Three-terminal Link, *CIGRÉ Session, Paris*, 1994, paper reference 14–107.
[62] Québec – New England | Hitachi Energy
[63] People's Daily: The world's first multi-terminal flexible DC transmission project was put into operation – Media attention – 2013 – China Southern Power Grid News – China Southern Power Grid (archive.org)
[64] Fairley P. High-Voltage DC Power Roars Ashore in Europe. *IEEE Spectrum*, 2024, pp. 28–31.
[65] RXHK: WuDongDe UHVDC Scheme Commercial Operation
[66] New Zealand Inter-Island HVDC | Hitachi Energy
[67] Ainsworth JD. The Phase Locked Oscillator – A New Control System for Controlled Static Converters. *IEEE Transactions on Power Apparatus and Systems*, vol. PAS-87, no. 3, 1968, pp. 859–865.
[68] Changji-Guquan UHVDC link | Hitachi Energy
[69] Hitachi Energy hands over North Sea Link, the first interconnector between Norway and the UK, exchanging renewable energy
[70] HVDC PROJECTS LISTING Prepared for the HVDC and Flexible AC Transmission Subcommittee of the IEEE Transmission and Distribution Committee. Wayback Machine (archive.org)
[71] Morocco-UK Power Project – Xlinks
[72] Whitlock R. Sun Cable Announces Global Expert Team to Deliver the Australia–Asia PowerLink Project. *Renewable Energy Magazine*. 23 Nov. 2022.
[73] Renz BA, Keri A, Mehraban AS, *et al.* AEP Unified Power Flow Controller Performance. *IEEE Transactions on Power Delivery*, vol. 14 no. 4, 1999, pp. 1374–1381.
[74] Nilsson SL, Xu S, Lei B, Deng Z, and Andersen BR. Applications Examples of UPFC and Its Variants. *Flexible AC Transmission Systems*, Paris: CIGRE-Springer, 2020, Chap. 15, pp. 645–703.
[75] Chang BH, Kim SY, Yoon JS, Choo JB, Lima LTG, and Feltes JW. Study of UPFC Operation Strategies in KEPCO System (II)-Dynamic Characteristics. *IFAC Proceedings Volumes*, vol. 36, no. 20, 2003, pp. 1087–1095.
[76] Fardanesh B and Schuff A. Dynamic Studies of the NYS Transmission System with the Marcy CSC in the UPFC and IPFC Configurations. *Proceedings of the IEEE/PES Transmission & Distribution Conference & Exhibition*, Dallas, TX, USA, Sep. 2003, pp. 1126–1130.

Chapter 3

Current status of power semiconductor valves

Norberto Garcia-Barriga[1]

The pioneering work on solid-state electronic devices in the Bell Laboratories in the 1940s and 1950s spearheaded the commercial developments of the Silicon-Controlled Rectifier (SCR) in GE and other manufacturers of power semiconductor devices in the USA and Europe in the following decades. The SCR found a natural niche application in the nascent area of HVDC power transmission, displacing the bulkier mercury-arc valves originally employed in such installations. Over the decades, the SCR or thyristor – as it is known more generally – has widened its use in the global electricity supply industry very substantially. Silicon (Si), as a semiconductor material, is ubiquitous in today's processes relating to electrical energy: industrial processes, all forms of transport, power generation, power transport, distribution, and its usage. Si has served very well the needs of the electricity supply industry so far. Still, as the growing electricity demand continues unabated, coupled with a wider social awareness of being supplied with electricity produced with a minimum ecological footmark and more competitive prices, the need for increased efficiencies and economies is becoming imperative in the industry. Green electricity is being transported over thousands of kilometers from remote sites to large conurbation areas to fulfill such requirements, using state-of-the-art power transmission technologies; however, these may be approaching their physical limits. This chapter examines the Si material and most popular Si-based power electronics devices. Then, it draws attention to what may become, in the fullness of time, the natural replacement of Si devices and valves in the handling of larger volumes of electrical power, namely, a range of wide bandgap semiconductor materials, which today are in their experimental and prototype stage.

3.1 Introduction

Conventional power grids use AC overhead transmission lines and large centralized power generation stations and, by in large, electromechanical servomechanisms for control purposes. Power semiconductor devices are found in relatively few but key applications, such as exciter systems for synchronous generators, flexible AC

[1]Faculty of Electrical Engineering, University of Michoacan, Campus Universitario S/N, México

transmission systems (FACTS), and HVDC (High Voltage Direct Current) power transmission. However, this scenario is changing rapidly; power semiconductor-based circuits are essential for linking today's renewable power generation centers to the power grid. Examples are the large-capacity offshore wind farms in the North Sea and the huge solar PV power plants in northern China. In the future, it may be the widespread use of marine power plants in the North Atlantic and the South Pacific oceans. Smart Grids will enable the coexistence of smart, self-controlling power grids with large, centralized power plants, the widespread use of embedded generation, energy storage, custom power solutions, and flexible AC and DC transmission and distribution systems. Large-scale inverter-based renewable energy sources will be a key component of the smart grid concept to reduce fossil fuel dependency and emissions of greenhouse gases [1].

The energy shift from conventional fossil fuels usage to electricity in combination with economic growth and relentless population increases, are harbingers for the continuous increase of electricity consumption. As the electricity supply industry moves to future scenarios of even larger amounts of electric energy demand, the utmost efficient generation, transportation means, and distribution practices of this vital fluid become a matter of greater strategic importance than today. In such a scenario, power electronics is a key technology to increase the energy efficiency of power grids and to enable the grid integration of renewable energy sources, energy storage systems, and electric vehicle recharging stations. There are already several power electronics solutions that are being used to underpin such objectives, but the full potential of solid-state power electronics applications in electrical power networks is yet to be unleashed [2].

Until now, the power electronics industry has been dominated by semiconductor valves and bridges using silicon (Si) and Germanium (Ge) materials. These are both mature technologies with cost-efficient and high-volume production. However, in applications of power electronics usage in the kW range and above, this has been limited to Si. During the past four decades, the drive has been in the development of wide bandgap (WBG) semiconductor materials based on silicon carbide (SiC) and gallium nitride (GaN). Both development initiatives have advanced closer to maturity, with good commercialization prospects in different low and medium-power applications, owing to their superior performance in higher energy handling, power density, and energy efficiency [3]. Nonetheless, the development of SiC and GaN materials suitable for use in the MW range is still in its infancy. In recent years, ultra-wide bandgap (UWBG) semiconductor materials such as Gallium Trioxide (Ga_2O_3), hexagonal boron nitride (h-BN), Aluminium Nitride (AIN), and Diamond have received research attention. They exhibit improved characteristics, such as high critical electric fields and extra-large bandgaps. Despite their great potential, research on UWBG semiconductor materials started only recently and represents a challenging research endeavor in the field of power electronics.

These developments align well with the current social, economic, and environmental concerns of today's global community. Power semiconductor devices are key technological components, directly impacting the sustainable development goals (SDG) put forward by the United Nations (UN). Power semiconductors enable the creation of economic progress in harmony with the environment and society, at least on the following counts [4,5]:

- UN-SDG-7 (affordable and clean energy): Transform renewable energy sources to electric energy, transport and distribute renewable energy to consumers, and ensure high energy efficiency.
- UN-SDG-8 (decent work and economic growth): Enable new industries and business models based on emerging power electronics solutions such as electric vehicles and green hydrogen.
- UN-SDG-9 (industry, innovation, and infrastructure): Development of environmentally friendly technologies to address society's demands, foster innovation, and promote sustainable industry and infrastructure.
- UN-SDG-12 (responsible consumption and production): Reduction of dependency on fossil fuels by the application of renewable energy, electric vehicles, and green hydrogen.

3.2 Historical developments and first applications

A timeline of the development of semiconductor devices is presented in Figure 3.1, which summarizes milestones in power electronics valves. Further, key applications of power semiconductor devices are summarized in this figure. The mercury arc rectifier (MAR) was the first power electronic device developed in 1902 [6]. The first Silicon Controlled Rectifier (SCR) was conceptualized and manufactured in the USA in 1957, in the Bell Telephone Laboratories and in General Electric, respectively [7]. The SCR was invented to replace mercury-arc valves and once it had reached production lines, was quickly adopted in controlled rectifier circuits in industry and city transport and opened doors to wider applications using phase control devices. However, its inability to be turned off proved to be a problem for applications beyond the line-commutated mode, and emphasized the need for solid-state devices which could be turned off.

Ten years earlier, in 1948, Bardeen and Brattain had developed the point-contact transistor (PCT) at the Bell Telephone Laboratories [8]. The PCT was the

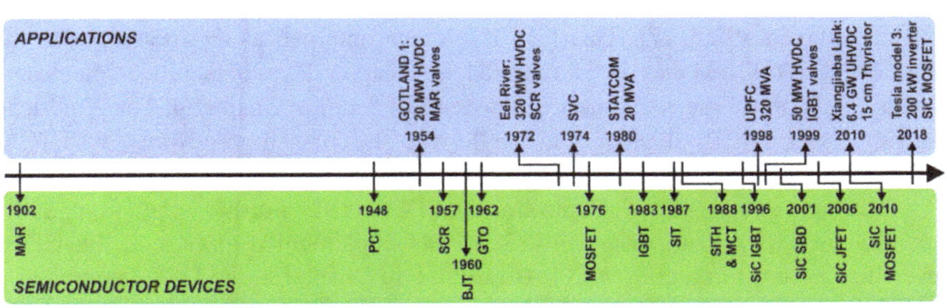

Figure 3.1 Timeline of the development of modern semiconductor devices and first applications

first solid-state device based on modern single-crystal semiconductor materials. With the advent of the metal oxide semiconductor (MOS) technology for digital circuits, a new class of power devices, named Metal Oxide Semiconductor Field Effect Transistor (MOSFET), was developed in the 1970s for high-frequency applications and low operating voltages. A prominent member of the transistor family is the static induction transistor (SIT), which was introduced in 1987 [9], just a year before the introduction of the static induction thyristor (SITH) and the MOS-controlled thyristor (MCT) in 1988 [10]. Another class of silicon devices evolved in the 1980s, where the insulated gate bipolar transistor (IGBT) represents the most successful power transistor in this class of devices [11]. IGBTs have been applied in medium and high-power applications because of their good power density and simple interface. As early as 2001, the first SiC Schottky barrier diode (SiC SBD) was produced with outstanding characteristics, such as high blocking voltage, better thermal stability, and very short reverse recovery time [12]. The SiC junction field-effect transistor (JFET) was developed since the 1990s and the first commercial SiC JFET device was available in around 2006. Similarly, the SiC MOSFET, the most market-oriented SiC device at this stage, was developed in 1987 and the first commercial devices were produced around 2010 [12]. Although SiC IGBTs have not been commercialized yet, the first SiC IGBT was fabricated in 1996 [13].

Over the years, electrical power systems have witnessed the introduction of pioneering projects based on state-of-the-art power converter technology. Figure 3.1 outlines ground-breaking developments in power electronics applications. The Gotland 1 project was the first commercial HVDC system in the world. It was a submarine link using mercury-arc bridges and phase control. The Gotland HVDC link began operation in 1954 and had a rating of 20 MW, 100 kV [14]. The first HVDC project to incorporate thyristor valves was the Eel River project, commissioned in Brunswick, Canada, in 1972. This HVDC link was a back-to-back scheme with a capacity of 320 MW and operating at 80 kV [15]. These HVDC installations have been described in Chapter 2. The Static VAR Compensator (SVC) is a shunt-connected FACTS device that was first demonstrated in 1974 by General Electric [16]. As already outlined in Chapter 2 and addressed in detail in Chapter 4, the SVC comprises a thyristor-controlled reactor, a thyristor-switched capacitor, and filters. The first prototype of the static synchronous compensator (STATCOM) was developed in Japan in 1980 [17]. This FACTS device emerged as an alternative to the SVC. The STATCOM has been outlined in Chapter 2 and its power electronic structure and control mechanisms are addressed in detail in Chapter 4. The American Electric Power (AEP) deployed in 1998 was the first high-power STATCOM installation. This was part of a larger FACTS project involving the full development of a Unified Power Flow Controller (UPFC) [18]. Further, the first Static Synchronous Series Compensator (SSSC), was implemented as part of the UPFC power flow control project by AEP. As outlined in Chapter 2, the UPFC combines a STATCOM and a series of voltage source converters working under a unified control scheme to provide an independent power flow control of active and reactive power, as well as voltage magnitude control. The Swedish island of Gotland is the site of the world's first commercial HVDC transmission based on voltage source converter (VSC) technology and IGBT valves, which was rated 50 MW and commissioned in 1999 [15]. The first UHVDC transmission link connecting Xiangjiaba-Shangai started commercial service in 2010 [19]. This 6.4 GW transmission scheme operates at ±800 kV and represents the first application of 15 cm diameter power thyristors. In 2018, Tesla introduced the first commercial electric powertrain with a 200 kW inverter using SiC MOSFETs [20].

3.3 Si power devices

Modern power electronic devices can be classified in several ways: based on the number of *pn* junctions, their degree of controllability, their bidirectional capability, their gate signal requirements, or the number of terminals. A semiconductor power device is the kernel element of the power electronics system, which is used as a rectifier, an inverter, or a switch. Considering the number of *pn* junctions, power electronic devices are broadly divided into three groups, namely: 1-junction/2-layer, 2-junction/3-layer, and 3-junction/4-layer. As it can be observed in Figure 3.2, the diode is the sole occupant of the 1-junction or 2-layer group. The 2-junction or 3-layer group consists of transistors, while the 3-junction or 4-layer group contains transistors and the thyristor family of power devices.

3.3.1 Diode

A diode is the simplest unidirectional electronic switch, which is formed from two layers of semiconductor materials and one *pn* junction. The symbol and structure of the semiconductor diode are depicted in Figure 3.3. This two-terminal device is an

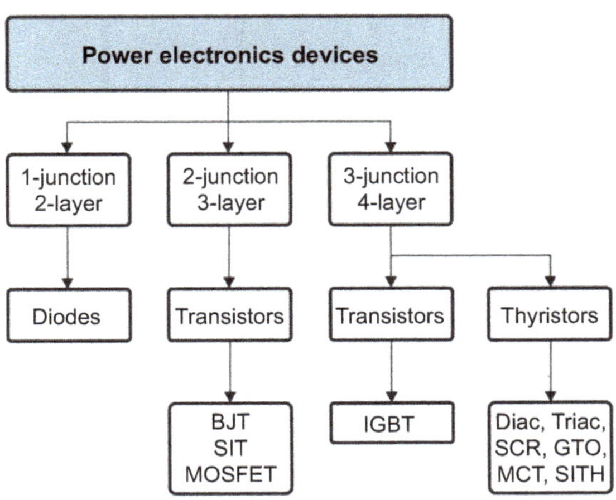

Figure 3.2 Classification of Si power electronic devices according to the number of pn junctions

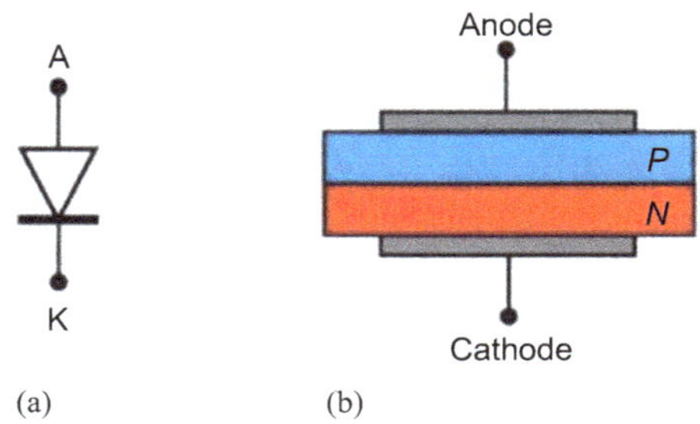

Figure 3.3 Semiconductor diode: (a) symbol and (b) composition

uncontrolled switch because the turn-on and turn-off are determined solely by voltages and currents in the power circuit. The ideal diode behaves as a short circuit and as an open circuit when it is forward-biased and reversed-biased, respectively. In the nonideal case, the diode shows a reverse recovery current characteristic. Diodes based on WBG materials, such as SiC, have very little reverse recovery, a characteristic that makes them good candidates for high-power applications [21].

The most important parameters of the diode are the following:

- Breakdown voltage: It is the reverse voltage that makes the diode conduct when it is reverse-biased.
- Reverse current: It is the current for a given voltage below the breakdown voltage.
- Forward voltage: It represents the voltage drop when the diode is forward-biased.
- Reverse recovery time: It is the time needed to stop conducting when the diode is reverse-biased.
- Forward recovery time: It is the time required to conduct when the diode is switched from reverse to forward bias.

Diodes are classified into six major categories, namely:

- Schottky diode: This diode is formed from a metal-*n* semiconductor junction instead of the conventional *pn* junction. It shows a low on-state voltage drop and very fast switching action. However, its breakdown voltage is much lower than that of the power diode.
- Zener diode: This type of diode operates in the reverse-biased mode and is used in voltage regulation applications. When it is reversed-biased above a breakdown voltage known as Zener voltage, a current called Zener current conducts through it.
- Small signal diode: These diodes have low reverse breakdown voltage and low forward current capabilities. It is suitable for low-power applications.
- Power diodes. There are two types of power diodes: line-frequency diodes with slow recovery times and fast recovery diodes. While the line-frequency diodes

are optimized for line-frequency AC–DC power rectifiers requiring high voltage and high currents, fast recovery diodes are utilized in high-frequency applications.

- Fast recovery diodes: These diodes are suitable for high-frequency switching applications due to their small turn-on and turn-off times, which are in the range of nanoseconds and microseconds, respectively. Nevertheless, they have high conduction losses.
- Fast-switching diodes: These diodes are optimized for high-frequency applications. They have very small recovery times in the order of a few microseconds to a few nanoseconds.

3.3.2 Thyristor family

The thyristor family of semiconductor switching devices is shown in Figure 3.4. As can be seen, these power electronic switches are four-layer semiconductor devices with alternating *pn* junctions. Some of these devices offer controlled turn-on of currents but not their interruption, while others are fully controllable. Some of the members of the thyristor family are the diode AC switch (DIAC), the triode AC switch

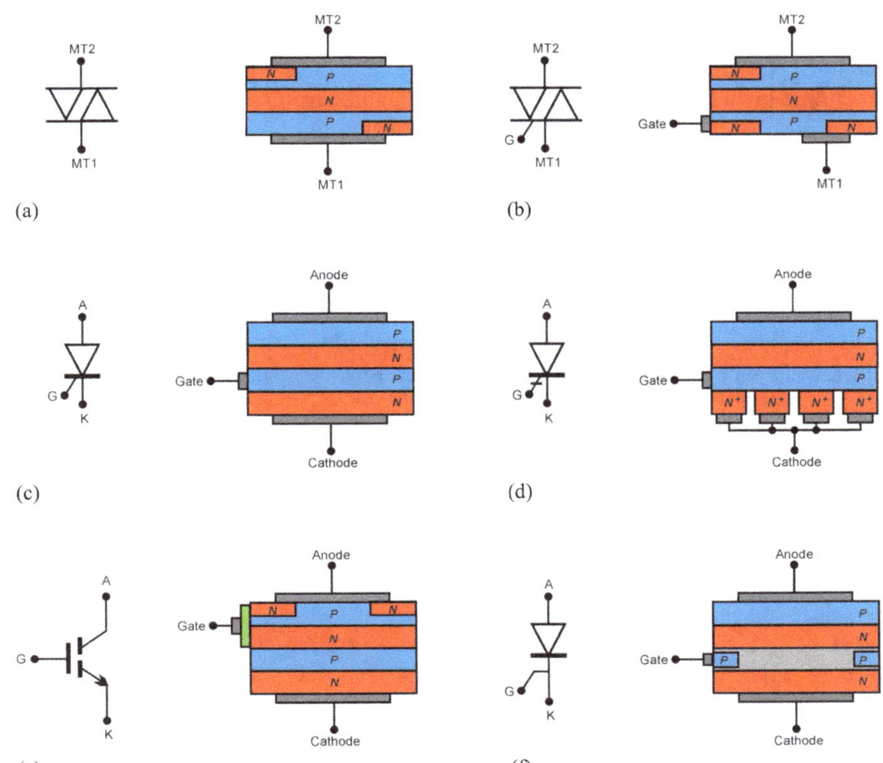

Figure 3.4 Thyristor family semiconductor switching devices: (a) Diac, (b) triac, (c) SCR, (d) GTO, (e) MCT, and (f) SITH

(TRIAC), the silicon-controlled rectifier (SCR), gate turn-off thyristor (GTO), the MOS-controlled thyristor (MCT), and the static induction thyristor (SITH). The thyristor family devices are used to switch AC currents with high power requirements [22].

3.3.2.1 DIAC

The diode AC switch (DIAC) is an uncontrolled switch whose turn-on and turn-off cannot be controlled. Figure 3.4(a) depicts the schematic symbol and cross-section of a DIAC. The structure of the DIAC is very similar to that of a TRIAC, except that it possesses no gate.

3.3.2.2 TRIAC

The triode AC switch (TRIAC) is a semi-controllable switch that can be turned on but cannot be turned off directly from the gate. It can be observed from Figure 3.4(b) that the TRIAC consists basically of two thyristors in an antiparallel configuration. It consists of three terminal layers, with the anode layer connected to the main terminals MT1 and MT2, and the cathode connected to the gate terminal G.

3.3.2.3 SCR

A silicon-controlled rectifier (SCR) is a four-layer three-terminal semiconductor device, which consists of alternating *p*-type and *n*-type semiconductor materials (see Figure 3.4(c)). While a *p*-type layer acts as an anode and an *n*-type layer acts as a cathode, the *p*-layer closer to the cathode functions as a gate. An SCR is a unidirectional switch like a diode but possesses a gate terminal to control its turn-on. The turn-off of the SCR is determined by the external power circuit.

3.3.2.4 GTO

The Gate Turn-Off Thyristor (GTO) of Figure 3.4(d), like the SCR, can be turned on by a gate current when the anode-to-cathode voltage is positive. Nevertheless, unlike the SCR, the GTO can be turned off using a negative gate current. While the duration of the negative gate turn-off current is brief, its magnitude must be very large compared to the turn-on current. Therefore, the GTO is a suitable device for applications where control of both turn-on and turn-off capability is required.

3.3.2.5 MCT

The MOS-controlled thyristor (MCT) is a device that combines the characteristics of a MOSFET and a SCR. The MCT shows a similar functionality to the GTO, but it requires a lower turn-off gate current. However, its main drawback is its low voltage blocking ability. Figure 3.4(e) depicts the structure of the MCT.

3.3.2.6 SITH

The static induction thyristor (SITH) is a fully controlled device that can be turned off in a similar way to the GTO (see Figure 3.4(f)). However, a SITH power device shows a higher conduction drop compared to the GTO.

3.3.3 Transistors

Figure 3.5 compares the structures of different types of Si power transistors.

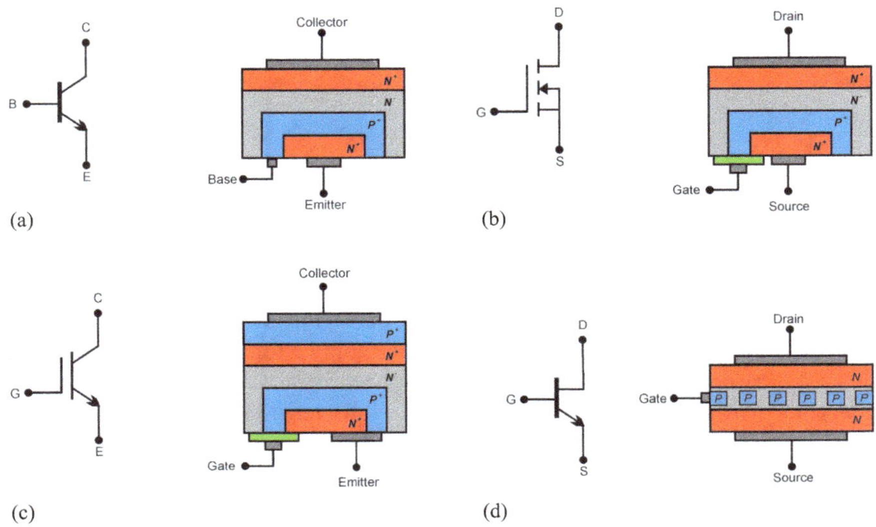

Figure 3.5 Simplified structure of transistors: (a) BJT, (b) MOSFET, (c) IGBT, and (d) SIT

3.3.3.1 Bipolar Junction Transistor (BJT)

Bipolar Junction Transistor (BJT) is a three-layer device that uses both electrons and holes as the charge carriers. Figure 3.5(a) depicts the schematic symbol and the structural diagram of an NPN BJT. In the NPN transistor, the current flows from the collector to the emitter once the base terminal is supplied with a positive voltage. Since electrons are majority carriers and holes are minority carriers in an NPN transistor, the BJT is turned on when the electrons enter the base region.

3.3.3.2 MOSFET

Metal-oxide-semiconductor field-effect Transistor (MOSFET) is a voltage-controlled unipolar device that only uses the majority of carriers during conduction. Since only the majority of carriers contribute to the current flow, MOSFETs have the highest switching speed. Figure 3.5(b) shows the cross-sectional view for a power MOSFET. The drain and source metal contacts are connected directly to N+ layers. The *n*-drift region provides the forward voltage-blocking property, where the lower doping N- is needed to provide higher voltage-blocking capabilities.

3.3.3.3 IGBT

IGBTs are voltage-controlled devices that combine gating characteristics taken from the MOSFET with the high current-carrying capability of the BJT. Further, this device is characterized by short switching transients and low conduction voltages. The vertical cross-section of an *n*-channel IGBT is shown in Figure 3.5(c). The IGBT's structure is like a power MOSFET except for an additional P+ layer at the top. This collector metal contact is directly connected to this P+ layer. Therefore, IGBTs are four-layer bipolar devices that use both electron and hole carriers.

3.3.3.4 SIT

The static induction transistor (SIT) is depicted in Figure 3.5(d) [9]. One of the main advantages of the SIT relies on a higher breakdown voltage and, therefore, it can operate at high voltage and current.

3.4 Semiconductor materials

The silicon atom contains 14 electrons, four at its most outer orbit. Germanium has 32 electrons around the nucleus with four valence electrons on the outmost orbit. Figure 3.6(a) presents the atomic representation of the Si element. The development of semiconductor products requires the application of single Si crystals. Figure 3.6(b) depicts a semiconductor crystal where all atoms make a covalent bond with their nearest neighborhood atoms. However, this pure monocrystalline silicon cannot conduct electricity. To overcome this problem, the conductive property of the silicon crystal is modified by doping it with doping impurities. These impurity atoms replace some of the Si semiconductor atoms in the crystal and occupy their positions. The semiconductor is categorized as a *n*-type or a *p*-type depending on the type of impurities that are used for the doping. In an *n*-type semiconductor, the pure semiconductor crystal is doped with pentavalent impurities, such as arsenic, antimony, or phosphorus (see Figure 3.6(c)). Similarly, boron is used to produce *p*-type semiconductor material by doping pure Si semiconductor crystals with Boron, which has three valence electrons, as shown in Figure 3.6(d).

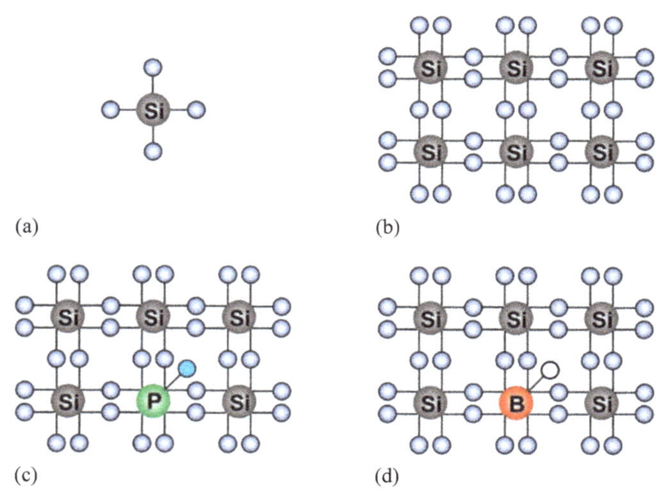

Figure 3.6 Silicon semiconductor material: (a) silicon element, (b) silicon crystal, (c) n-type semiconductor, and (d) p-type semi-conductor

3.5 Power electronics devices applications

Modern power electronics applications span a wide range of power capacities and switching frequencies, from home appliances to very high-power utility infrastructure. This scenario, in terms of power level versus switching frequency, is illustrated in Figure 3.7. A high switching frequency leads to smaller passive components and system downsizing. However, this is at the expense of increased switching losses of the power semiconductor devices. Hence, a suitable compromise between switching frequency and switching power losses ought to be made at the design stage of the equipment. In Figure 3.7, Si power device constraints are outlined against the SiC and GaN power devices capability limits of potential, future devices. The properties and potential benefits of wider bandgap semiconductors are surmised in Section 3.8.

- **Low power:** Si power Triacs, IGBTs, and MOSFETs are preferred for home appliance applications such as air conditioning, refrigeration, and washing machines. For low-power and high switching frequency applications, Si-power MOSFETs are used in switching power supplies and automotive applications.
- **Industrial applications:** In the medium power area, IGBT transistor dominates in applications such as robots and electric car power trains.
- **Power distribution:** IGBTs are the major device of choice for the development of renewable energies and microgrids. The most popular custom power devices in distribution systems for increased reliability and power quality improvement are STATCOM, DVR, and UPFC, which use IGBT power valves.

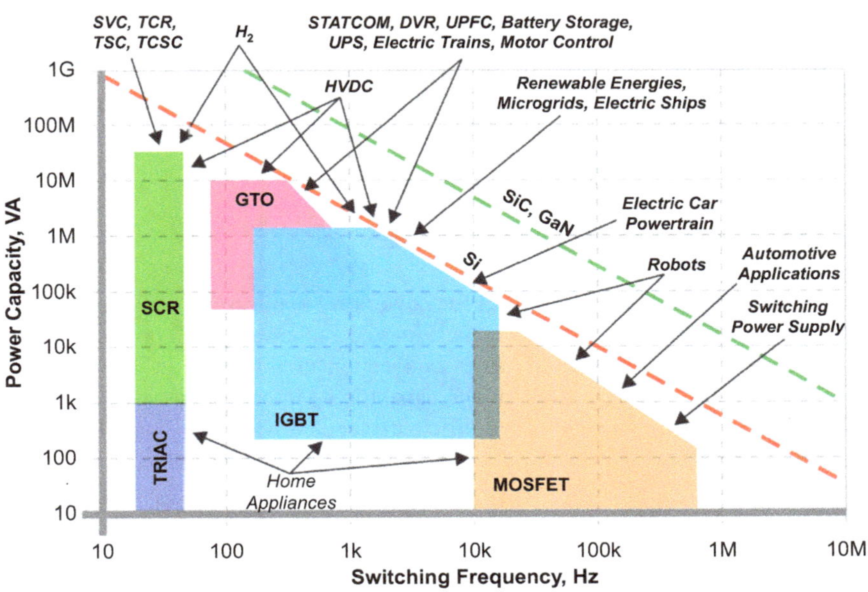

Figure 3.7 Applications of power semiconductor devices

- **Power transmission:** Thyristors and transistors have been employed in power transmission and distribution applications, providing numerous benefits that result in cost-effective solutions. Early applications have been in high-voltage DC transmission (HVDC) and FACTS. HVDC transmission has exploited line-commutated power electronics such as SCR for the past five decades, while self-commutated power electronics based on GTOs and IGBTs were introduced at the transmission level until the 1990's decade. High power static var compensators such as SVC, TCR, TSC, and TCSC have relied on phase control thyristors, which offer the highest power ratings and, therefore, the most cost-effective.
- **Large power traction:** In the field of electric trains, GTOs and IGBTs have demonstrated their capabilities to handle energy concerns. Similarly, the development of electric ships has been favored by the application of state-of-the-art IGBTs.
- **Energy storage:** Bi-directional power converters with GTO and IGBT technologies are applied in battery energy storage systems ranging from a few kilowatts to several hundreds of megawatts.
- **Renewable energies:** PV inverters based on Si IGBTs became the industry standard in residential, commercial, and large PV installations with power ratings ranging from several kilowatts to several megawatts. Onshore and offshore wind parks based on DFIG and directly driven turbines rely on durable and cost-effective power semiconductors for grid connection. IGBT-based solutions are used in today's multi-megawatt wind power converters. The next generation of wind turbines is set to surpass 15 MW, and power semiconductors with increased blocking voltages will be required to achieve sufficient high-power densities.
- **Green hydrogen generation:** Green hydrogen will play a key role in the green energy portfolio easing the great current onus put on energy storage systems based on batteries and, in the long-term, providing the preferred option for energy storage. At present, large-scale electrolyzers are integrated into the power grid via thyristor or IGBT-based converters, delivering maximum powers in the range of 10 MW.

3.6 Mechanical and thermal stress

State-of-the-art power valves are ideal for power grid applications with high-current and high-blocking voltage requirements. Figure 3.8 shows a high-power thyristor, 15 cm diameter silicon wafer. A series connection of thyristors is used to create composite valves with very high voltage capability. Moreover, the stacks of press-pack high-power semiconductors are efficiently cooled down. Press-pack thyristor and IGBT are mature technologies used in HVDC and FACTS applications.

Thyristors and IGBTs in press pack housing are subjected to thermal and mechanical stresses. As can be seen in Figure 3.9, a mounting force is applied from a single point above the center of the device. Mechanical stress can occur either due to mechanical tolerances or heat expansion and contraction during changes in the electrical load [23]. Voltage drops during conduction of an HVDC thyristor are in

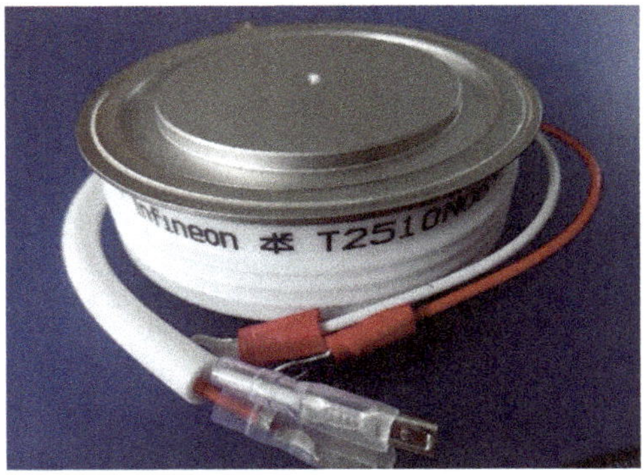

Figure 3.8 High-power thyristor for a 15 cm diameter silicon wafer

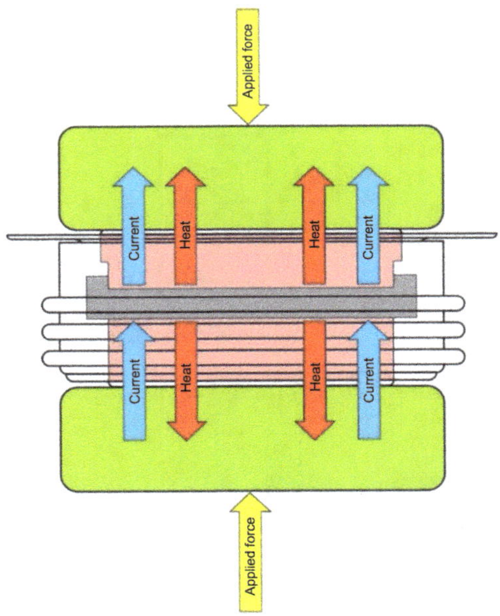

Figure 3.9 Mechanical and thermal stresses on press pack high-power devices

the order of two to three volts for typical currents of several kiloamperes. Therefore, considerable power losses are dissipated during conduction [19].

The operating temperatures of power electronics devices must be kept at reasonable levels using closed-loop liquid cooling systems, which ensure heat removal from critical components to limit thermal stress [24]. The cooling system is depicted schematically in Figure 3.10, which consists of the following subsystems:

- Main cooling circuit: It comprises a liquid coolant pump, which is used to circulate deionized water in a closed loop at a constant flow rate. For sites

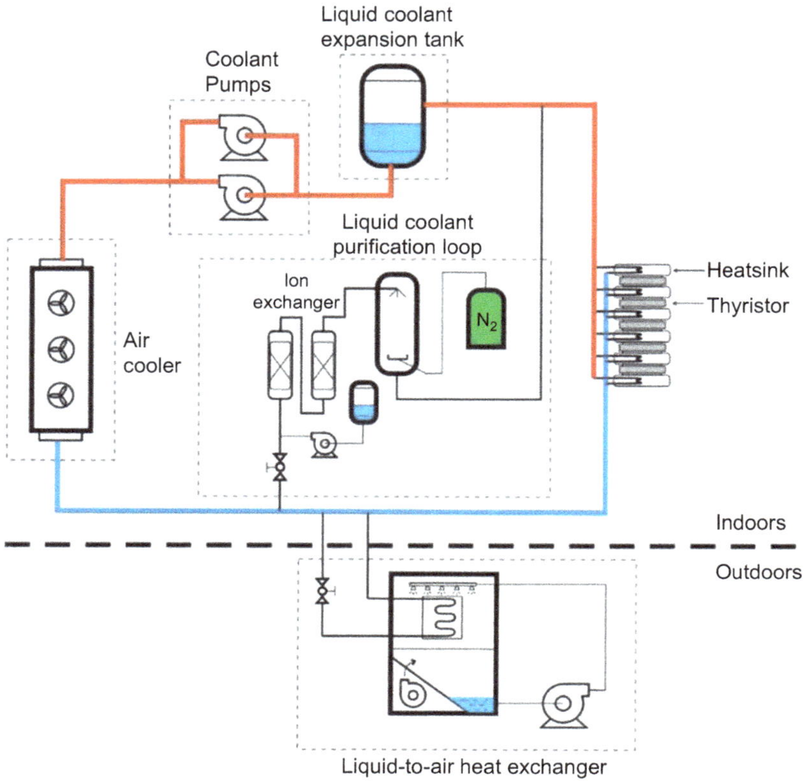

Figure 3.10 Cooling system diagram

with low-temperature climates, pure glycol is added to the water. An extra pump for redundancy is added to the main circuit. Further, it includes a liquid coolant expansion tank and interconnecting piping. The extracted heat of power devices is transferred to the surrounding air by using a forced-air cooler.

- Liquid coolant purification loop: The main circulating cooling circuit operates in parallel with a liquid coolant purification loop to reduce leakage current in high-voltage environments. The cooling liquid flows through two ion exchangers connected in series and an expansion vessel.
- Liquid-to-air heat exchanger: If the ambient temperature is close to or higher than the inlet water temperature of the valve, then the air cooler will not be able to remove the losses dissipated in the power electronics valves. Therefore, an external cooling circuit system consisting of a liquid-to-air heat exchanger is incorporated into the converter valve cooling system. It can be seen from Figure 3.10 that after passing through the air cooler, the water/glycol solution is recirculated through a liquid-to-air heat exchanger [25].

3.7 Switching characteristics

Switching characteristics encapsulate the dynamic performance of semiconductor devices when changing from conduction to blocking states and vice versa. Figure 3.11 depicts an elementary switching circuit, which is used to describe conduction and switching power losses of fully controlled semiconductor valves. It comprises a semiconductor switch, a freewheeling diode, and an inductor with a high inductance value, where the inductor is represented as a constant current source I_o. The semiconductor switch connects and disconnects the source and the load at a fast rate. Figure 3.12 illustrates a switching transition interval, the switching control signal (q), the on-state voltage drops (v_T), current (i_T), and power losses (p_T). Furthermore, Figure 3.12 describes on-off transitions and associated power losses. It

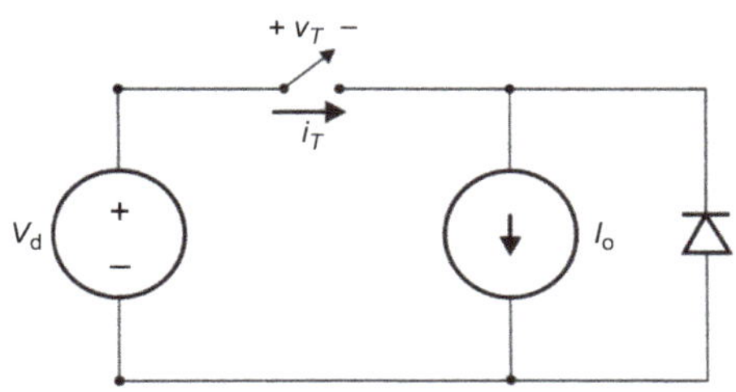

Figure 3.11 Elementary switching circuit

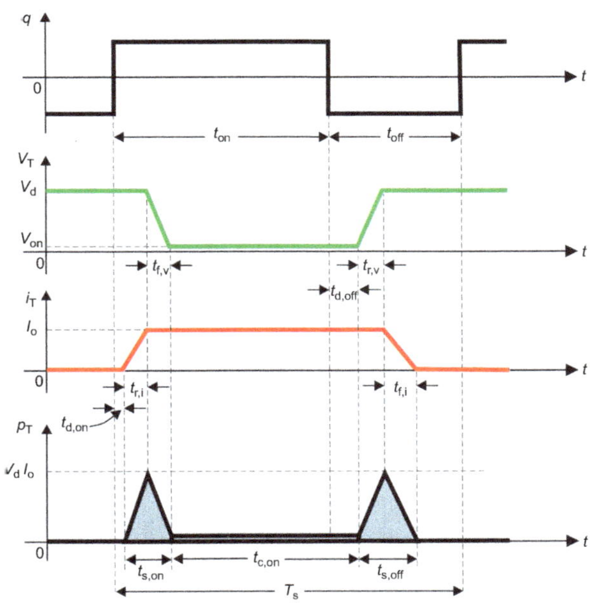

Figure 3.12 Instantaneous voltage, current, and power for a fully controlled semiconductor device

can be appreciated that changes in voltage and current are linear and do not occur simultaneously, except for the short overlap periods.

The energy losses in a switching transition interval are computed as the area under the power curve, as follows,

$$E_T = \int_0^{T_s} p(t)dt \tag{3.1}$$

The average power losses are defined as,

$$P_T = \frac{1}{T_s}\int_0^{T_s} p_T(t)dt = \frac{1}{T_s}\int_0^{T_s} v_T(t)i_T(t)dt = \frac{E_T}{T_s} = E_T \cdot f_s \tag{3.2}$$

where f_s is the switching frequency. Total power losses are:

$$P_T = P_c + P_s \tag{3.3}$$

where P_c and P_s are the conduction and switching power losses, respectively.

3.7.1 MOSFET

The conduction and switching losses of the MOSFET can be calculated as follows:

3.7.1.1 Conduction loss

According to Figure 3.12, the conduction losses are:

$$P_c = R_{DS} I_{d,RMS}^2 \tag{3.4}$$

where R_{DS} is the Drain-Source on-state resistance, in Ω, and $I_{d,RMS}$ is the RMS value of the drain current in A.

In (3.4), $I_{d,RMS}$ is calculated as follows:

$$I_{d,RMS} = \sqrt{\frac{1}{T_s}\int_0^{T_s} {i_T}^2 dt} \simeq \sqrt{\frac{1}{T_s}\int_0^{t_{on}} {I_o}^2 dt} \simeq I_o\sqrt{\frac{t_{on}}{T_s}} = I_o\sqrt{D}$$

where D is the duty cycle.

3.7.1.2 Switching loss

Switching losses basically depend on the energy losses during on and off states. The on-state power loss can be calculated approximately as follows:

$$P_{s,on} \simeq \frac{1}{2} V_d I_o t_{s,on} f_s \tag{3.5}$$

and

$$t_{s,on} = t_{r,i} + t_{f,v} \tag{3.6}$$

where $t_{r,i}$ is the rise time of current i_T, and , and $t_{f,v}$ is the fall time of voltage v_T.

Similarly, the off-state loss can be expressed as,

$$P_{s,off} \simeq \frac{1}{2} V_d I_o t_{s,off} f_s \tag{3.7}$$

and

$$t_{s,off} = t_{r,v} + t_{f,i} \tag{3.8}$$

where $t_{r,v}$ is the rise time of voltage v_T, and $t_{f,i}$ is the fall time of current i_T.

Rising and falling times for current and voltages can be computed from data-sheet parameters of power devices as follows,

$$t_{r,i} = R_G C_{iss} \ln\left(\frac{V_{GS} - V_{TH}}{V_{GS} - V_{gp}}\right)$$

$$t_{f,v} = R_G C_{gd} \frac{V_{DS}}{(V_{GS} - V_{gp})}$$

$$t_{r,v} = R_G C_{gd} \frac{V_{DS}}{V_{gp}}$$

$$t_{f,i} = R_G C_{iss} \ln\left(\frac{V_{gp}}{V_{TH}}\right)$$

and

$$R_G = R_g + R_{gext}$$

$$C_{gd} = \frac{Q_{GD(D)}}{V_{DS(D)}}$$

where V_{DS} is the drain-source voltage, V_{GS} is the gate-source voltage, V_{TH} is the gate threshold voltage, V_{gp} is the gate plateau voltage, R_g is the internal gate resistance, R_{gext} is the external gate resistance, $Q_{GD(D)}$ is the gate-drain charge at a given test conditions, $V_{DS(D)}$ is the drain-source voltage at a given test conditions, C_{iss} is the input capacitance.

The switching losses due to the gate switching control circuit are:

$$P_{GS} = \frac{1}{2} C_{iss} V_{GG}^2 f_s \tag{3.9}$$

where C_{iss} is the input capacitance and V_{GG} is the gate voltage.

The total power switching losses are defined as,

$$P_s = P_{s,on} - P_{s,off} + P_{GS} \tag{3.10}$$

3.7.2 IGBT

Conduction and switching losses for the IGBT are calculated in a similar way as the conduction and switching losses for the MOSFET.

3.7.2.1 Conduction loss

The total conduction losses over a period of switching transitions are:

$$P_c = \left(V_{CE,on} I_C t_{c,on}\right) \cdot f_s \tag{3.11}$$

where $V_{CE,on}$ is the collector-to-emitter voltage, I_C is the collector current and $t_{c,on}$ is defined as,

$$t_{c,on} = t_{on} - t_{s,on} \tag{3.12}$$

Substituting (3.12) in (3.11) yields,

$$P_c = V_{CE,on} I_C \left(t_{on} - t_{s,on}\right) \cdot f_s \tag{3.13}$$

Since $t_{on} \ggg \mathrm{t}_{s,on}$ then (3.13) can be approximated to,

$$P_c \simeq V_{CE,on} I_C t_{on} f_s = V_{CEon} I_C D \tag{3.14}$$

where D is the duty cycle.

3.7.2.2 Switching loss

The IGBT exhibits a residual current called tail current after being switched off. The tail current appears when the voltage across the IGBT is at its highest level. Therefore, tail current results in additional switching losses, which must be properly taken into account. Switching losses for an IGBT are computed using energy losses instead of applying switching times. From the IGBT datasheet, energy losses can be obtained for two collector current operating conditions. Figure 3.13 depicts

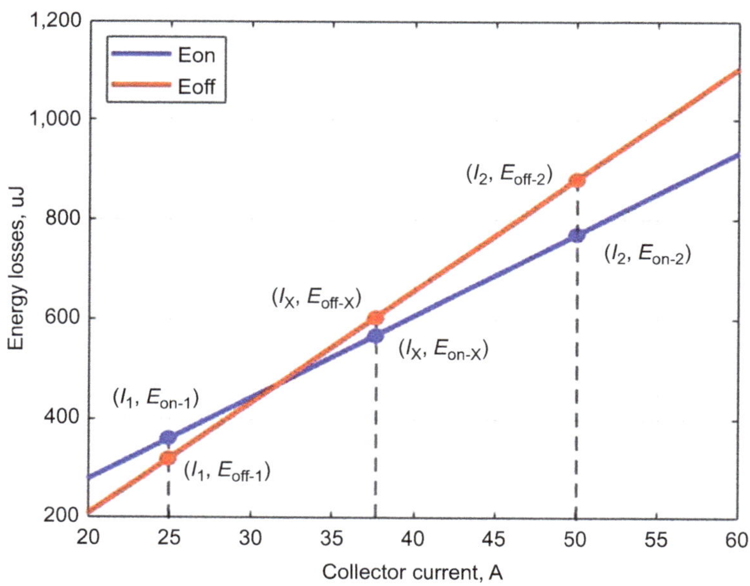

Figure 3.13 Energy losses during turn-on and turn-off for the IGBT

the energy lost during turn-on and turn-off as a function of the collector current. Then, the energy losses during turn-on E_{on-X} and turn-off E_{off-X} at a given collector current I_X can be determined using the equations below:

$$E_{on-X} = E_{on-1} + \left(\frac{E_{on-1} - E_{on-2}}{I_1 - I_2}\right)(I_X - I_1) \tag{3.15}$$

$$E_{off-X} = E_{off-1} + \left(\frac{E_{off-1} - E_{off-2}}{I_1 - I_2}\right)(I_X - I_1) \tag{3.16}$$

The turn-on loss can be calculated by multiplying the turn-on energy loss E_{on} by the switching frequency f_s as follows:

$$P_{s,on} = E_{on-X} f_s \tag{3.17}$$

Similarly, the off-state loss can be expressed as:

$$P_{s,off} = E_{off-X} f_s \tag{3.18}$$

where E_{off} is the turn-off energy loss. Both E_{on} and E_{off} are reported by manufacturers in the technical datasheets for specific IGBT power devices.

The switching losses due to the gate driver circuit are calculated as,

$$P_{GC} = \frac{1}{2} C_{ies} V_{GG}^2 f_s \tag{3.19}$$

where C_{ies} is the input capacitance.

The total power switching losses are defined as:

$$P_s = P_{s,on} + P_{s,off} + P_{GC} \tag{3.20}$$

3.7.3 Thermal model

The ambient temperature, the maximum junction temperature, and the heat dissipation conditions determine the permissible power dissipation of a power device. The thermal energy flow of a power device in steady-state may be modeled by using an electric circuit analogy. Figure 3.14 depicts the simplified thermal model of the power devices if no heat sink is used. When thermal energy is produced at the junction due to a power dissipation P_T, a resulting temperature T_j will appear at the junction. Considering the electrical analogies of the thermal variables, the power dissipation in the power device is represented as a current source, and the

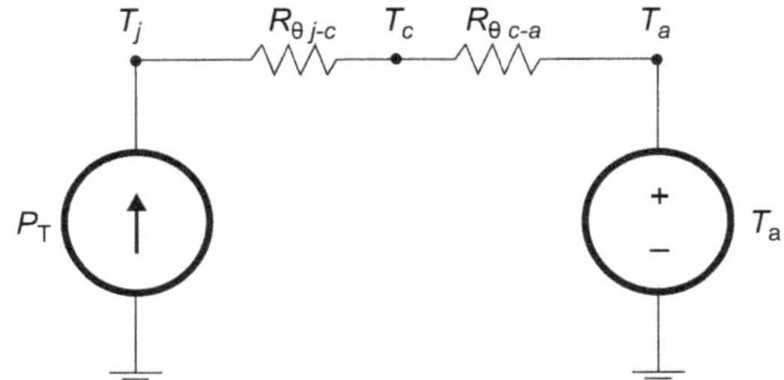

Figure 3.14 Simplified thermal model for power devices

temperature level is expressed as a voltage source. Furthermore, the thermal energy flow from the junction of a power device to the ambient air is derived from thermal resistances. Thermal resistance is a characteristic that indicates the resistance of a semiconductor material to a heat flow.

The junction-to-ambient thermal resistance $R_{\theta\, j-a}$ is simplified to,

$$R_{\theta\, j-a} = R_{\theta\, j-c} + R_{\theta\, c-a} \tag{3.21}$$

where $R_{\theta\, j-c}$ is the junction-to-case thermal resistance and $R_{\theta\, c-a}$ is the case-to-ambient thermal resistance.

From the equivalent circuit of Figure 3.14, the resulting junction temperature T_j can be calculated as follows [26],

$$T_j = R_{\theta\, j-a} \cdot P_T + T_a \tag{3.22}$$

where P_T is the power consumption of the power device, in W, and T_a ambient temperature, in °C.

3.8 Future trends

There is common agreement within the power electronics community that wide-bandgap (WBG) semiconductor materials, such as silicon carbide (SiC) and gallium-nitride (GaN), exhibit superior physical properties to the conventional semiconductor materials used in today's mass market applications, such as Si, enabling WBG devices to operate at higher voltages, switching frequencies, and temperatures than the conventional Si technology. The bandgap is a measure of the strength of the electric field of a semiconductor material. The critical electric field is its maximum breakdown voltage capability. Table 3.1 summarizes the properties of key semiconductor materials [27]: Silicon (Si), 4H crystal structure for silicon carbide (4H-SiC), 2H crystal structure for gallium nitride (2H-GaN), diamond, and 2H crystal structure for Aluminum Nitride (2H-AlN).

3.8.1 Wide bandgap

Figure 3.15 depicts relevant properties of semiconductor materials. Figure 3.15(a) indicates that SiC and GaN have higher energy bandgaps, higher breakdown electric field strengths, higher thermal conductivities, and higher carrier saturation drift velocities than Si. It should be noted that the critical electric field of SiC and GaN is ten times fold that of Si. Hence, for power devices of comparative voltage ratings, SiC and GaN devices would be designed using ten times thinner drift layers and one hundred times higher dopant densities [5]. This property of WBG materials will result in more than several hundred times lower specific on-resistance than Si devices. This can be better appreciated by looking at the specific on-resistance for the drift-region drop expression defined by the equation below [28]:

$$R_{ONsp} = \frac{4}{a^3 \varepsilon_0} \frac{V_B^2}{\mu_e \varepsilon_r E_G^{3n}} \tag{3.23}$$

where V_B is the voltage breakdown, E_G is the energy bandgap, μ_e is the electron mobility, ε_r is the relative permittivity, and ε_0 vacuum permittivity.

Table 3.1 Physical properties of key semiconductor materials

Material	**Energy bandgap** $E_g(eV)$	**Relative dielectric constant** ε_r	**Electron mobility** $\mu_e\left(\frac{cm^2}{V-s}\right)$	**Critical electric field** $E_c\left(\frac{MW}{cm}\right)$	**Saturated electron drift velocity** $v_{sat}\left(10^7\,\frac{cm}{s}\right)$	**Thermal conductivity** $\lambda\left(\frac{W}{cm-K}\right)$	**Gap type**	**Technology level**
Si	1.12	11.8	1,350	0.25	1.0	1.5	Indirect	Mass production
4H-SiC	3.26	10	720	2.0	2.0	4.5	Indirect	Partial mass production
2H-GaN	3.39	9.9	1,000	3.3	2.5	2.5	Direct	Partial mass production
Diamond	5.45	5.5	2,800	10	2.7	22	Indirect	Research stage
2H-AlN	6.2	8.5	300	12	1.7	2.85	Direct	Research stage

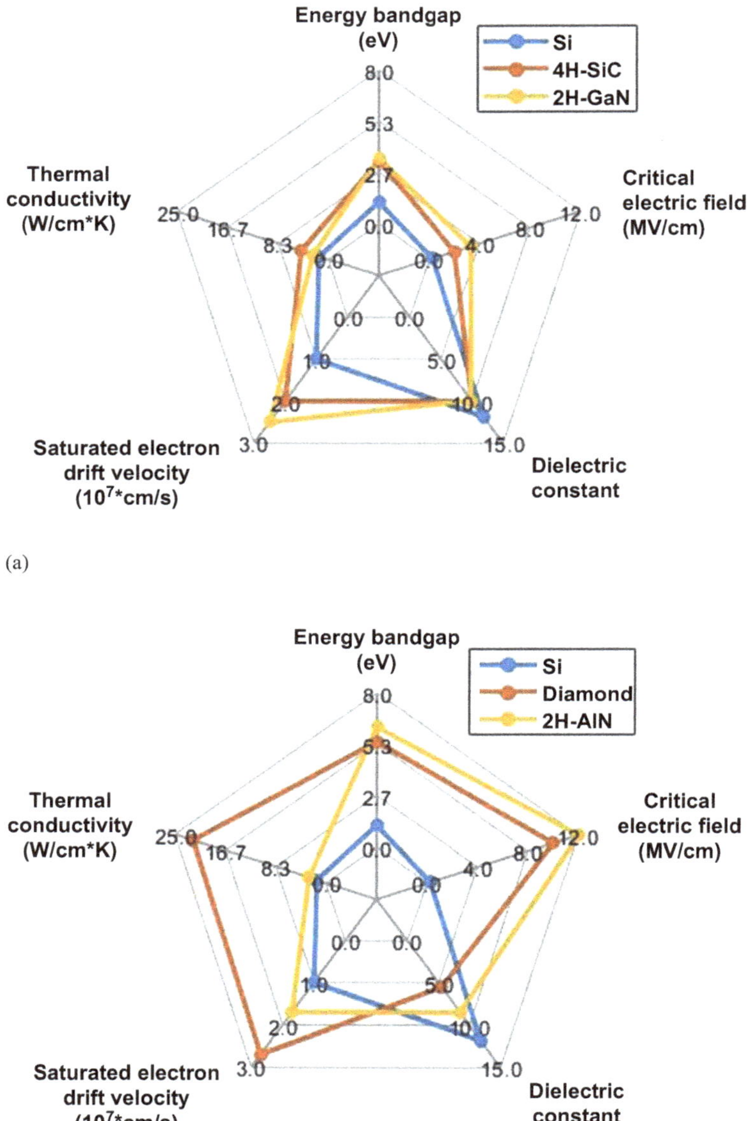

(b)

Figure 3.15 Relevant properties of semiconductor materials for (a) WBG and (b) UWBG

It is observed from (3.23) that the on-state resistance is inversely proportional to μ_e and ε_r. More importantly, the cubic dependence of the on-state dependence on E_G and, therefore, on the critical electric field E_c favors semiconductor materials such as SiC and GaN. Hence, conduction losses essentially depend on properties such as electron mobility, relative permittivity, and energy bandgap. The breakdown voltage can be computed as follows [28]:

$$V_B = 2.76 \times 10^5 \frac{a^2 \varepsilon_r E_G^{2n}}{N_B} \tag{3.24}$$

where N_B is the impurity doping concentration.

The parameters a and n are calculated according to the relationship between the energy bandgap E_G and the critical electric field E_c. In [26], the parameter values a and n are reported for different types of semiconductor materials.

Figure 3.16 shows the specific on-state resistance as a function of the maximum supported voltage for Si and WBG semiconductor materials. It can be appreciated that for an on-state resistance of 0.1 Ωcm^2, which yields reasonable conduction losses, the maximum supported voltage for Si semiconductor material is 0.66 kV. In contrast, WBG materials can support a few tens of kilovolts. The SiC semiconductor material possesses a breakdown voltage of 19.33 kV, whereas the breakdown voltage for GaN is approximately 26.14 kV. Furthermore, the specific on-resistance values for 4H-SiC and 2H-GaN are 900 and 1,600 times smaller than the corresponding specific on-resistance provided by the Si material for a given breakdown voltage.

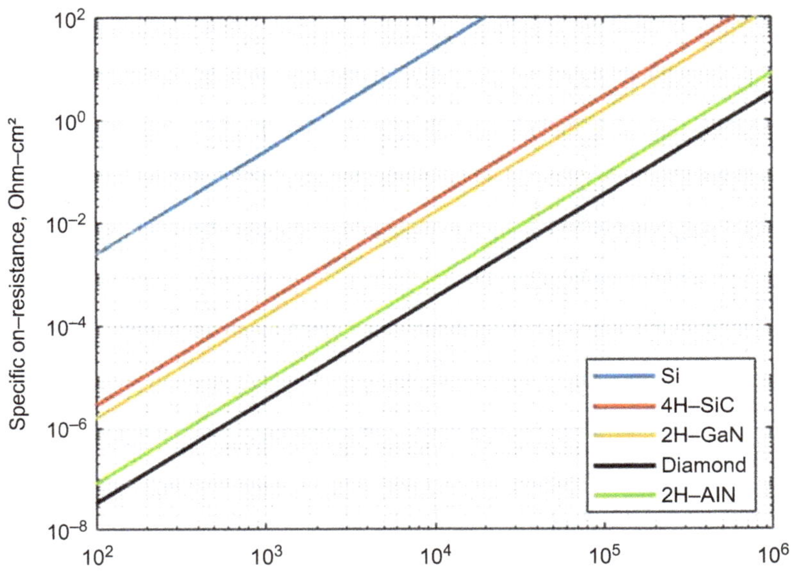

Figure 3.16 Specific on-state resistance of drift regions in WBG and UWBG semiconductor materials as a function of the maximum supported voltage

The properties exhibited by WBG materials have encouraged the development of new power devices, such as SiC Schottky diodes, SiC MOSFET, and GaN HEMT. In general, GaN has greater figures of merit than SiC, but with current techniques available, this semiconductor material is also more difficult to crystallize and process than SiC.

The improvements in drift region resistance for SiC and GaN, compared to Si, are largely due to their larger energy band gap and critical electric field breakdown. This would enable the formulation of an informed surmise that WBG materials contribute to significantly lower energy consumption, allowing designers to develop more environmentally friendly solutions and helping to reduce CO_2 emissions.

3.8.1.1 WBG power devices

- *SiC IGBT*
  Over the last few decades, numerous SiC IGBT prototypes have been demonstrated, with increasing blocking voltages, improved on-state performance, and high breakdown voltage [29]. Nevertheless, no commercial solutions for IGBT devices based on WBG materials, such as SiC and GaN, seem to exist today. However, semiconductor manufacturers do offer SiC MOSFETS and GaN HEMTs.
- *GaN HEMT*
  Gallium Nitride High Electron Mobility Transistor (GaN HEMT) is a voltage-driven power device. The properties of the GaN allow the development of power devices with high voltage breakdown, high current, and high switching frequency. These unique attributes lead to greater efficiency, significant reduction in size and weight, and enhanced thermal performance.

3.8.2 Ultrawide-bandgap semiconductors

With their unique properties, UWBG semiconductors represent a new class and promising material platform for developing a new set of high-performance devices for a wide range of applications. Nevertheless, important research efforts are still needed to understand better their basic properties and develop appropriate processing techniques. Ultrawide bandgap materials such as Ga_2O_3, h-BN, AIN, and Diamond are semiconductor materials that possess excellent properties, including very large bandgap, high critical electric field, high carrier mobility, and chemical inertness [30]. Owing to their very attractive properties, UWBG semiconductor materials are promising candidates for developing high-performance devices for high voltage, high-frequency switching, and high-temperature power electronics applications. According to Figure 3.15(b), Diamond and 2H-AIN materials possess higher energy bandgap, higher breakdown electric field strength, higher thermal conductivity, and higher carrier saturation drift velocity than SiC and GaN.

The specific on-resistance of drift regions in UWBG semiconductor materials is compared with that of Si material in Figure 3.16. The 2H-AIN and diamond materials can perform beyond the limits of Si. For instance, the breakdown voltage for Si material for a specific on-resistance of 0.1 Ωcm^2 is only 0.66 kV, while 2H-AIN and Diamond have breakdown voltages of 113.85 kV and 172.92 kV, respectively. For any breakdown voltage, the specific on-resistance values for

2H-AIN and diamond are 30,000 and 70,000 times smaller than the specific on-resistance of the Si material.

3.8.3 Numerical example

The elementary switching circuit of Figure 3.11 is given in the following parameters: V_d = 400 V, I_o = 15 A, D = 0.6, V_{GG} = 12 V, f_s = 20 kHz, and T_a = 25 °C. The details of the power semiconductor devices such as Si MOSFET, Si IGBT, SiC MOSFET, and GaN HEMT are given in Table 3.2.

The theoretical prediction of conduction losses, switching losses, gate circuit losses, total power losses, and the junction temperature are shown in Table 3.3 for

Table 3.2 Details of devices used in the numerical example: Si MOSFET, SiC MOSFET; GaN HEMT; Si IGBT

Device type	Si MOSFET	SiC MOSFET	GaN HEMT	**Device type**	Si IGBT
Manufacturer	RHOM	Wolfspeed	Infineon	**Manufacturer**	Infineon
Model number	R6042JNZ4	C3M0065090J	GS-065-030-6-LL	**Model number**	IKZ50N65ES5
V_{DS} **(V)**	600.0	900.0	700.0	V_{CE} **(V)**	650
I_D **(A)**	42.0	35.0	40.0	I_C **(A)**	50
P_{TOT} **(W)**	495.0	113.0	–	P_{TOT} **(W)**	274
$R_{\theta\ j\text{-}c}$ **(°C/W)**	0.25	1.1	0.46	$R_{\theta\ j\text{-}c}$ **(°C/W)**	0.55
$R_{\theta\ j\text{-}a}$ **(°C/W)**	30.0	40.0	29.5	$R_{\theta\ j\text{-}a}$ **(°C/W)**	40
R_{DS} **(Ω)**	0.08	0.065	0.04	$V_{CE,on}$ **(V)**	1.35
C_{iss} **(nF)**	3.5	0.760	0.235	C_{ies} **(nF)**	3.1
R_g **(Ω)**	1.2	3.5	1.3	E_{on} **@ 50 A (mJ)**	0.77
$Q_{GD(D)}$ **(nC)**	35.0	9.0	2.3	E_{off} **@ 50 A (mJ)**	0.88
$V_{DS(D)}$ **(V)**	300.0	400	400	E_{on} **@ 25 A (mJ)**	0.36
V_{gp} **(V)**	10.0	8.0	2.5	E_{off} **@ 25 A (mJ)**	0.32
V_{TH} **(V)**	6.0	2.1	1.7		

Table 3.3 Comparison of power losses and junction temperature for Si, SiC, and GaN technologies

Device	**Conduction losses (W)**	**Switching losses (W)**	**Gate circuit losses (mW)**	**Total power losses (W)**	**Junction temperature (°C)**
Si MOSFET	10.80	2.42	5.04	13.23	421.80
Si IGBT	12.15	5.84	4.46	18.00	744.79
SiC MOSFET	8.78	1.07	1.09	9.84	418.72
GaN HEMT	5.40	0.10	0.33	5.50	187.24

different power semiconductor devices. It was found that Si IGBT incurs maximum conduction losses ≈12.15 W followed by the Si MOSFET (10.8 W), SiC MOSFET (8.78 W) and GaN HEMT (5.40 W). Comparing switching loss, Si IGBT has a maximum of ≈5.84 W, followed by Si MOSFET ≈2.42 W, SiC MOSFET ≈1.07 W and GaN HEMT ≈0.10 W. Gate circuit losses are found to be approximately 5 mW, 4.5 mW, 1.1 mW, and 0.3 mW for Si MOSFET, Si IGBT, SiC MOSFET and GaN HEMT, respectively.

Total power losses for all the devices are given in Table 3.3. Total power losses are higher for Si IGBT (18.00 W), followed by Si MOSFET (13.23 W), SiC MOSFET (9.84 W) and GaN HEMT (5.50 W). These are ≈45% and 69% fewer total power losses in SiC MOSFET and GaN HEMT compared to Si IGBT. Similarly, total power losses for SiC MOSFET and GaN HEMT are reduced by 25.6% and 58.4%, respectively, compared to Si MOSFET.

The junction temperatures are shown in the last column for all devices. The highest junction temperatures are 744.79 °C and 421.80 °C for Si IGBT, and Si MOSFET, respectively, whereas the junction temperature is 418.72 °C for SiC MOSFET and only 187.24 °C for GaN HEMT. The junction temperature for GaN HEMT is 74.9% lower than Si IGBT.

3.9 Conclusions

The existence of Si-based power electronics devices is fast approaching ¾ of a century; a period marked for their continuous innovation and an ever-expanding range of uses, encompassing military and civil applications. A wide range of Si-based devices have been developed among which the most popular are the following: diodes, thyristor, GTO, IGBT, and MOSFET. The functionality of these devices increases with the complexity of their physical structures, namely, the number of layers and terminals employed. The Si material is very well characterized, and it is largely the dominant material used in power electronics applications. However, its physical limits, such as its energy bandgap, may be about to come into consideration as the electricity supply industry moves into the use of ultra-high-power transmission installations.

Key characteristics of these semiconductor devices are the following: power capacity, switching frequency, power losses, heat dissipation, size, and their flexibility to make up series/parallel combinations. Their applications range from home appliances to ultra-high-power utility infrastructure. A high switching frequency leads to smaller components and system downsizing. However, this is at the expense of increased switching losses, thermal stress, and a decrease in power handling capacity. Si power devices exhibit higher power loss and are subject to higher thermal stress compared to wide bandgap semiconductor materials, such as SiC and GaN, owing to their much lower bandgap. This chapter has conjectured what power levels were to be attained if current Si-based devices were to be replaced with wider bandgap materials. However, such devices only exist at the prototype stage, with limited scope for commercial exploitation at the present time.

References

[1] Kang C, Kirschen D, and Green TC. The evolution of smart grids. *Proceedings of the IEEE*, vol. 111, no. 7, pp. 691–693, 2023.

[2] Fairley P. High-voltage DC power roars ashore in Europe. *IEEE Spectrum*, vol. 61, no. 1, pp. 28–31, 2024.

[3] Baliga BJ. *Wide Bandgap Semiconductor Power Devices: Materials, Physics, Design and Applications*. Amsterdam: Elsevier, 2018.

[4] United Nations. *The Sustainable Development Goals Report*. New York: United Nations, 2023.

[5] IEC. *Power Semiconductors for an Energy-Wise Society*. Geneva: International Electrotechnical Commission, 2023.

[6] Prince DC. Mercury arc rectifier phenomena. *Journal of the AIEE*, vol. 46, no. 7, pp. 667–674, 1927.

[7] Owen EL. SCR is 50 years. *IEEE Industry Applications Magazine*, vol. 13, no. 6, pp. 6–10, 2007.

[8] Bardeen J, and Brattain WH. The transistor: A semi-conductor triode. *Physical Review Journal*, vol. 74, no. 2, pp. 230–231, 1948.

[9] Nishizawa J, and Yamamoto K. High-frequency high-power static induction transistor. *IEEE Transaction on Electron Devices*, vol. 25, no. 3, pp. 314–322, 1978.

[10] Fisher CA, Paxman DH, and Slatter JAG. Design and performance of a new static induction thyristor—The gated V-groove p–i–n diode. *IEEE Transactions on Electron Devices*, vol. 31, no. 9, pp. 1299–1308, 1984.

[11] Baliga BJ. *The IGBT Device: Physics, Design, and Applications of the Insulated Gate Bipolar Transistor*. Amsterdam: Elsevier, 2015.

[12] Guo X, Xun Q, Li Z, and Du S. Silicon carbide converters and MEMS devices for high-temperature power electronics: A critical review. *Micromachines*, vol. 10, no. 6, pp. 1–26, 2019.

[13] Ramungul N, Chow TP, Ghezzo M, Kretchmer J, and Hennessy W. A fully planarized, 6H-SiC UMOS insulated-gate bipolar transistor. *Proceedings of the Annual Device Research Conference Digest*, pp. 56–57, 1996.

[14] Korytowski M. Uno Lamm: The father of HVDC transmission [history]. *IEEE Power and Energy Magazine*, vol. 15, no. 5, pp. 92–102, 2017.

[15] Peake O. The history of high voltage direct current transmission. *Australian Journal of Multi-Disciplinary Engineering*, vol. 8, no. 1, pp. 47–55, 2010.

[16] Hingorani NG, and Gyugyi L. *Understanding FACTS: Concepts and Technology of Flexible AC Transmission Systems*. New York: Wiley-IEEE Press, 1999.

[17] Sumi Y, Harumoto Y, Hasegawa T, Yano M, Ikeda K, and Matsuura T. New static var control using force-commutated inverters. *IEEE Transactions on Power Apparatus and Systems*, vol. 100, no. 9, pp. 4216–4224, 1981.

[18] Renz BA. AEP unified power flow controller performance. *IEEE Transactions on Power Delivery*, vol. 14, no. 4, pp. 1374–1381, 1999.

[19] Huang H, Uder M, Barthelmess R, and Dorn J. Application of high power thyristors in HVDC and FACTS systems. *Proceedings of the 17th Conference of the Electric Power Supply Industry (CEPSI)*, pp. 1–8, 2008.

[20] Morya AK. Wide bandgap devices in AC electric drives: Opportunities and challenges. *IEEE Transactions on Transportation Electrification*, vol. 5, no. 1, pp. 3–20, 2019.

[21] Hart DW. *Power Electronics*. McGraw Hill: New York, 2011.

[22] Sugandhi RK, and Sugandhi KK. *Thyristors: Theory and Applications*. Hoboken, NJ: Wiley, 1981.

[23] Hitachi. *Mechanical Clamping of Press-Pack High Power Semiconductors*. Lenzburg: Hitachi Energy Switzerland Ltd. Semiconductors, 2016.

[24] Schauder C. Development of a ±100 MVAr static condenser for voltage control of transmission systems. *IEEE Transactions on Power Delivery*, vol. 10, no. 3, pp. 1486–1496, 1995.

[25] Miller TJE. *Reactive Power Control in Electric Systems*. New York: Wiley-Interscience, 1982.

[26] Rohm Application Note. *What Is a Thermal Model*. Kyoto: Rohm Co. Ltd., 2016.

[27] Chow TP, Omura I, Higashiwaki M, Kawarada H, and Pala V. Smart power devices and ICs using GaAs and wide and extreme bandgap semiconductors. *IEEE Transactions on Electron Devices*, vol. 64, no. 3, pp. 856–873, 2017.

[28] Hudgins JL, Simin GS, Santi E, and Khan MA. An assessment of wide bandgap semiconductors for power devices. *IEEE Transactions on Power Electronics*, vol. 18, no. 3, pp. 907–914, 2003.

[29] Zhang Q. Design and characterization of high-voltage 4H-SiC p-IGBTs. *IEEE Transactions on Electron Devices*, vol. 55, no. 8, pp. 1912–1919, 2008.

[30] Xu M. A review of ultrawide bandgap materials: properties, synthesis and devices. *Oxford Open Materials Science*, vol. 2, no. 1, pp. 1–31, 2022.

Chapter 4
Working principles of power electronic circuits and power converters

Pedro Roncero-Sánchez[1]

The evolution of solid-state power electronics over the last 50 years has enabled the development of modern HVDC systems. Line-Commutated Converters (LCCs) using thyristor valves were widely used in the early stages of HVDC, having superseded an earlier form of power electronics based on the use of mercury-arc power rectifiers, i.e., non-solid power electronics. However, at the turn of the millennium, the advent of Voltage-Source Converters (VSCs) aimed at MW (megawatt)-level applications, coupled with ease of control, increased power capability, and the development of new converter topologies, has facilitated the construction of new HVDC systems, coexisting side-by-side with LCC HVDC systems, each having their own application market niche. Besides, LCC and VSC converters are also used in AC power grid applications relating to grid power control, enhanced reliability of supply, improved power quality, and power generation; this equipment and operational applications in AC power grids are termed FACTS. This chapter is dedicated to both LCCs and VSCs, analysing their operational principles, and the primary topologies used, and explaining the key features of both types of technologies.

4.1 Introduction

High Voltage Direct Current (HVDC) power transmission systems are pivotal in the modern electrical power grid, facilitating long-distance and undersea power transmission, the interconnection of asynchronous power systems, and the evacuation of wind and solar power generation. Two primary technologies dominate the HVDC landscape: line-commutated converters (LCCs) and voltage-source converters (VSCs) [1]. An intrinsic characteristic of LCC-based HVDC systems is that the DC current does not change direction; it flows from the rectifier to the inverter station, and it is assumed to be constant. On its AC side, the LCC converter exhibits a current source behavior, injecting fundamental frequency and harmonic

[1]School of Industrial Engineering, University of Castilla-La Mancha, Campus Universitario S/N, Spain

currents into the AC power grid. Hence, LCC-based converters are also known as current-source converters (CSC), and the whole DC link: CSC-HVDC link. The two terms LCC-HVDC and CSC-HVDC, are used interchangeably throughout the book. Concerning VSCs, these converters impress fundamental and harmonic frequency voltages on their AC bus. Accordingly, the HVDC links using VSCs are termed VSC-HVDC links. Examples of both kinds of HVDC installations around the globe are detailed in Chapter 2.

Flexible AC Transmission Systems (FACTS) is an area of power electronics applied to AC power transmission systems that seek to maximise power throughputs with utmost operational reliability. Many FACTS equipment suitable for alleviating one or more operational problems in the neighbourhood of its installation is available from all the major manufacturers of electrical power equipment. The most popular FACTS equipment and installations are detailed in Chapter 2. LCC and VSC converters are the core components of FACTS equipment.

Likewise, power electronic converters make up the heart of an HVDC power system: the rectifier and the inverter stations are made up each of one or more converters suitably connected to make up a monopolar, a bipolar, or a multiterminal HVDC system of a given rating. The evolution of these power electronic circuits, particularly VSCs, has been quite rapid, marked by significant technological advancements, leading to enhanced performance and reliability. They have progressed from being two-level and three-level structures at the turn of the century to today's multilevel structures of high ratings, low power losses, and near-sinusoidal voltage outputs. In contrast, the development of HVDC systems using LCC converters represents quite a mature technology where additional technological breakthroughs may not be seen as pressing as in VSC-HVDC [2,3]. Si thyristor valves are one of the most robust components of modern electrical power grids; they incur relatively low power losses and conduct rather large currents and withstand large blocking voltages. They are easily combined in series and parallel to make up bridges and converter stations of ultra-high capacity, enabling HVDC links spanning distances of more than 3,000 km. The great strength of classic HVDC systems – using LCC converters – is their ability to transport efficiently large volumes of power between two points. In contrast, their fundamental and practical limitation is their difficulty in transporting power between more than two points, and three points following a tandem structure.

The VSC technology, underpinned by the modular multilevel converter (MMC) design appeared in 2010, revolutionising HVDC systems. MMCs are renowned for their modularity, scalability, and ability to integrate offshore renewable energy sources seamlessly into the power grid. As seen from the HVDC installations described in Chapter 2, although the rating powers and transmission distances of VSC-HVDC systems are increasing at a healthy rate, they are still much lower than those of LCC-HVDC systems. It is foreseen that a new generation of semiconductor valves using wide bandgap materials, such as SiC and GaN, would result in higher capacity transistor valves, with reduced power losses, as surmised in Chapter 3. This would increase the capabilities of MMC equipment well beyond their current ratings.

LCCs have served as the foundation technology from which HVDC systems have been developed. However, new HVDC link designers are beginning to favour the use of the VSC technology over the CSC technology, provided the megawatt requirements of the new DC link are within the current capacities of the VSC technology.

The operating principles of LCC and VSC converters, together with their control strategies, are addressed in this chapter. The first part analyses the fundamental principles of LCCs and their various applications, including HVDC and FACTS systems. The second part of the chapter focuses on VSCs, studying several multilevel topologies and comparing them with the conventional two-level topology. The chapter concludes with a comparison of the features of LCC and VSC in HVDC systems.

4.2 Line-commutated converters

The LCC converter has been extensively utilised in HVDC systems owing to its great power handling capacity and relatively low power losses.

As discussed in Section 2.2.5.2, modern LCC-HVDC systems employ thyristors valves and even though their operation requires in-situ reactive-power support and have limited voltage regulation, they constitute a mature and proven power transmission technology when employed for bulk power transmission over long distances, spanning thousands of kilometres. It should be noted that Load Tap Changing (LTC) transformers are employed with LCC-HVDC systems to give them some additional voltage regulation support, but the regulation afforded by LTCs is not fast-acting.

This section outlines the LCC converter's basic operating principles, discusses key LCC-HVDC features, and delves into applications of this type of converter.

4.2.1 Principle of operation

The basic schematic structure of a three-phase, six-pulse, full-bridge LCC is depicted in Figure 4.1. This converter topology is termed the Graetz bridge. It comprises three legs, with each leg containing two series-connected thyristors. The cathodes of the upper thyristors are interconnected, and so are the anodes of the lower thyristors. The midpoint of each leg is connected to a voltage source (an equivalent representing the AC power grid). The LCC output voltage is obtained between the interconnected cathodes of the upper thyristors and the interconnected anodes of the lower thyristors. Normally, the LCC's load exhibits a current source behaviour [2].

The output AC voltage of the LCC is controlled by adjusting the firing delay angle, α, applied to the thyristor valves. This angle determines the point in time at which current pulses are applied through the gates of the thyristors to turn them on. Considering ideal rectification, there are always two thyristor valves conducting at any one time, one upper valve and one lower valve, hence, each thyristor valve conducts for 120°. During normal but non-ideal rectifying conditions, three valves

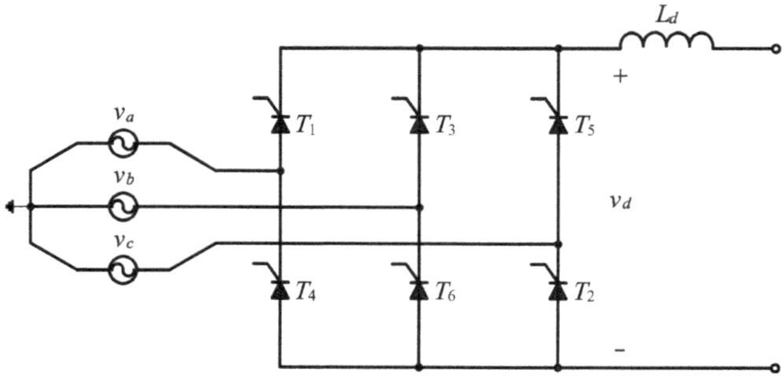

Figure 4.1 Scheme of the full-bridge LCC

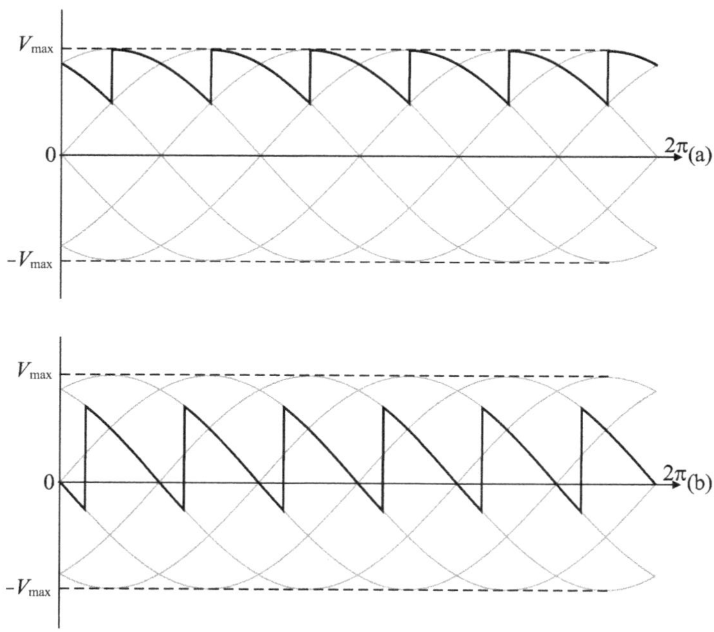

Figure 4.2 Voltage at the load for two firing delay angles: (a) $\alpha = 30°$, and (b) $\alpha = 75°$

will conduct simultaneously during the so-called overlap or commutation period. During faulted operation, such as commutation failure, more than three valves will be conducting at the same time, giving rise to anomalous voltages and currents in the AC and DC sides of the CSC converter [1]. Let us consider first the case of ideal commutation.

Figure 4.2 shows the output voltage of the Graetz bridge for two values of the firing delay angle α: In Figure 4.2(a) α is 30°, while in Figure 4.2(b) a value of 75°

has been used for α. The mean voltage, V_d, can be calculated by averaging the output voltage over the repetition period [4]:

$$V_d = \frac{1}{2\pi}\int_{C}^{2\pi} v_d d(\omega t) = \frac{6}{2\pi}\int_{-\frac{\pi}{6}+\alpha}^{\frac{\pi}{6}+\alpha} V_{\max}\cos(\omega t)d(\omega t) = \frac{3}{\pi}V_{\max}\cos(\alpha) \tag{4.1}$$

where $V_{\max} = \sqrt{2}V_{LL}$, and V_{LL} is the root-mean-square value of the line-to-line grid voltage.

Equation (4.1) shows that the mean voltage can take either positive or negative values depending on the firing delay angle α. It is observed that for α values lower than 90°, the mean voltage is positive, which implies that the direction of power flow is from the AC grid to the DC side of the LCC. This operating mode is known as rectifier mode. Furthermore, when the firing delay angle is greater than 90°, the LCC mean voltage becomes negative and, since the direction of current through the thyristors cannot reverse, the direction of power flow is from the DC side of the LCC to the AC grid. In this case, the LCC operates in inverter mode. Note that in reference to Figure 4.1, such an operating mode would imply having an active power source on the DC side of the circuit. In practice this would come from a second LCC converter, making up an HVDC link. Both operation modes are plotted in Figure 4.3, where the firing delay angle takes values from 0 rad to π rad. In principle at least, if $\alpha = 90°$, the mean value of the output voltage is zero, which means that there is no power transfer between the AC grid and the DC side of the LCC.

In practice, the rectification/inversion process of a Graetz bridge is non-ideal, and it is achieved at the expense of drawing reactive power and producing harmonic current distortion on its AC side and harmonic voltage distortion on its DC side. These harmonic currents are termed characteristic harmonics, namely, 5th, 7th, 11th, 13th, 17th, 19th, 23rd, and 25th. If the LCC converter is connected to an unbalanced AC power grid, then non-characteristic harmonics may be produced. Normally, shunt, tuned filters, and high-pass filters are used to prevent these harmonics from reaching the AC power grid.

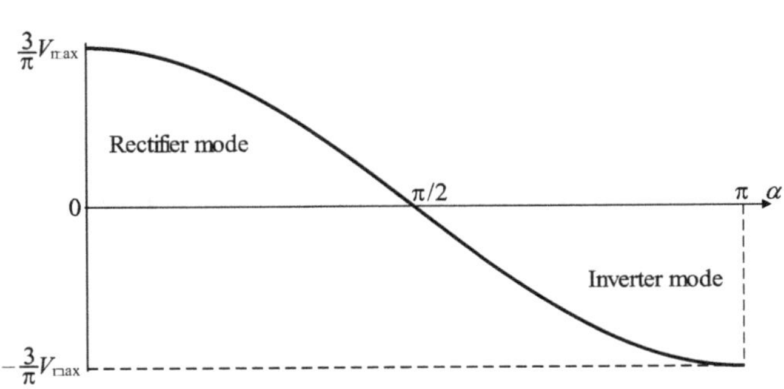

Figure 4.3 Variation of the mean value of the load voltage with firing delay angle in the range $0 < \alpha < \pi$

4.2.2 Effect of commutation on DC voltage

In practice, rectification is non-ideal, i.e., the output DC voltage suffers an area loss, and the current contains ripple. This ripple has a harmonic frequency of six times the main operating frequency (i.e., 50 Hz or 60 Hz) owing to the six-pulse nature of the rectifier circuit. Each thyristor valve conducts the full current for 1/3 of the cycle, and two valves conduct in series at any one time.

Note that the equivalent voltage source representing the AC power grid in Figure 4.1, was drawn with no inductance; thus, assuming an infinitely strong power grid exhibiting a zero-source impedance. In practice, this situation does not exist. Note that even if the AC power grid were to be an infinitely strong system, for practical reasons there is an interfacing transformer sitting between the power converter and the AC power grid and the transformer contains leakage reactance which contributes to the non-ideal process of rectification of the AC waveform using a Graetz bridge.

The main effect of a reactance linking the AC voltage source and the rectifier converter is to reduce the rate of change of current from the outgoing valve into the incoming valve, resulting in a longer commutation time.

During commutation, a self-clearing, short-circuit of rather short duration is established between the phase that is ceasing to conduct and the incoming phase. The magnetic energy stored in the reactance of the former phase is transferred to the reactance of the latter phase. The DC current and the line inductance define the amount of energy being swapped between any given pair of phases undergoing commutation.

The commutation phenomenon affects the mean voltage value (4.2), given a modified expression:

$$V_d = \frac{3}{\pi} V_{\max} \cos(\alpha) - \frac{3X_c}{\pi} I_d \tag{4.2}$$

where X_c is the commutation reactance and I_d is the DC current.

4.2.3 Applications of LCC systems

LCCs are used in both HVDC and FACTS equipment, exploiting different control aspects of LCC converters. As discussed above, in HVDC converters, LCCs are used to delay the transfer of current from one phase to another, i.e., from the outgoing to the incoming phase. In this application, the firing angle takes values between 0 and $\pi/2$ for the case of rectification and between $\pi/2$ and π for the case of inversion. These values apply after the natural $\pi/6$ crossing between the outgoing and incoming valves. In thyristor controlled reactor (TCR) circuits, LCC is used to ignite the current flow through an air-core reactor. In this application, the firing angle takes values between $\pi/2$ and π. This principle applies to the TCR circuits employed in the SVC and the thyristor-controlled series compensator (TCSC) [5]. There is a simpler variant of the LCC, which reduces its action to fire at $\pi/2$, acting effectively as a switch. This finds application in thyristor-switched capacitor circuits. In this section, some of these applications are briefly explained.

4.2.3.1 Thyristor controlled reactor

The TCR is a shunt converter with an inductive load whose reactance can be modified by varying the firing angle of the thyristors. The TCR is employed to modify the reactive power by controlling the load current [6].

The most basic scheme of a TCR is shown in Figure 4.4. Two thyristors placed in anti-parallel are connected in series with the inductor. For firing angles between 90° and 180°, the fundamental harmonic of the current reduces as the firing angle increases. The achieved effect is a net increase in the reactor inductance, and with it, decreasing the reactive power [7].

When the firing angle is $\alpha = 90°$, the load voltage and the load current are continuous with the natural phase-shift of 90° between them, as shown in Figure 4.5. Nevertheless, when the firing angle increases, both, the load voltage and the load current are discontinuous as can be seen in Figures 4.6 and 4.7, where values for the firing angle equal to 120° and 135° have been employed,

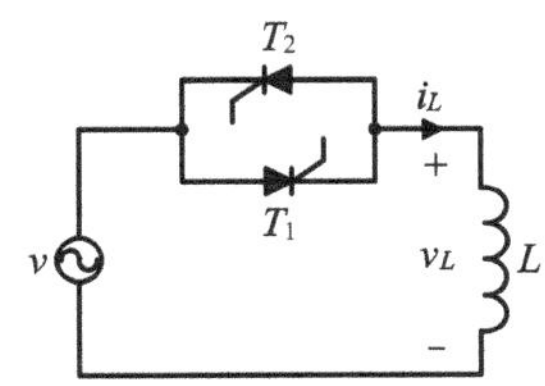

Figure 4.4 Basic scheme of a TCR

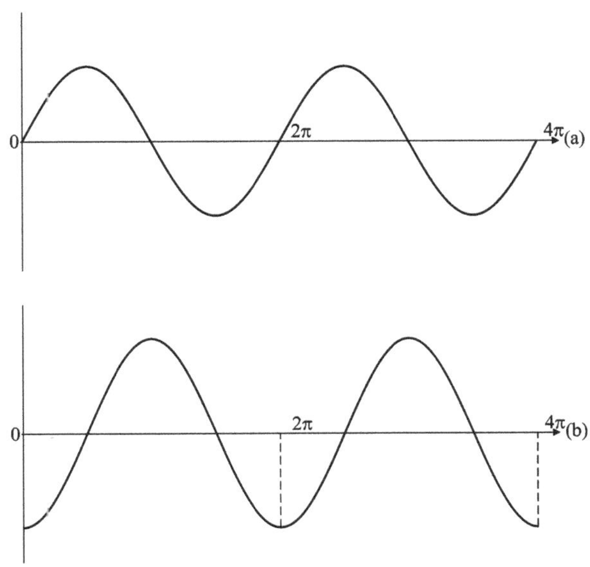

Figure 4.5 Waveforms of the TCR obtained for α = 90°: (a) load voltage, and (b) load current

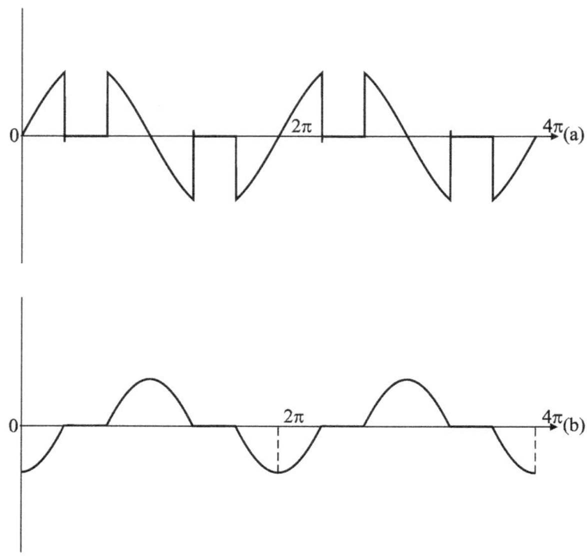

Figure 4.6 Waveforms of the TCR obtained for α = 120°: (a) load voltage, and (b) load current

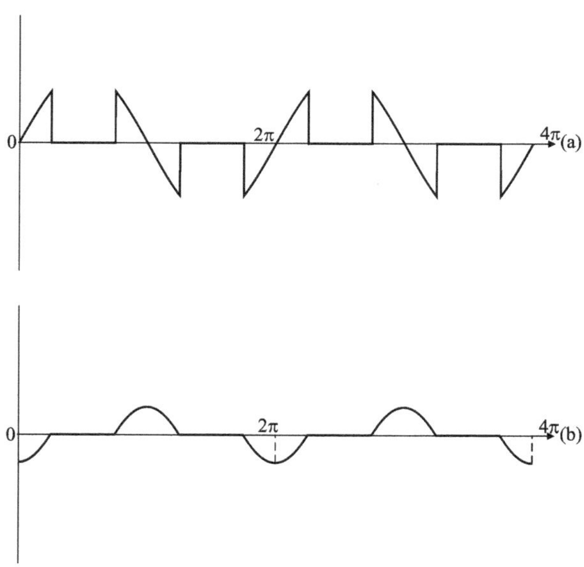

Figure 4.7 Waveforms of the TCR obtained for α = 135°: (a) load voltage, and (b) load current

respectively. As these results show, the fundamental harmonic of the current decreases as the firing angle increases. However, the harmonic content of the current also increases. To overcome this side-effect, different solutions based on multi-stage topologies may be used as proposed in [6].

Apart from the condition shown in Figure 4.5 where a firing angle of 90° was applied – corresponding to a full current conduction – all other firing angles selected will yield non-sinusoidal waveforms. The harmonic currents generated by a single-phase TCR are 3rd, 5th, 7th, 9th, and 11th, and their magnitudes will be a function of the thyristors' firing angles.

In power transmission applications, the normal TCR circuit comprises three single-phase TCRs connected in a delta. It is termed a six-pulse TCR, and the delta connection confines the 3rd harmonic current within the delta, avoiding the need to install a large, tuned filter to prevent this harmonic current from reaching the network. Tuned filters though are needed for filtering out the 5th and 7th harmonics, and a high-pass filter for the 11th and higher harmonic frequency currents.

An alternative arrangement to the six-pulse TCR is the twelve-pulse TCR, where two half-size TCR units are connected in parallel employing a three-phase transformer having two identical secondary windings, except for their connection: one is connected in star and the other in delta. Each TCR unit is connected to one of the secondary windings and owing to the delta connection there is a natural elimination of the 5th and 7th harmonic currents. The only significant harmonic currents left to filter out are the 11th and 13th. This TCR arrangement reduces substantially the filtering requirements but increases the complexity of the whole VAR support installation, particularly in terms of transformer requirements. It should be remarked that the TCR only makes one-half of the Static Var Compensation (SVC) plant; the other half is the thyristor switched capacitor (TSC), as outlined in Section 2.2.1.1 of Chapter 2.

4.2.3.2 Thyristor switched capacitor (TSC)

The TSC is a shunt converter designed to supply reactive power. The scheme of the TSC is depicted in Figure 4.8. It comprises two thyristors connected in an anti-parallel arrangement, like the TCR, along with a capacitor, an inductor, and a small resistance. The inductor and the resistance are employed to limit the inrush current under unexpected operating conditions [8]. For three-phase systems, the TSC is connected in delta [9].

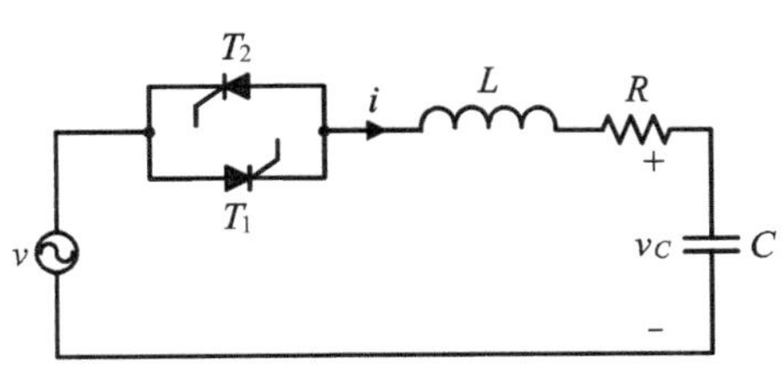

Figure 4.8 Basic scheme of a TSC

Unlike the TCR, the TSC is operated by an ON/OFF control scheme with a firing angle α= 90° [10]. In order to reduce the current transient response, the capacitor should be charged to [11]:

$$v_{c0} = \pm \frac{V_M}{1 - \omega^2 LC} \tag{4.3}$$

where V_M is the amplitude of the grid voltage. Moreover, the thyristors must be switched on at maximum grid voltage, and the voltage at the capacitor must have the same polarity as the grid voltage. If these two conditions are fulfilled, the current only exhibits steady-state behaviour [11]:

$$i = I_M \cos\left(\omega t + \frac{\pi}{2}\right) \tag{4.4}$$

The natural frequency of the TSC is selected to be within the range of [125, 250] Hz for 60-Hz power systems, while for 50-Hz systems, the frequency values are within the range of [120, 210] Hz [5].

Figure 4.9 shows the TSC waveforms. The grid voltage is plotted in Figure 4.9(a), while the capacitor voltage is shown in Figure 4.9(b). The voltage

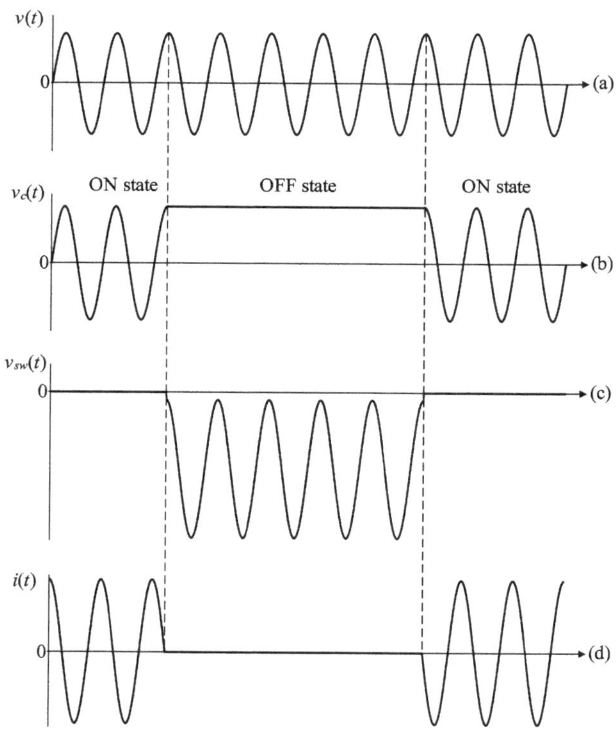

Figure 4.9 Waveforms of the TSC obtained with ON/OFF control: (a) grid voltage, (b) capacitor voltage, (c) thyristors voltage, and (d) current through the capacitor

across the thyristors is plotted in Figure 4.9(c), and the current flowing through the capacitor is plotted in Figure 4.9(d). The results show the ON and OFF states of the TSC. When the TSC is in the ON state, the current leads the capacitor voltage with a phase shift of 90°. When the TSC is in the OFF state, the capacitor voltage fulfills (4.3), and the current is zero. Moreover, the voltage across the thyristors is the difference between the grid voltage and the capacitor voltage, resulting in a negative value. For the connection of the TSC, the thyristors are switched on when the grid voltage reaches the positive peak value. As shown in Figure 4.9(d), there is neither overshoot nor transient in the current, achieving a sinusoidal waveform in consonance with (4.4).

Since the TSC is operated by an ON/OFF control scheme, this switch-like operation does not introduce harmonic distortion.

4.2.3.3 Thyristor-switched series capacitor

The thyristor-switched series capacitor (TSSC) is a capacitive series compensator that can be used to reduce the series reactive impedance of the compensated transmission line. It increases the transient stability limit and provides damping of power oscillations and sub-synchronous oscillations [8].

The basic structure of the TSSC is depicted in Figure 4.10. The TSSC comprises several capacitors in series, with each capacitor being shunted by a bypass switch consisting of two antiparallel-connected thyristors. When the thyristor switches are turned off, the capacitors are inserted into the main circuit [5].

A capacitor can be inserted into the main circuit at zero crossing of the current, as the thyristor switches off when the current crosses zero [8]. Following this action, the capacitor charges, and a full half cycle of current in the opposite direction is required to discharge the capacitor to zero voltage. This introduces a DC offset in the voltage across the capacitor. Accordingly, capacitor insertion should be carried out when the capacitor voltage is zero to mitigate the inrush current [12].

A TSSC with two identical capacitors has been simulated. The obtained results, as waveforms, are shown in Figure 4.11. The TSSC current is plotted in Figure 4.11(a), while the voltages across capacitors C_1 and C_2 are shown in Figure 4.11(b) and (c), respectively. The voltage across both capacitors is plotted in Figure 4.11(d). Initially, both capacitors are bypassed with $v_{C1} = v_{C2} = 0$. This is followed when capacitor C_1 is inserted into the circuit at zero crossing of the current, while capacitor C_2 is inserted after three fundamental periods of the insertion of C_1. Figure 4.11(b) and (c) shows the DC offset of the voltages across the capacitors. Finally, both capacitors are again bypassed when their voltages are equal to

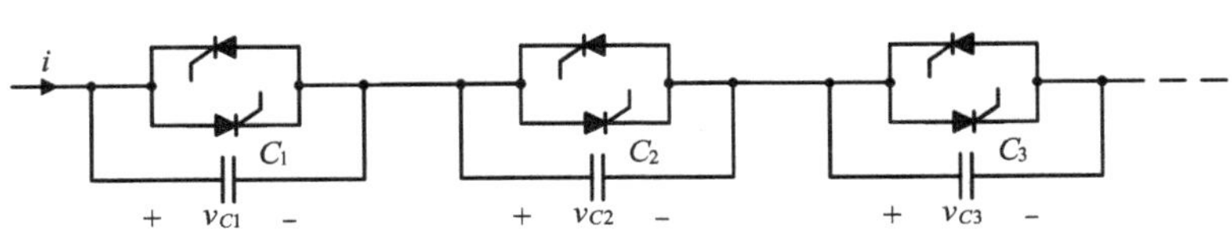

Figure 4.10 Basic scheme of a TSSC

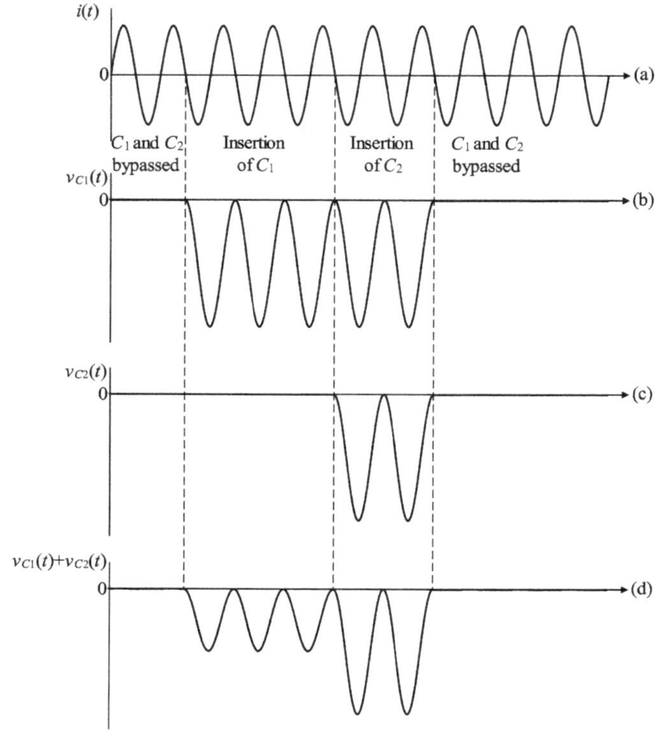

Figure 4.11 Waveforms of a TSSC scheme with two capacitors: (a) current waveform, (b) voltage at capacitor 1, (c) voltage at capacitor 2, and (d) total voltage across both capacitors

zero. Figure 4.11(a) shows that there is no inrush current in the circuit when this commutation is carried out.

4.2.3.4 Thyristor-controlled series compensator

The TCSC is designed to vary the electrical length of a compensated transmission line on an almost instantaneous basis while increasing the stability margins of the power grid [13]. It is a series compensator that increases or decreases the impedance of the compensated transmission line, according to operational requirements [8]. The change in impedance is continuous and with little delay.

The basic scheme of the TCSC is depicted in Figure 4.12: The TCSC comprises a capacitor that is shunted by a switch (SW) consisting of two antiparallel-connected thyristors together with a series inductor. As discussed in Section 2.2.2.1 of Chapter 2, the switch can be operated in vernier mode by controlling the phase of the gate signals of the thyristors by means of a firing angle α [14]. By modifying this firing angle, a variable series reactance $X_{TCSC}(\alpha)$ that exhibits either capacitive or inductive behaviour is achieved. In accordance with [15], the thyristors should

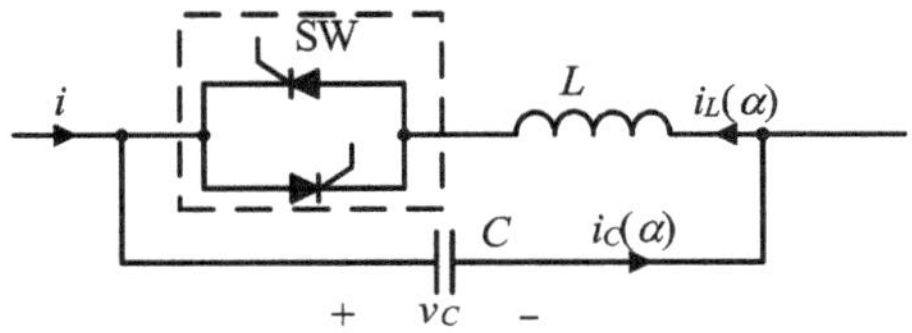

Figure 4.12 General scheme of a TCSC device

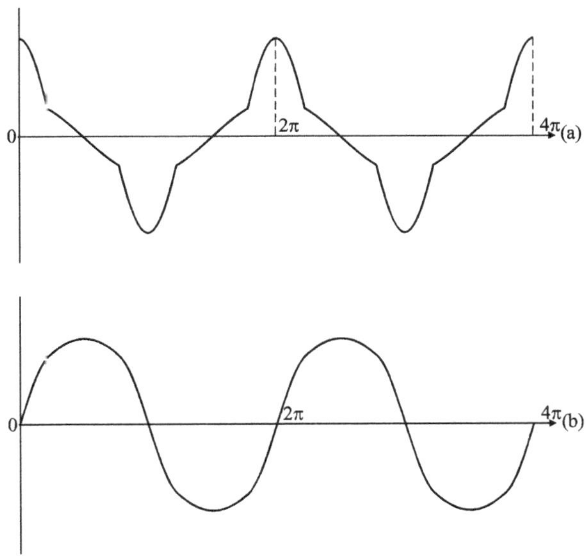

Figure 4.13 Waveforms of the TCSC with α = 140°: (a) capacitor current, and (b) capacitor voltage

be fired when the capacitor voltage and the current have opposite directions, which implies a firing angle between 90° and 180°.

The TCSC voltage and current waveforms are non-sinusoidal, and their harmonic content is a function of the firing angle value. To show this point, simulations are carried out using PSCAD/EMTDC using the circuit in Figure 4.12 with the following parameters: $C = 200$ mF, $L = 12$ mH, and a firing angle $\alpha = 140°$. The results are plotted in Figures 4.13–4.15, where two cycles of the waveforms in steady-state operation, are shown. Figure 4.13(a) plots the current that flows through the capacitor, while the capacitor voltage is plotted in Figure 4.13(b). The complementary waveforms in the inductor are shown in Figure 4.14(a) for the current, and in Figure 4.14(b) for the voltage, respectively. Furthermore, the waveforms in the thyristors are shown in Figure 4.15, where the current is plotted in Figure 4.15(a), and the voltage is shown in Figure 4.15(b).

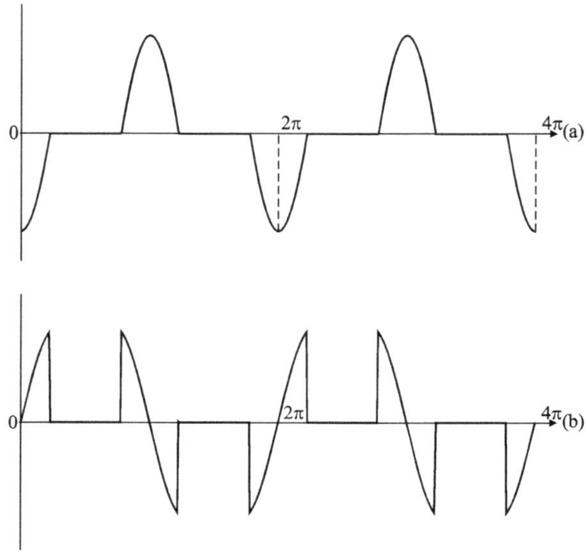

Figure 4.14 Waveforms of the TCSC with α = 140°: (a) inductor current, and (b) inductor voltage

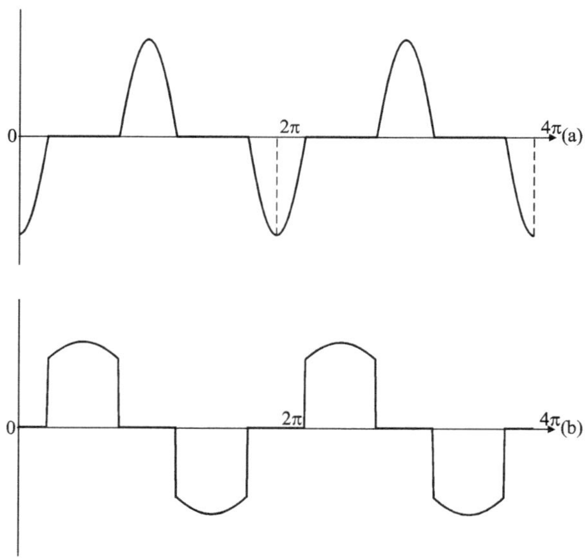

Figure 4.15 Waveforms of the TCSC with α = 140°: (a) switch current, and (b) switch voltage

These results show that all the TCSC voltage and current waveforms are non-sinusoidal. Admittedly, these results are for the specific case when the thyristor firing angle takes a value of $\alpha = 140°$, but to a greater or lesser extent, this feature applies to values of α in the range: $90° < \alpha < 180°$.

To assess the harmonic content of these waveforms as a function of the firing angle, Figure 4.16 shows the root-mean-square (RMS) value of the third harmonic capacitor voltage, normalised to the RMS value of the fundamental voltage. This is presented as a function of the firing angle, which ranges between 90° and 180°. The normalised value of the third harmonic is almost constant and equal to 0.16 pu for α less than 126°. For higher values of α, the third harmonic component decreases to zero in an almost linear fashion.

The corresponding values for the 5th, 7th, 9th, and 11th harmonic frequencies are plotted in Figure 4.17. Comparing the results in Figures 4.16 and 4.17 reveals that the third harmonic is the most significative component in the capacitor voltage for any value of α in the range: $90° < \alpha < 180°$.

Furthermore, the 5th harmonic exhibits a constant value of 0.0037 pu from $\alpha = 90°$ to $\alpha = 126°$; after this point, its value increases, reaching a maximum of 0.023 pu at $\alpha = 143°$, after it decreases to zero in an almost linear fashion, as shown in Figure 4.17(a). The 7th harmonic maintains a maximum value of 0.01 pu, which remains constant up to $\alpha = 126°$. Moreover, the 7th harmonic is zero for $\alpha = 142.5°$ and $\alpha = 180°$, as shown in Figure 4.17(b). Figure 4.17(c) and (d) reveals that the values of the 9th and 11th harmonics, respectively, are smaller than the other harmonics. In addition, the 9th harmonic is zero for $\alpha = 130°$, $\alpha = 151°$ and $\alpha = 180°$, while the 11th component is zero for $\alpha = 139°$, $\alpha = 156.5°$, and $\alpha = 180°$.

As discussed in Section 2.2.2.1 of Chapter 2, the TCSC exhibits an inductive and a capacitive range of operation. For steady-state operation, there is no incentive for operating the TCSC in the inductive region since this will effectively make

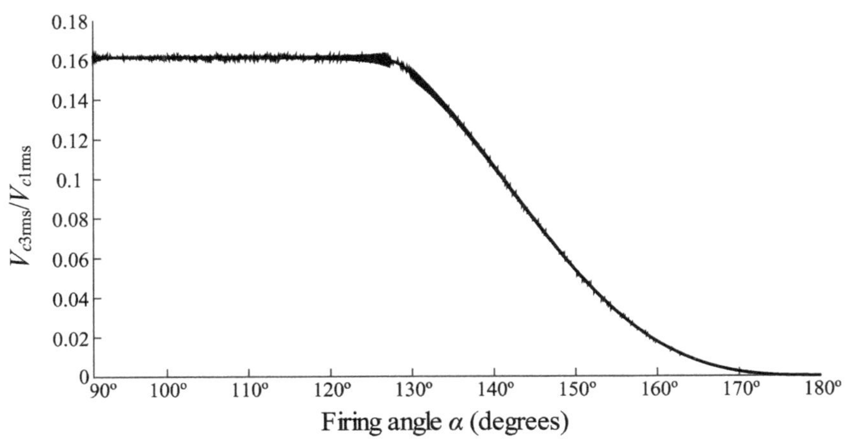

Figure 4.16 Normalised RMS value of 3rd harmonic capacitor voltage as a function of firing angle α

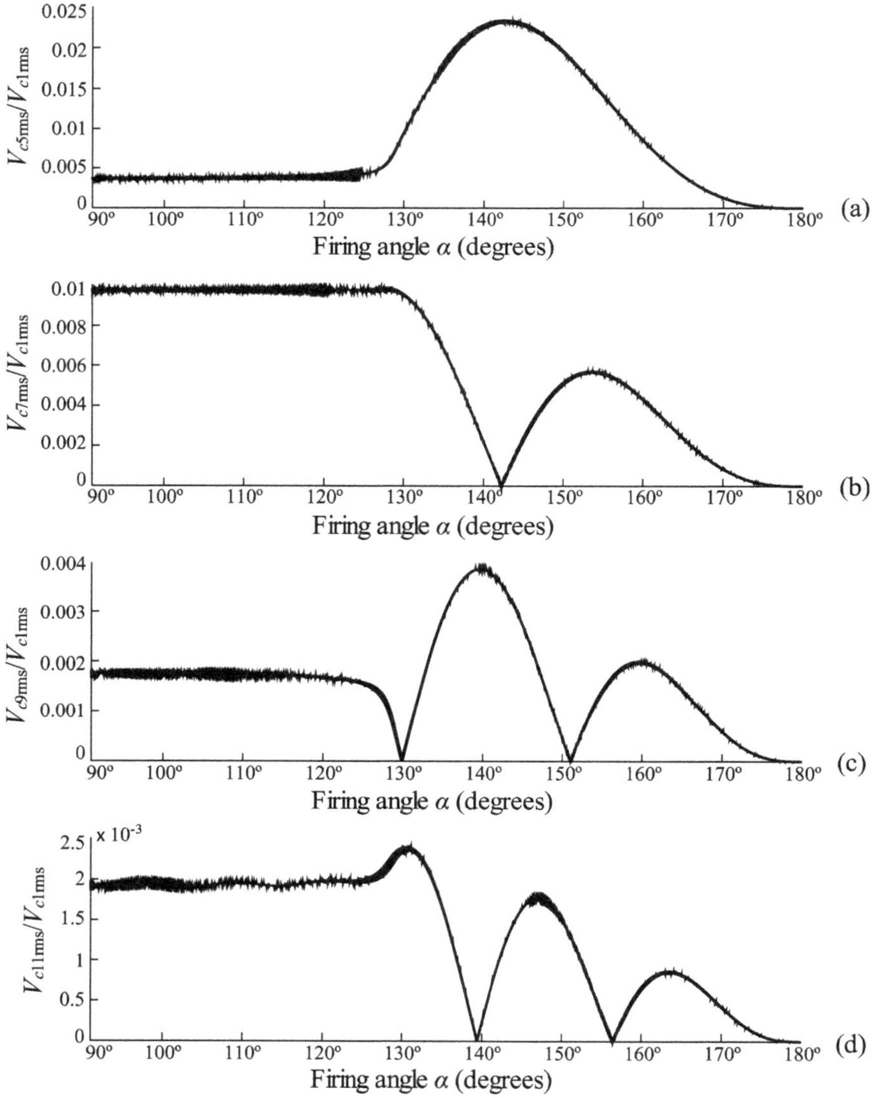

Figure 4.17 Normalised RMS values as a function of the firing angle α for the following components of the voltage capacitor: (a) 5th harmonic, (b) 7th harmonic, (c) 9th harmonic, and (d) 11th harmonic

longer the electrical distance of the compensated transmission line. For the test case simulated here, the range of TCSC capacitive operation would be in the region of 150° to 170°, which would yield acceptable maximum levels of harmonic voltage distortion, i.e., 5% of 3rd harmonic, 2% of 5th harmonic, 0.5% of 7th, 0.2% of 9th, and 0.15% of 11th.

4.2.4 LCC in HVDC systems

Various topologies are available for HVDC systems. Using LCC converters, the simplest HVDC system is the monopolar DC link, as depicted in Figure 4.18. It comprises two Graetz bridges, where one of them operates as a rectifier and the other one works as an inverter. Smoothing reactors and filters are present on the DC side of both the rectifier and inverter bridges. This ancillary equipment is essential for the smooth operation of the HVDC link since the rectification/inversion processes are non-ideal, i.e., these are achieved at the expense of producing harmonic distortion on the DC voltage. Moreover, the rectifier and converter stations act as sinks of reactive power, requiring *in-situ* reactive power compensation to discourage the import of reactive power from the connecting AC power grids. The use of filters prevents the injection of harmonic currents into the connecting AC power grids.

The connecting transformer at the rectifier station is a step-down, load-tap-changing (LTC) transformer. Conversely, the connecting transformer at the inverter station is a step-up LTC transformer. The tap mechanisms (T_r and T_i) at both transformers act as control variables when the HVDC link operates in constant current (power) mode, which is its normal operating mode.

A drawback of this configuration is that the ground is used as a current return path, but it is more economical, as it only uses one metallic conductor. It corresponds to the early HVDC links that existed in the late 1950s and early 1960s. Nowadays, it is used in some submarine cable connections. A simplified representation of the monopolar LCC-HVDC system is shown in Figure 4.19.

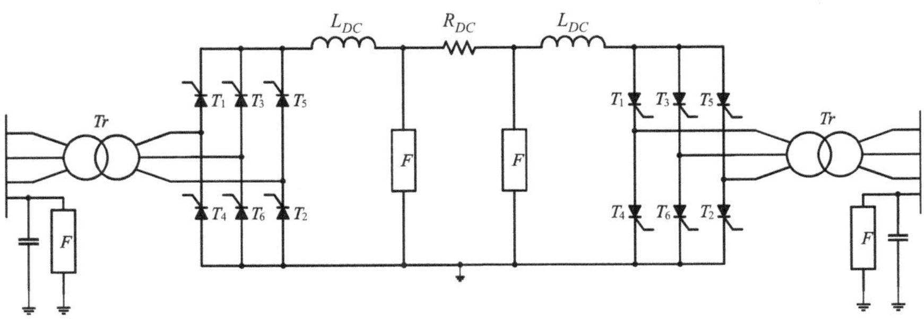

Figure 4.18 Scheme of a monopolar HVDC-based LCC

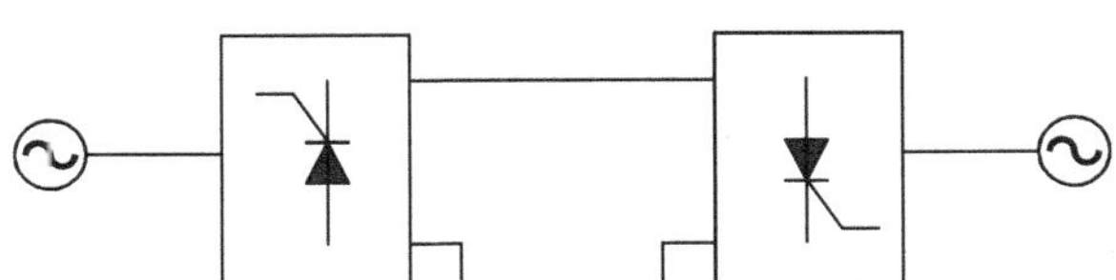

Figure 4.19 Simplified scheme of a monopolar LCC-HVDC system

The structures of modern HVDC links are more complex than the monopolar structure. Standard point-to-point HVDC schemes are homopolar or bipolar, as opposed to monopolar and the converter stations are quadrupoles; this is in the interest of higher power transfer yields and increased reliability. This is exemplified in Figures 4.20 and 4.21 in a simplified manner.

The homopolar configuration comprises two or more DC conductors of the same polarity, which is normally negative to reduce the corona effect. The poles operate in parallel, and the ground is used as the return path. In case of a fault in one conductor, the healthy conductor continues to supply power. The main advantage of this topology is that it has lower installation costs than other HVDC configurations [16].

The bipolar configuration is by far the most popular HVDC configuration. It is made up of two identical monopolar links, one at positive and one at negative polarity with respect to ground. Each monopolar side can operate on its own with ground return; but when the two poles operate with equal currents, they cancel each other out at ground return. The ground path is used for limited periods in an emergency when one pole is out of service [17]. The cost of this topology is reduced compared to two monopolar systems since it employs a common ground path [18].

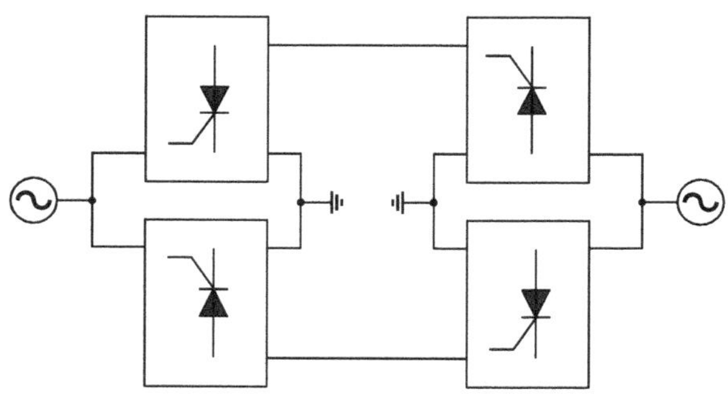

Figure 4.20 Homopolar LCC-HVDC system

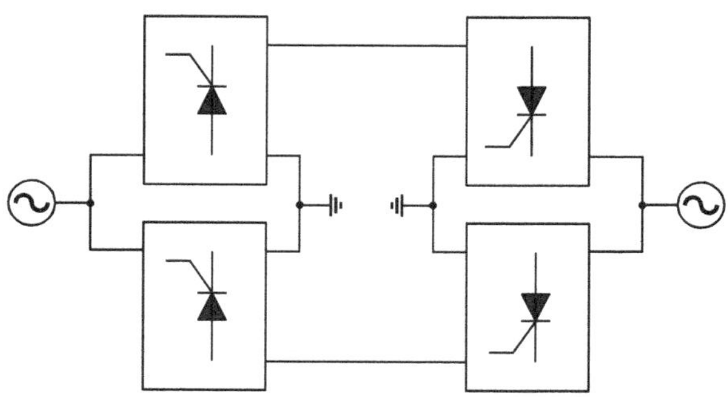

Figure 4.21 Bipolar LCC-HVDC configuration

More than two terminals can be connected in tandem to build a so-called multiterminal configuration. For instance, in Figure 4.22, three terminals are used: the terminal placed on the right side operates in inverter mode whereas the other two terminals work as rectifiers. Alternatively, with an automatic reversal of voltage and firing-angle control, other operating combinations are possible.

A rather comprehensive list of LCC-HVDC installations is given in Section 2.3.4 of Chapter 2. It is seen that these systems continue to be installed even though Modular Multilevel Converter (MMC)-HVDC systems are available from all the major manufacturers of HVDC equipment. One reason may be that the LCC-HVDC option may be cheaper than the equivalent VSC-HVDC option and that the technical requirements are sufficiently met by the former alternative. A case in point is Estlink 2 between Estonia and Finland, commissioned in 2014; it is a submarine LCC-HVDC system and contrasts with the VSC-HVDC Estlink 1 commissioned in 2006. However, the main reason for the continuous installation of LCC-HVDC systems is the capacity of the link. To date, the highest rating VSC-HVDC system is 1,400 MW and ±525 kV, which is insufficient to meet the largest power transmission requirements. Table 4.1 lists some of these projects [17].

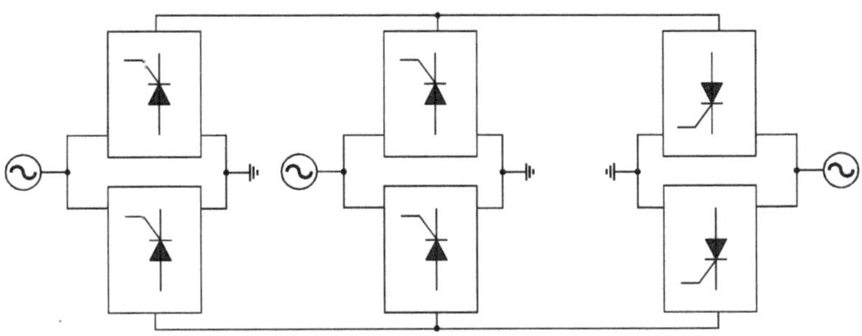

Figure 4.22 Multiterminal LCC-HVDC topology

Table 4.1 Examples of LCC-HVDC facilities

Installation name	Rated power (MW)	Rated voltage (kV)	Year
Jinpin – Sunan	7,200	±800	2012
Mundra – Haryana	2,500	±500	2012
Rio – Madeira	7,100	±600	2013
Xiluodu – Guangdong	6,400	±500	2013
Nuozhadu – Guangdong	5,000	±800	2013
Southern Hami – Zhengzhou	8,000	±800	2014
Biswanath – Agra	6,000	±800	2014
Xiluodu – Zhejiang	8,000	±800	2014
Zhu Ndong – Sichuan	10,000	±1,100	2015
Xingu – Rio	4,000	±800	2019
Changji – Guquan	12,000	±1,100	2019

4.3 Voltage-source converters

4.3.1 GTO- and IGBT-based VSCs

Voltage Source Converters (VSCs) used in HVDC systems employ two types of semiconductors: Si Gate Turn-Off Thyristors (GTOs), and Si Insulated Gate Bipolar Transistors (IGBTs). The main characteristics of these valves have been addressed in Chapter 3 mainly from the physics of semiconductors. Hence, additional characteristics are discussed here to explain their great success in HVDC systems.

At this point in time, IGBTs with anti-parallel diodes are preferred over GTOs [19] and are employed in state-of-the-art VSCs [20]. They offer several advantages over GTOs, such as a faster switching frequency, a lower power consumption of the control circuit, and the control operation of the device by using a voltage signal. Moreover, GTOs have several disadvantages like complex gate drive circuits to achieve the turn-off of the device, the high power required to control the device, and high switching power losses. However, they possess some very interesting features: high current density during conduction, while being able to block high voltages when in switched-off mode. Moreover, GTOs withstand large voltage gradients and can integrate an inverse diode [21]. These features make GTOs a useful option for very high-power HVDC systems, even though conventional thyristors are still the preferred option in the gigawatt range, as exemplified by the ultra-high voltage DC links in Table 4.1.

4.3.2 Three-phase two-level VSCs

The three-phase two-level VSC is the most fundamental converter topology utilised in HVDC systems: As depicted in Figure 4.23, this VSC comprises three independent legs, each corresponding to one phase. Each leg is equipped with two switches, i.e., IGBTs, and each switch has a diode connected in antiparallel, facilitating the flow of reverse current. The operation of this VSC can be carried out using either a Pulse-Width-Modulation (PWM) scheme or the square-wave mode.

In terms of conduction, the two switches of the same leg cannot be turned on simultaneously in order to avoid short circuits. Similarly, both switches cannot be turned off at the same time to ensure that the output voltage remains referenced.

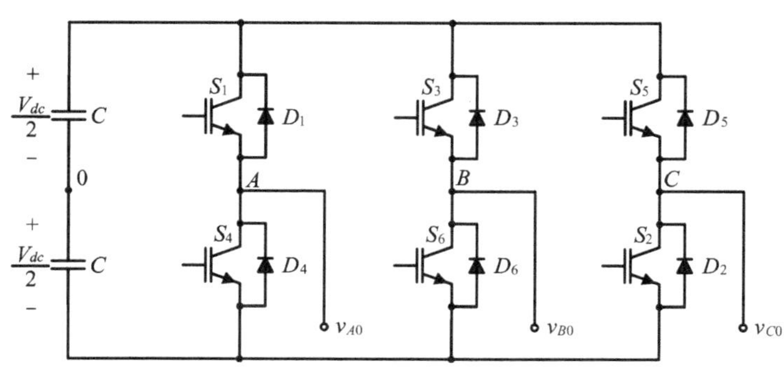

Figure 4.23 Scheme of a three-phase two-level VSC

If the VSC is controlled by a PWM scheme, two possibilities are available: Space Vector Pulse Width Modulation (SVPWM) and Sinusoidal Pulse Width Modulation (SPWM).

In accordance with [22] the SVPWM technique provides a higher output voltage for the same DC-bus voltage, lower switching losses, and better harmonic performance in comparison with the SPWM method. Although traditionally SVPWM is employed in the control of electric drives [23], research work in the field of HVDC systems and FACTS has been carried out using the SVPWM technique [24].

In the SVPWM, the voltage generated by the VSC is expressed in a static reference frame using the Clarke transformation, i.e., in α–β coordinates. Only eight modes are possible according to the restrictions of the switches of one leg. These modes are shown in Table 4.2 based on the states of the three upper switches (S1, S3, and S5 – refer to Figure 4.23).

Depending on the states of the switches, an output voltage is obtained in each branch of the inverter. The voltages per branch, for a balanced star load, are shown in Table 4.3.

Figure 4.24 shows the voltages in α–β coordinates for each mode of the VSC when a star load is used. Each point signifies the spatial voltage vector generated in the corresponding mode, and the lines connecting them represent the adjacent modes. Adjacent modes are those where only a state change of one leg is necessary to transition from one mode to another. In modes 7 and 8, the voltage is zero, and these are referred to as inactive modes, while the rest are referred to as active modes.

In order to generate the vector, $\mathbf{u}_s$, the VSC must be switched between various modes during a switching period t_s, such that the average value of the voltage vector in the stationary axes generated during t_s coincides with the magnitude of the vector $\mathbf{u}_s$. To achieve this, there are infinite combinations of the inverter modes. In

Table 4.2 Different VSC modes

Switch/Mode	1	2	3	4	5	6	7	8
S1	ON	ON	ON	OFF	OFF	OFF	ON	OFF
S3	OFF	OFF	ON	ON	ON	OFF	ON	OFF
S5	OFF	ON	OFF	ON	OFF	ON	ON	OFF

Table 4.3 Line-to-neutral voltages for a star connected load

Voltage/Mode	1	2	3	4	5	6	7	8
v_{an}	$\frac{2}{3}V_{dc}$	$\frac{1}{3}V_{dc}$	$\frac{1}{3}V_{dc}$	$\frac{-2}{3}V_{dc}$	$\frac{-1}{3}V_{dc}$	$\frac{-1}{3}V_{dc}$	0	0
v_{bn}	$\frac{-1}{3}V_{dc}$	$\frac{-2}{3}V_{dc}$	$\frac{1}{3}V_{dc}$	$\frac{1}{3}V_{dc}$	$\frac{2}{3}V_{dc}$	$\frac{-1}{3}V_{dc}$	0	0
v_{cn}	$\frac{-1}{3}V_{dc}$	$\frac{1}{3}V_{dc}$	$\frac{-2}{3}V_{dc}$	$\frac{1}{3}V_{dc}$	$\frac{-1}{3}V_{dc}$	$\frac{2}{3}V_{dc}$	0	0

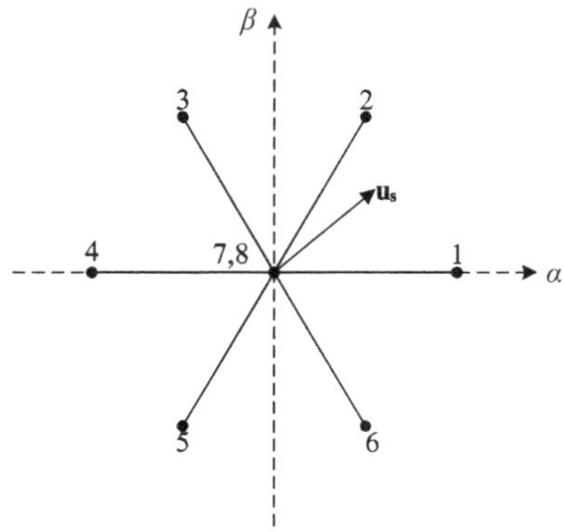

Figure 4.24 Voltages generated in α–β coordinates

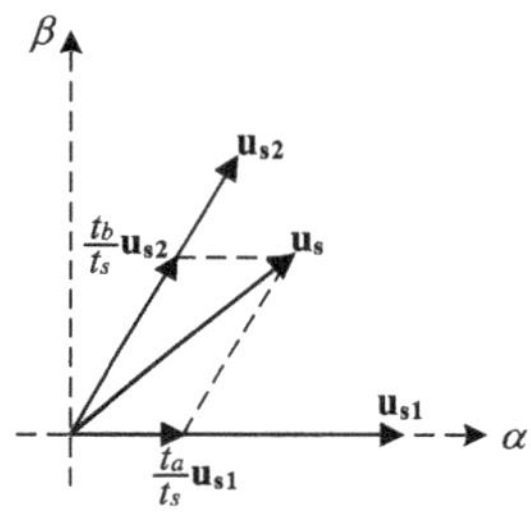

Figure 4.25 Decomposition of the voltage vector in two adjacent modes

[25], a method is presented that minimises the number of commutations during t_s: every generated vector is always located between two adjacent active modes a and b (modes 1 and 2 in Figure 4.25). The vector $\mathbf{u_s}$ can be generated by applying mode a for a time t_a (thus applying the voltage $\mathbf{u_{sa}}$ to the load) and mode b for a time t_b (in this case the applied voltage is $\mathbf{u_{sb}}$), such that the following equation is fulfilled:

$$t_s \cdot \mathbf{u_s} = t_a \cdot \mathbf{u_{sa}} + t_b \cdot \mathbf{u_{sb}} \tag{4.5}$$

In general, t_a+t_b will be less than t_s, therefore the remaining time (t_s-t_a-t_b) is distributed between the inactive modes 7 and 8.

To minimise the number of switch commutations, transitions are always between adjacent modes. Therefore, a switching period always starts and ends with an inactive mode. Furthermore, the time that the VSC is in modes 7 and 8 is evenly distributed between both modes:

$$t_7 = t_8 = \frac{t_s - t_a - t_b}{2} \tag{4.6}$$

When the SPWM scheme is employed to control the VSC, the firing signals of the switches are determined by the intersections of a sinusoidal reference signal, i.e., a modulating signal, of fundamental frequency and given amplitude in per-unit, with a triangular signal called carrier signal, whose frequency is the switching frequency of the VSC. There are three modulating signals, one per leg of the VSC, with a 120° phase shift between them in order to achieve a balanced set of three-phase voltages. The ratio between the amplitude of the modulation signal and the amplitude of the carrier wave is the amplitude modulation index (m_a), while the ratio between the frequency of the carrier wave and the frequency of the modulating signal is the frequency modulation index (m_f) [4,26].

Figure 4.26 shows the principle of the SPWM process, where a value of 9 has been chosen for m_f, and $m_a<1$. In this case, only one modulating signal has been used for the sake of simplicity to depict the switching pattern of the corresponding leg. When the modulating waveform is higher than the carrier signal, the upper switch of one leg of the VSC in Figure 4.23 is turned on, the output voltage of that leg is, therefore, $V_{dc}/2$ with regard to the midpoint of the DC side of the VSC. On the other hand, when the carrier signal is higher than the modulating waveform, the lower switch of the leg is turned on and the output voltage is $-V_{dc}/2$.

It is desirable for the frequency modulation index to be an integer, odd, and a multiple of three in order to ensure that the line-to-line output voltage of the inverter does not contain even harmonics or harmonics that are multiples of three. This exploits the fact that the modulating wave has half-wave and quarter-wave symmetries with respect to the carrier wave [27]. For high values of the m_f index, the harmonics of the output voltage appear in sidebands around the switching frequency, i.e., the frequency of the carrier signal, and its multiples, according to (4.7),

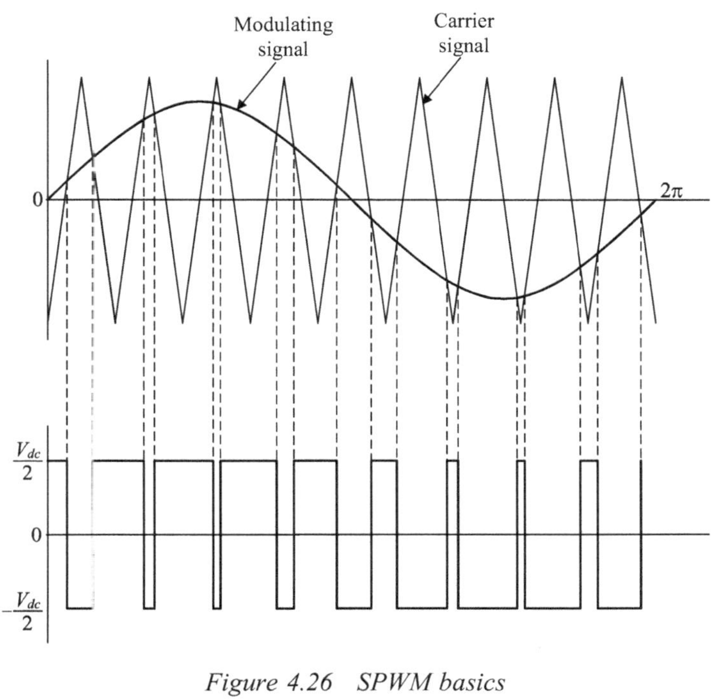

Figure 4.26 SPWM basics

where f_1 is the fundamental frequency, i.e., the frequency of the modulating signal, index i takes even values for odd values of index j, while i is odd when j takes even values [4]:

$$f_h = (i \cdot m_f \pm j) \cdot f_1 \tag{4.7}$$

Therefore, to achieve currents with low harmonic distortion, it is desirable to work with high values of the frequency modulation index.

Furthermore, if the m_a index is lower or equal to one, the amplitude of the fundamental harmonic of the line-to-neutral voltage of the VSC can be obtained as:

$$\widehat{V}_{Ln_1} = m_a \cdot \frac{V_{dc}}{2} \tag{4.8}$$

The root-mean-square (RMS) value of the fundamental component of the line-to-line output voltage is derived from (4.8) as follows:

$$V_{LL_{1\text{rms}}} = \sqrt{\frac{3}{2}} \cdot m_a \cdot \frac{V_{dc}}{2} = 0.6123 \cdot m_a \cdot V_{dc} \tag{4.9}$$

Equation (4.9) shows that the fundamental harmonic of the output voltage of the VSC can be changed linearly by modifying the amplitude modulation index. Nevertheless, the m_a index can be higher than 1 and, under these circumstances, the VSC operates in square-wave mode: The fundamental component of the output voltage has constant amplitude, and all switches operate at a 50% duty cycle. Moreover, the switching frequency corresponds to the fundamental frequency.

Figure 4.27 illustrates the waveforms of the VSC output voltage in square-wave mode: Figure 4.27(a) depicts the line-to-neutral voltage referenced to the

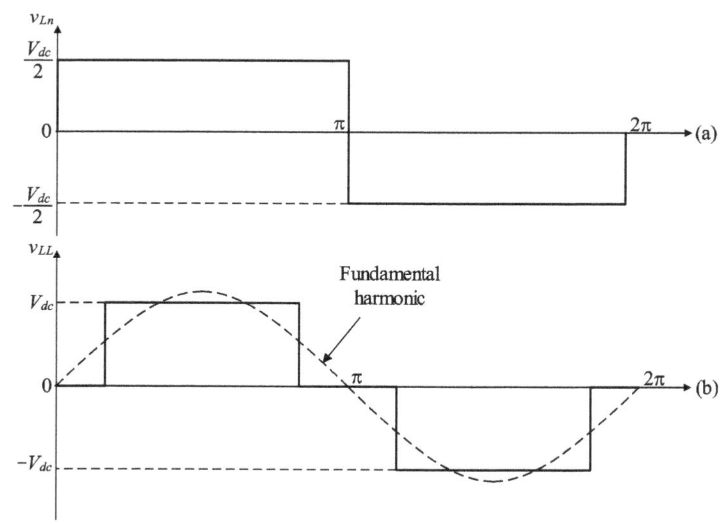

Figure 4.27 Square wave mode in the VSC: (a) line-to-neutral output voltage, and (b) line-to-line output voltage

midpoint of the DC side. This voltage is a square waveform whose amplitude is $V_{dc}/2$. Moreover, the line-to-line output voltage is plotted in Figure 4.27(b) along with its fundamental component. The harmonics present in the line-to-line voltage are odd components excluding those that are multiple of 3.

4.3.3 Three-phase multilevel VSCs

The conventional three-phase, two-level VSCs face some challenges when applied to HVDC systems, such as high switching losses, high voltage-derivative stress, as well as difficulties in connecting many IGBTs in series, owing to the very high operating voltages required and ensuing electromagnetic interference complications.

To address these issues, multilevel converters have emerged as a powerful solution for HVDC applications. Multilevel converters can generate output voltages with multiple levels by using several DC sources or capacitors in series or parallel. This results in lower switching losses, lower common-mode voltages, lower voltage-derivative stress, higher voltage and power ratings, and better harmonic performance [28]. Furthermore, multilevel converters offer higher efficiency, modularity, scalability, and fault tolerance [29].

Among the various topologies of multilevel converters, three types have been used in HVDC applications: the neutral point clamped (NPC), the flying capacitor (FC), and the very successful modular multilevel converter (MMC). The current multilevel converters exhibit a great deal of reliability and efficiency as proven by the many submarine DC installations used to evacuate offshore wind power where MMC is the *de facto* power converter technology. Nonetheless, the search for improvements is never-ending [30].

4.3.3.1 NPC topology

The NPC converter is one of the earliest and most popular multilevel converters. It consists of several levels of clamping diodes connected to a midpoint of two DC capacitors. The NPC converter can generate three or more levels of output voltage by switching different combinations of devices. The devices are controlled to produce a staircase waveform that approximates a sinusoidal waveform. The number of levels determines the resolution and the quality of the output voltage.

The NPC converter employs clamping diodes to ensure the proper distribution of voltage across the power switches. This converter was first proposed by Baker and Nabae *et al.* in 1981 [31] and has been widely used in high-power applications.

The main features of NPC converters are:

- They produce three output voltage levels: positive, negative, and zero, which improves the quality of the output waveform and reduces harmonic distortion.
- They operate at higher voltages than two-level converters since the power switches are exposed to only half of the DC-link voltage.
- They operate with reduced switching power losses and increased efficiency by using different modulation techniques.

The basic structure of a three-level NPC converter is shown in Figure 4.28. It consists of three legs, each with four power switches (usually IGBTs with

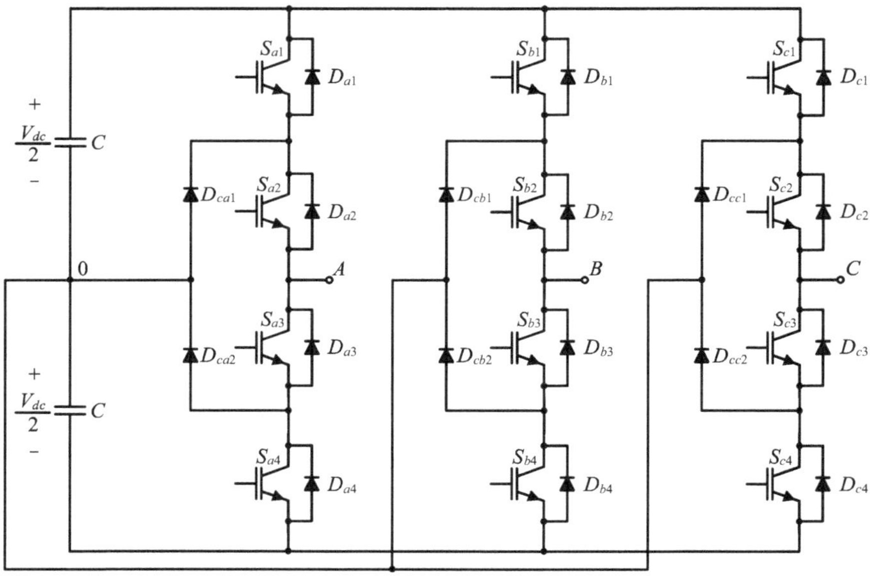

Figure 4.28 Scheme of a three-level NPC VSC

Table 4.4 Switching states in leg A

Switch/voltage v_{A0}	**$\frac{V_{dc}}{2}$, P**	**0, 0**	**$-\frac{V_{dc}}{2}$, N**
S_{a1}	ON	OFF	OFF
S_{a2}	ON	ON	OFF
S_{a3}	OFF	ON	ON
S_{a4}	OFF	OFF	ON

anti-parallel diodes) and two clamping diodes. The DC bus comprises two capacitors, with their midpoint connecting to the clamping diodes. The output voltage of each leg can take positive, negative, or zero values, depending on the switching states of the power switches.

To control the NPC converter, different modulation techniques can be used, sinusoidal PWM or space vector PWM (SVM). These techniques determine the switching patterns of the switches to generate the desired output voltage. A significant challenge in this converter topology is to balance the voltages of the two DC capacitors.

Table 4.4 shows the possible switching states and output voltages for leg A of the NPC converter. There are three kinds of switching states: P (output voltage $V_{dc}/2$), 0 (zero output voltage), and N (output voltage $-V_{dc}/2$) in each leg.

The space vector diagram for a three-level NPC converter in α–β coordinates is shown in Figure 4.29. It consists of 19 vectors: one zero vector ($\mathbf{u}_0$), six small

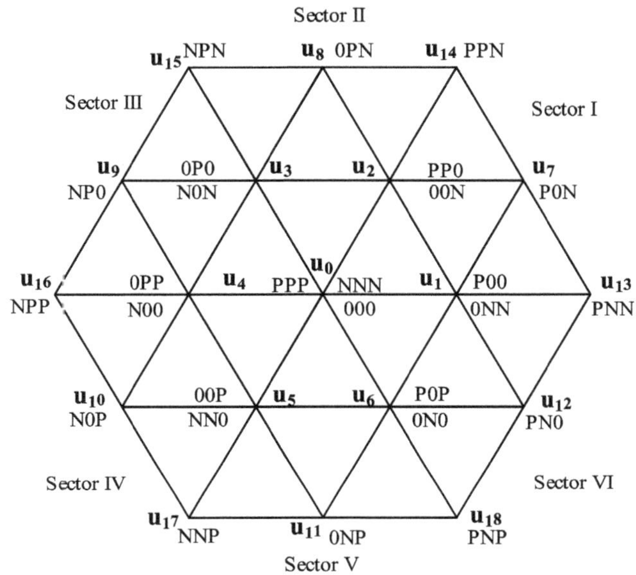

Figure 4.29 Space-vector diagram for a three-level NPC converter

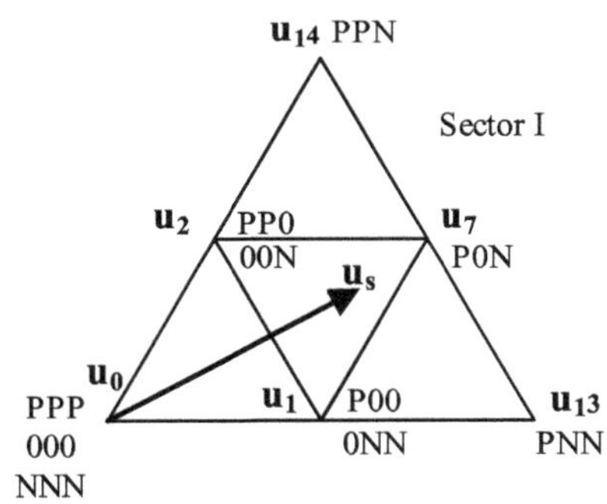

Figure 4.30 Synthesis of vector $\mathbf{u}_s$ in sector I

vectors ($\mathbf{u}_1$–$\mathbf{u}_6$), six medium vectors ($\mathbf{u}_7$–$\mathbf{u}_{12}$), and six large vectors ($\mathbf{u}_{13}$–$\mathbf{u}_{18}$). Each vector corresponds to a switching state of the converter, as shown in Table 4.4 [32].

The SVM technique consists of synthesising a desired output voltage vector $\mathbf{u}_s$ by applying a combination of adjacent vectors for a certain duration. The duration of each vector is proportional to its distance from $\mathbf{u}_s$. The principle of volt-second balance is employed to select the closest three space vectors. The selection of these three proximate vectors is aimed at minimising the harmonic distortion in the voltage output of the converter. For example, in accordance with [32], if $\mathbf{u}_s$ is located in sector I as shown in Figure 4.30, it can be synthesised by fulfiling

(4.10) and (4.11):

$$\mathbf{u}_1 \cdot t_1 + \mathbf{u}_2 \cdot t_2 + \mathbf{u}_7 \cdot t_7 = t_s \cdot \mathbf{u}_\mathbf{s} \tag{4.10}$$

$$t_1 + t_2 + t_7 = t_s \tag{4.11}$$

where t_1, t_2 and t_3 are the times that vectors $\mathbf{u}_1$, $\mathbf{u}_2$ and $\mathbf{u}_3$ are applied, respectively.

One of the main issues to consider in the NPC converters is the balance of the neutral point to avoid overvoltage or undervoltage stress on the switches and diodes. In accordance with [33], the neutral point voltage depends on three main factors: the switching states of the converter, the uneven shunt losses of the clamping diodes, and the capacitance errors of the DC link capacitors. The switching states determine how the converter connects the output phases to the positive, negative, or neutral points of the DC link. Depending on the load current and power factor, different switching states can cause a charge or discharge of the capacitors, resulting in a fluctuation of the neutral point voltage. The uneven shunt losses are caused by the different forward voltages and reverse recovery currents of the clamping diodes, which also affect the charge balance of the capacitors. The capacitance errors are due to the manufacturing tolerances and aging effects of the capacitors, which make them have different values and equivalent series resistances. Different methods are available to achieve the neutral-point balance; some of them are detailed in [34,35].

An alternative to the SVM applied to the NPC converter is the use of the phase-shifted SPWM technique, which is based on comparing a sinusoidal reference signal with triangular carrier signals to generate the switching signals for each phase. The carrier signals have a constant frequency, which determines the switching frequency of the NPC converter. The number of carrier signals n is calculated as $n = m - 1$ [36], with m being the number of levels of the NPC converter. For example, for a three-level, three-phase NPC converter, two carrier signals are used for each phase, as shown in Figure 4.31. The two carrier waveforms along with the three modulating waveforms corresponding to each of the three legs of the converter are plotted in Figure 4.31(a). The output voltage of phase A with respect to the neutral point is shown in Figure 4.31(b) and exhibits three levels, while the line-to-line output voltage contains five levels and is plotted in Figure 4.31(c).

Furthermore, the NPC converter can be operated in square-wave mode. In this mode, the converter switches at the fundamental frequency of the output voltage and produces a square waveform with an amplitude of $\pm V_{dc}/2$. The fundamental component of the output voltage can be controlled by modifying the phase-shift angle α shown in Figure 4.32(a) and (b) in accordance with [4,37].

For a three-level NPC converter, the line-to-neutral voltages contain three levels, while the line-to-line voltage exhibits five levels, as shown in Figure 4.32(c).

The square wave mode has the advantage of low switching losses, but the disadvantages of a higher harmonic distortion and a lower voltage-control capability.

These three methods of controlling the output voltage in an NPC converter can be also applied to the operation of FC multilevel converters.

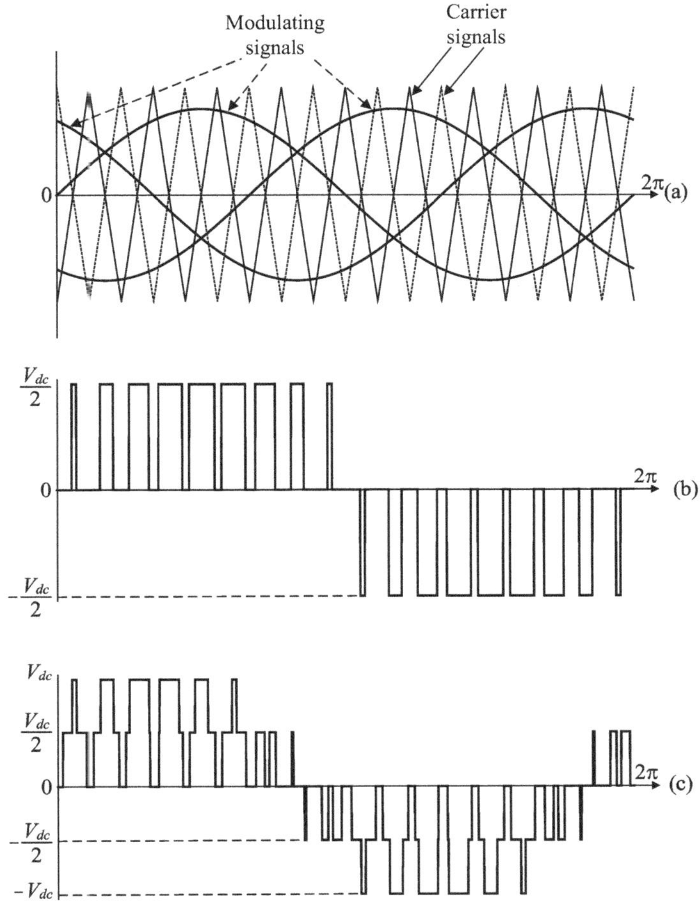

Figure 4.31 Sinusoidal PWM scheme for a three-level converter: (a) carrier and modulating signals; (b) line-to-neutral voltage, and (c) line-to-line voltage

4.3.3.2 FC topology

The FC converter is a multilevel topology where the clamp diodes of the NPC converter are substituted by capacitors. The principle of FC converters is the use of capacitors charged to different voltage levels to synthesise the output voltage waveform. This approach offers several advantages, including reduced harmonic distortion and lower switching power losses. The most basic structure of an FC converter is depicted in Figure 4.33, where a three-level converter is shown. The flying capacitors C_{FCa}, C_{FCb}, and C_{FCc} are charged to $V_{dc}/2$, and a suitable combination of the switches provides three possible values for the output voltage of the line-to-neutral point, i.e., $V_{dc}/2$, 0 and $-V_{dc}/2$.

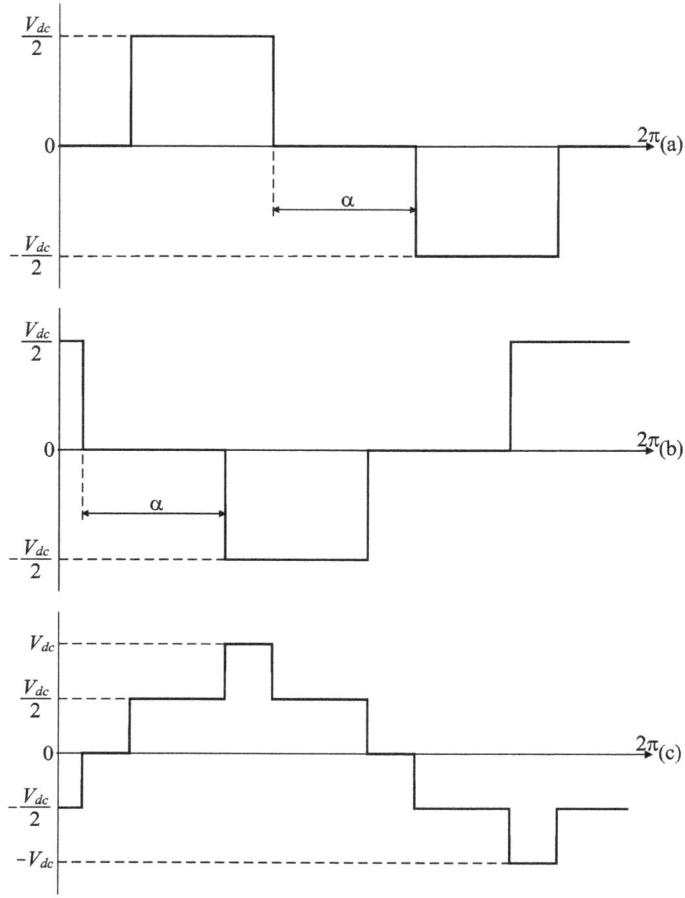

Figure 4.32 Square wave operation in a three-level converter: (a) output voltage v_{A0}; (b) output voltage v_{B0}, and (c) line-to-line output voltage v_{AB}

Table 4.5 shows the possible switching states and output voltages for leg A of the FC converter. Like in the case of the NPC converter, three kinds of switching states are possible: Output voltage equal to $V_{dc}/2$, zero output voltage, and output voltage equal to $-V_{dc}/2$. The zero-output voltage can be obtained by switching off devices S_{a1} and S_{a3} while turning on devices S_{a2} and S_{a4} or, alternatively, by switching on devices S_{a1} and S_{a3} while turning off devices S_{a2} and S_{a4}. This redundancy is an advantage to maintaining balanced voltage at the flying capacitors. Moreover, the gate signals of devices S_{a1} and S_{a4} are complementary, and so are the gate signals of switches S_{a2} and S_{a3} [38].

The FC topology has the advantage of presenting a modular configuration that allows to achieve VSCs with a higher number of levels than those obtained with the

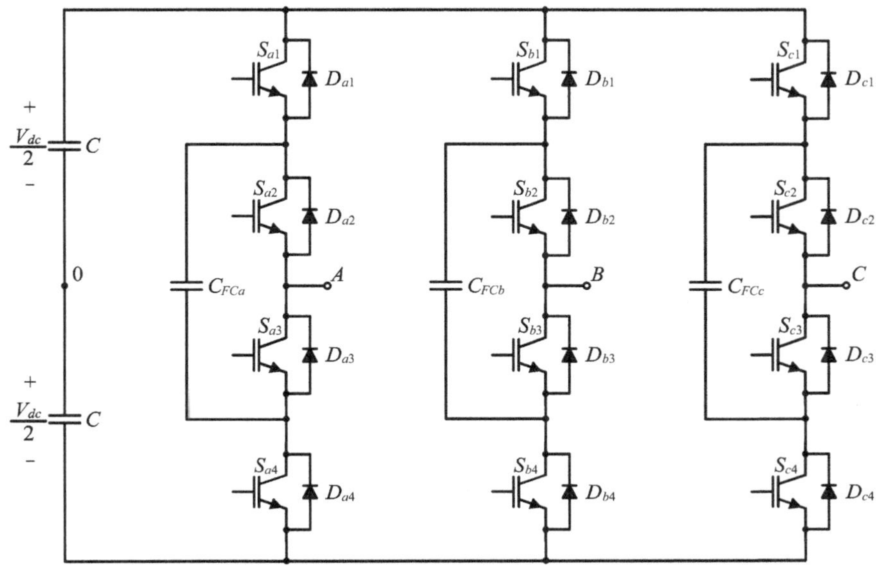

Figure 4.33 Scheme of a three-level FC VSC

Table 4.5 FC converter. Switching states in leg A.

Switch/voltage v_{A0}	$\frac{V_{dc}}{2}$	0	0	$-\frac{V_{dc}}{2}$
S_{a1}	ON	OFF	ON	OFF
S_{a2}	ON	ON	OFF	OFF
S_{a3}	OFF	OFF	ON	ON
S_{a4}	OFF	ON	OFF	ON

NPC topology [39]. Furthermore, the flying capacitor voltages can be controlled by selecting the redundant states that are inherent to this topology [40].

However, the FC converters do have the following drawbacks [41]:

- Many flying capacitors are required when the number of levels is increased. Furthermore, additional pre-charge circuits must be included to charge the capacitor to the suitable voltage level.
- Complex control algorithms are employed to keep balanced the flying capacitor voltages.
- A switching frequency in the range of 1 kHz and 2 kHz is required to obtain a suitable voltage balance. This value might not be feasible in high-power applications that employ switching frequencies up to 700 Hz [30].

As explained in Section 4.3.3.1, the FC VSCs can be operated using the techniques applied to NPC converters, i.e., SVM and SPWM strategies, as well as the square-wave operation.

4.3.3.3 Modular multilevel VSCs

A newer multilevel converter topology aimed at HVDC systems is the so-called modular multilevel converter (MMC) [42,43], developed at the start of the century [44]. The MMC topology addresses successfully some of the main drawbacks of the NPC and FC converters and offers numerous advantages over the conventional two-level converters such as modularity, scalability, and high efficiency, making it the de facto converter standard in modern power systems since the year 2010 [45]. MMCs generate a near-sinusoidal voltage waveform. As illustrated in Figures 2.29 and 2.30 of Chapter 2, this is achieved by incrementally switching each module in sequence, aiming to approximate a sinusoidal waveform as closely as feasible, depending on the number of modules, with switching frequencies in the range of 150 Hz to 200 Hz [26]. It should be noted that these switching frequency values are significantly lower than those used in FC converters.

An MMC comprises multiple power modules (PM) or sub-modules. Each PM comprises a bidirectional half-bridge converter which can be controlled independently, as shown in Figure 4.34. Two IGBTs are the controllable switches employed in the PM.

These power modules are connected in series to form a chain, which can then be used to synthesise the output voltage waveform. For example, the connection in a series of two PMs is shown in Figure 4.35. Given that the two switches of a module cannot be simultaneously on, if switches S_2 in both modules are off then the module is off, and the total output voltage is $v_{out} = 0$. In the case that switches S_1 of both modules are on, the total output voltage is $v_{out} = 2\ V_c$. Finally, if switch S_1 of one module and switch S_2 of the other module are both activated, the total output voltage is $v_{out} = V_c$.

The example shown in Figure 4.35 is a three-level MMC, as the output voltage can have three different values. To increase the number of levels and build a three-phase converter, more PMs can be connected in series as illustrated in Figure 4.36, where each leg of the MMC contains several PMs. In this scheme, the three legs of the three-phase MMC are connected in parallel to the same DC side. The inductances are employed to limit the current in case of faults developing in the converter.

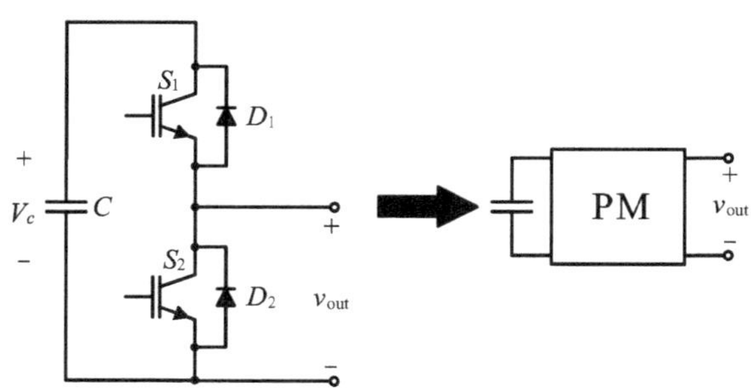

Figure 4.34 Scheme of a power module of an MMC

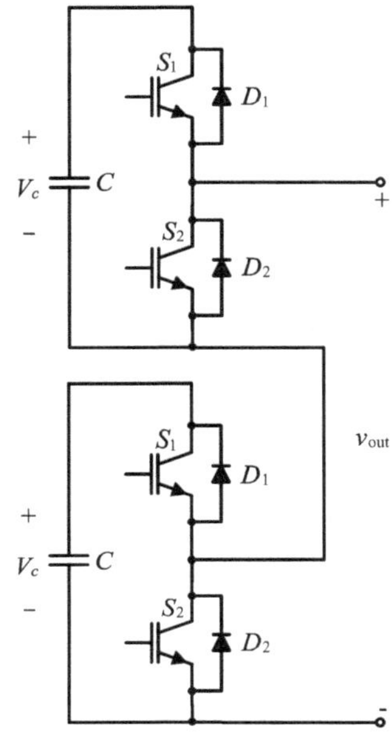

Figure 4.35 Series connection of two PMs

The main features and advantages of MMCs may be summarised as follows [46,47]:

- High-voltage capability: MMCs are capable of handling high voltage levels, making them ideal for HVDC transmission systems where high voltages are required for efficient long-distance power transmission.
- Superior performance: The modular structure of MMCs yields quasi-sinusoidal voltage waveforms and reduced power losses compared to conventional two-level converter topologies, resulting in enhanced system efficiency and reliability.
- Fault tolerance: The redundant nature of MMCs provides inherent fault tolerance capabilities. In the event of a PM failure, the converter can continue operating with minimal disruption, mitigating the risk of system downtime.
- Scalability: MMCs offer unparalleled scalability, allowing for easy expansion or contraction of the converter capacity by simply adding or removing PMs. This flexibility makes MMCs highly adaptable to evolving system requirements.
- Low harmonic distortion: By accurate control of the individual PMs, a near-sinusoidal MMC's output voltage can be achieved, which contains only a trickle of high-order voltage frequencies. These harmonics can be smoothed out using series inductive reactors; hence, ensuring compliance with existing stringent grid codes and harmonic standards.

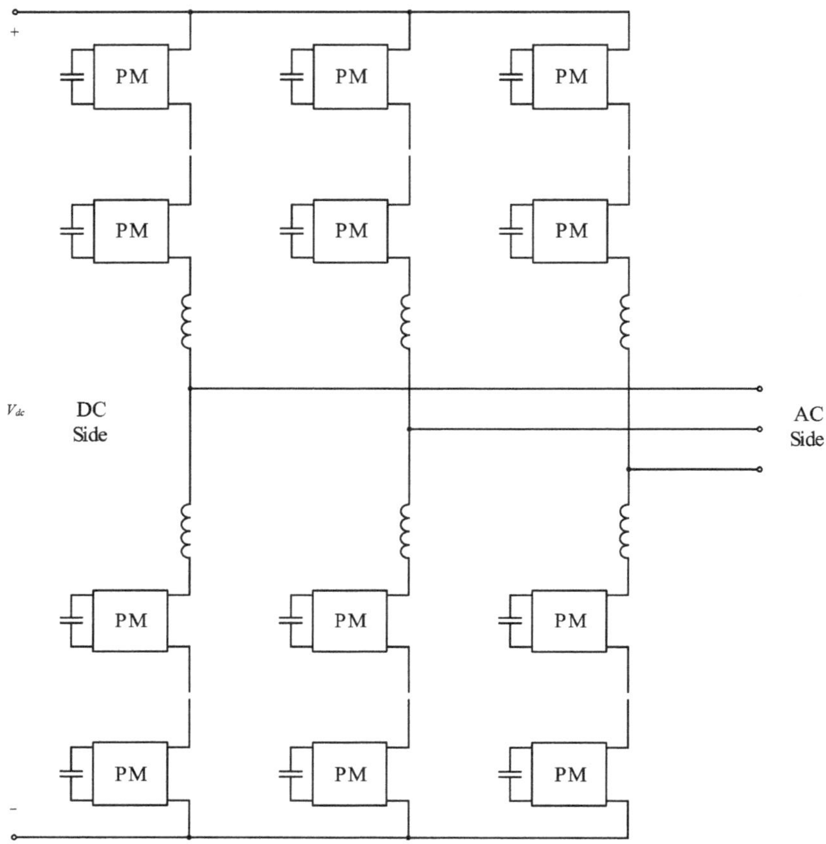

Figure 4.36 Scheme of a three-phase MMC

4.4 Comparison of features of LCC-HVDC and VSC-HVDC systems

LCC-HVDC systems are known for their robustness and cost-effectiveness. They are suitable for high-power bulk transmission over long distances. Conversely, VSC-HVDC systems are characterised by their flexibility and are ideal for applications requiring a fast dynamic response, such as grid stabilisation and integration of renewable energy sources, particularly the tapping of offshore wind power generation.

LCC-HVDC systems are known for their low power loss and comparatively lower construction expenses, particularly in connection with the thyristor valves. The power rating of such systems has been recorded to reach up to 12 GW and operate at $\pm 1{,}100$ kV. Despite these advantages, LCC-HVDC systems have several drawbacks, such as not being able to connect to passive networks, and may experience commutation failure. Furthermore, the reactive power consumption is very high in LCC-HVDC systems. On the other hand, VSC-HVDC systems do not experience commutation failure and can provide power to a passive network. Owing to recent developments in power electronics, VSC-HVDC systems can be used for power transmission up to ±525 kV. Nonetheless, VSC-HVDC systems have at present, higher installation costs, smaller power ratings, and higher power loss when compared to LCC-HVDC systems [48].

In accordance with [49], some of the most important features of LCC-HVDC and VSC-HVDC systems are:

- VSC-HVDC systems have independent active and reactive power control. This feature can be used to provide VAR support and voltage regulation to the connecting AC power grid. In contrast, LCC-HVDC systems are sinks of reactive power and *in-situ* reactive power is provided by means of switched capacitor banks or a suitable FACTS shunt compensator (e.g., SVC, STATCOM) but this latter option comes with a hefty price tag.
- In the case of VSC-HVDC systems, the direction of the active power flow can change rapidly. However, this is not the case in LCC-HVDC systems, where a change in the voltage polarity of the DC side is required to change the active power flow direction. This is a time-consuming operation since it involves switching off the LCC system to achieve depolarisation on the DC side.
- LCC-HVDC systems can operate without communication between terminals, which is an advantage, particularly for long-distance transmission. On the other hand, VSC-HVDC systems require communication between terminals of the HVDC link.
- VSC-HVDC systems can exchange active power under asymmetric disturbances. This operation cannot be performed when using LCC-HVDC systems.
- Unlike CSC-HVDC systems, VSC-HVDC systems can interact with passive power grids. Passive power grids are those that do not contain synchronous power sources.
- LCC-HVDC systems employ high-power capacitor banks on the AC side of each terminal, which can be prone to sudden voltage increases and resonance phenomena with the power grid.
- LCC-HVDC systems represent a mature technology, experience low power losses, and are surveyed to be more reliable than VSC-HVDC systems. It is expected that the current gap will narrow as future IGBT valves become sturdier with the availability of wide-bandgap materials, such as SiC and GaN.
- LCC-HVDC systems are unable to operate under voltage sags since these will induce commutation failures. On the contrary, VSC-HVDC systems will provide support to the AC power grid to ride through the voltage sag.
- LCC-HVDC systems employ tuned filters to mitigate the very substantial generation of harmonic currents, whereas VSC-HVDC systems do not typically require tuned filters, resulting in a smaller footprint and reduced costs. This is particularly the case when MMCs are used.

In summary, LCC-HVDC systems are better suited for point-to-point, bulk power transmission involving extra-long distances and power ratings not yet within the realm of VSC-HVDC systems. Also, LCC-HVDC systems have lower costs. On the other hand, VSC-HVDC systems, with their recent breakthroughs, are unassailable in applications such as offshore wind power [50], and in areas with limited space availability. They are increasingly being used for asynchronous interconnections. The availability of MMC converters from all the major manufacturers of HVDC systems has further enhanced their operational flexibility, efficiency, and reliability. Both LCC-HVDC and VSC-HVDC systems possess unique advantages, which make them suitable for different applications. Their selection should be based on specific project requirements, considering factors, such as power capacity, grid support, space availability, and cost efficiency. As the power electronics technology advances, the gap between these two power transmission options is narrowing, with hybrid solutions arising to leverage the strengths of both technologies.

4.5 Conclusions

The rapid evolution of megawatt-level power electronics in the last three decades has found a specialised marketplace badly in need of such technology. In the early 1990s the immediate requirements were to provide operational flexibility and reliability to the AC power grid and the area of FACTS was established. The equipment and operational experience in reactive power support encouraged the prototyping of VSC-HVDC links which far superseded expectations, giving rise to a new form of DC power transmission, which became established during the first decade of this century. However, it has been the arrival in circa 2010 of a new breed of VSC converters, termed MMC, that has made VSC-HVDC systems formidable contenders to replace LCC-HVDC systems in certain application areas, such as submarine power transmission for the tapping of renewable power generation and grid stabilisation. Nonetheless, extra-high-power transmission over extra-long distances is still the realm of LCC-HVDC.

References

[1] Mirsaeidi S, Tzelepis D, He J, Dong X, Said DM, and Booth C. A controllable thyristor-based commutation failure inhibitor for LCC-HVDC transmission systems. *IEEE Transactions on Power Electronics*. 2021;36 (4):3781–92.

[2] Li P, Adam GP, Finney SJ, and Holliday D. Operation analysis of thyristor-based front-to-front active-forced-commutated bridge DC transformer in LCC and VSC hybrid HVDC networks. *IEEE Journal of Emerging and Selected Topics in Power Electronics*. 2017;5(4):1657–69.

[3] Zhang T, Li C, and Liang J. A thyristor based series power flow control device for multi-terminal HVDC transmission. In: 2014 *49th International*

Universities Power Engineering Conference (UPEC). Piscataway, NJ: IEEE; 2014. p. 1–5.

[4] Mohan N, Undeland TM, and Robbins WP. *Power Electronics. Converters, Applications and Design*. 3rd edn. New York: Wiley; 2003.

[5] Mathur RM, and Varma RK. *Thyristor-based Facts Controllers for Electrical Transmission Systems*. New York: Wiley-IEEE Press; 2002.

[6] Asghar MSJ. Thyristor controlled reactors (TCR) and harmonic suppression by discontinuous phase controlled (DPC) switching. In: 2021 *IEEE 6th International Conference on Computing, Communication and Automation (ICCCA)*. Piscataway, NJ: IEEE; 2021. p. 738–43.

[7] Acha E, Agelidis VG, Anaya-Lara O, and Miller TJE. *Power Electronic Control in Electrical Systems*. Amsterdam: Elsevier; 2002.

[8] Hingorani NG, and Gyugyi L. Understanding FACTS [Internet]. *Understanding FACTS: Concepts and Technology of Flexible AC Transmission Systems*. Piscataway, NJ: IEEE; 1999. Available from: http://ieeexplore.ieee.org/xpl/bkabstractplus.jsp?bkn=5264253.

[9] Gelen A, and Yalcinoz T. The behavior of thyristor switched capacitor (TSC) installed in an infinite bus system. *In: IEEE EUROCON 2009*, EUROCON. 2009.

[10] Maza-Ortega JM, Acha E, García S, and Gómez-Expósito A. Overview of power electronics technology and applications in power generation transmission and distribution. *Journal of Modern Power Systems and Clean Energy*. 2017;5(4):499–514.

[11] Jianhua Z, Guangping D, Gang X, Jie Z, Hui Z, and Shuying W. Design of the control system for thyristor switched capacitor devices. In: *Proceedings of the IEEE Power Engineering Society Transmission and Distribution Conference*. 2003.

[12] Deotale BD, Paraskar SR. Transient stability improvement using Thyristor Switched Series Capacitor (TSSC) FACTS device. In: *2016 IEEE Students' Conference on Electrical, Electronics and Computer Science*, SCEECS 2016. 2016.

[13] Mahajan V. Thyristor controlled series compensator. In: 2006 *IEEE International Conference on Industrial Technology*. Piscataway, NJ: IEEE; 2006. p. 182–7.

[14] Larsen EV, Clark K, Miske SA, and Urbanek J. Characteristics and rating considerations of thyristor controlled series compensation. *IEEE Transactions on Power Delivery*. 1994;9(2):992–1000.

[15] Helbing SG, and Karady GG. Investigations of an advanced form of series compensation. *IEEE Transactions on Power Delivery*. 1994;9(2):939–47.

[16] Khan WA, Hamid S, Usman MR, *et al.* HVDC transmission line fault identification: A learning based UAV control strategy. *IEEE Access*. 2022; 10:121561–79.

[17] Oni OE, Davidson IE, and Mbangula KNI. A review of LCC-HVDC and VSC-HVDC technologies and applications. In: 2016 *IEEE 16th International Conference on Environment and Electrical Engineering (EEEIC)*. Piscataway, NJ: IEEE; 2016. p. 1–7.

[18] Okba MH, Saied MH, Mostafa MZ, and Abdel-Moneim TM. High voltage direct current transmission – A review, part I. In: 2012 *IEEE Energytech*. Piscataway, NJ: IEEE; 2012. p. 1–7.

[19] Chokhawala R, Danielsson B, and Ängquist L. Power semiconductors in transmission and distribution applications. In: *IEEE International Symposium on Power Semiconductor Devices and ICs (ISPSD)*. 2001.

[20] Vrana TK, and Energi S. Review of HVDC component ratings: XLPE cables and VSC converters. In: 2016 IEEE International Energy Conference (ENERGYCON). Piscataway, NJ: IEEE; 2016. p. 1–6.

[21] Bernet S. Recent developments of high power converters for industry and traction applications. *IEEE Transactions on Power Electronics*. 2000;15(6):1102–17.

[22] Shu Z, Tang J, Guo Y, and Lian J. An efficient SVPWM algorithm with low computational overhead for three-phase inverters. *IEEE Transactions on Power Electronics*. 2007;22(5):1797–805.

[23] Mohan N. *Advanced Electrical Drives*. Minneapolis, MN: Mnpere; 2001.

[24] Garcia-Gonzalez P, and Garcia-Cerrada A. Control system for a PWM-based STATCOM. *IEEE Transactions on Power Delivery*. 2000 Oct;15(4):1252–7.

[25] Bellini A, Figalli G, and Ulivi G. A microcomputer-based control system for an inverter supplying state feedback controlled induction motors. *IFAC Proceedings Volumes*. 1983;16(16):531–7.

[26] Acha E, Roncero-Sánchez P, De la Villa-Jaen A, Castro LM, and Kazemtabrizi B. *VSC-FACTS-HVDC: Analysis, Modelling and Simulation in Power Grids*. New York: Wiley; 2019.

[27] Arrillaga J, Liu YH, and Watson NR. *Flexible Power Transmission: The HVDC Options*. New York: Wiley; 2007.

[28] Rodriguez J, Franquelo LG, Kouro S, *et al.* Multilevel converters: An enabling technology for high-power applications. *Proceedings of the IEEE*. 2009;97(11):1786–817.

[29] Sharifabadi K, Harnefors L, Nee HP, Norrga S, and Teodorescu R. *Design, Control and Application of Modular Multilevel Converters for HVDC Transmission Systems*. New York: Wiley-IEEE Press; 2016.

[30] Kouro S, Malinowski M, Gopakumar K, *et al.* Recent advances and industrial applications of multilevel converters. *IEEE Transactions on Industrial Electronics*. 2010;57(8):2553–80.

[31] Nabae A, Takahashi I, and Akagi H. A new neutral-point-clamped PWM inverter. *IEEE Transactions on Industrial Applications*. 1981;IA-17(5):518–23.

[32] Shanmei Cheng, Yajun Liu, and Bing Wu. SVM algorithm of three-level NPC inverter. In: 2008 *3rd IEEE Conference on Industrial Electronics and Applications*. Piscataway, NJ: IEEE; 2008. p. 2194–8.

[33] Zhang M, Cui Y, Wang Q, *et al.* A study on neutral-point potential in three-level NPC converters. *Energies (Basel)*. 2019;12(17):3367.

[34] Ouerdani I, Ben Abdelghani H, Bennani Ben Abdelghani A, Montesinos-Miracle D, and Slama-Belkhodja I. Space vector modulation technique for 3-level NPC converter with constant switching frequency. *Advances in Power Electronics*. 2016; 2016.

[35] Rojas F, Kennel R, Cardenas R, Repenning R, Clare JC, and Diaz M. A new space-vector-modulation algorithm for a three-level four-leg NPC inverter. *IEEE Transactions on Energy Conversion*. 2017;32(1):23–35.
[36] Rodriguez J, Jih-Sheng Lai, and Fang Zheng Peng. Multilevel inverters: a survey of topologies, controls, and applications. *IEEE Transactions on Industrial Electronics*. 2002;49(4):724–38.
[37] Hart DW. *Power Eletronics*. New York: McGraw Hill; 2012.
[38] Rodríguez J, Franquelo LG, Kouro S, *et al.* Multilevel converters: An enabling technology for high-power applications. *Proceedings of the IEEE*. 2009;97 (11):1786–817.
[39] Klug RD, and Klaassen N. High power medium voltage drives – Innovations, portfolio, trends. In: 2005 *European Conference on Power Electronics and Applications*. Piscataway, NJ: IEEE; 2005. 10 pp.
[40] He L, and Cheng C. A flying-capacitor-clamped five-level inverter based on bridge modular switched-capacitor topology. *IEEE Transactions on Industrial Electronics*. 2016;63(12):7814–22.
[41] Wu B, and Narimani M. *High-Power Converters and AC Drives*. New York: Wiley; 2005.
[42] Gemmell BD, Dorn J, Retzmann D, and Soerangr D. Prospects of multilevel VSC Technologies for power transmission. In: Transmission and Distribution Exposition Conference: 2008 IEEE PES Powering Toward the Future, PIMS 2008; 2008.
[43] Jacobson B, Karlsson P, Asplund G, Harnefors L, and Jonsson T. VSC-HVDC transmission with cascaded two-level converters. *In: 43rd International Conference on Large High Voltage Electric Systems 2010*. Paris: CIGRE; 2010.
[44] Akagi H. Classification, terminology, and application of the modular multilevel cascade converter (MMCC). In: *2010 International Power Electronics Conference – ECCE Asia*, IPEC 2010. 2010.
[45] Raju MN, Sreedevi J, Mandi RP, and Meera KS. Modular multilevel converters technology: A comprehensive study on its topologies, modelling, control and applications. *IET Power Electronics*. 2019;12(2):149–69.
[46] Hollweg GV, Khan SA, Chaturvedi S, Fan Y, Wang M, and Su W. Grid-connected converters: A brief survey of topologies, output filters, current control, and weak grids operation. *Energies (Basel)*. 2023;16(9):3611.
[47] Jones PS, and Davidson CC. Calculation of power losses for MMC-based VSC HVDC stations. In: 2013 *15th European Conference on Power Electronics and Applications*, EPE 2013. 2013.
[48] Xiao H, Sun K, Pan J, and Liu Y. Operation and control of hybrid HVDC system with LCC and full-bridge MMC connected in parallel. *IET Generation, Transmission & Distribution*. 2020;14(7):1344–52.
[49] Rzepka P, Szablicki M, Halinka A, Kielak R, and Jurczyk D. Technological Conditions for VSC and LCC HVDC Systems in Reference to Network Code Requirements. In: 2019 *Modern Electric Power Systems (MEPS)*. Piscataway, NJ: IEEE; 2019. p. 1–6.
[50] Torres-Olguin RE, Molinas M, and Undeland T. Offshore wind farm grid integration by VSC technology with LCC-based HVDC transmission. *IEEE Trans Sustain Energy*. 2012;3(4):899–907.

Chapter 5

Power electronics in bulk power transmission

Ricardo Martínez-Parrales[1] *and Claudio R. Fuerte Esquivel*[2]

5.1 Introduction

The global electricity supply industry is facing new, very significant operational challenges owing to the relentless penetration of asynchronous and static solar PV generators, increased industrial production, and the deregulation of the electric sector. In this new operating paradigm, where the high-performance control of the power grid is mandatory, planners of the electric power system are motivated to explore new ways to maximize energy transfers using existing transmission facilities, while maintaining acceptable levels of grid reliability and stability. In the mid-1980s, an in-depth analysis of the options available to achieve such goals pointed in the direction of power electronics-based controllers [1,2]. After more than three decades of sustained equipment development, implementations, and vast experience gained in actual installations, these power electronics-based controllers have demonstrated to have been a viable alternative to conventional electromechanical technologies with slow response times and high maintenance costs [3,4].

Flexible Alternating Current Transmission Systems (FACTS) is a term used to describe AC transmission systems incorporating power electronics-based equipment into the high voltage side of the network to make it electronically controllable [2,5]. Many of the ideas making up FACTS may have been conceived during the late 1970s and early 1980s, but they may have first been articulated as a unified vision in a speech titled "High Power Electronics and Flexible AC Transmission System," delivered by N.G. Hingorani on 19 April 1988 at the American Power Conference's 50th Annual Meeting in Chicago [1]. FACTS, as an integrated philosophy, seeks to incorporate the leading-edge developments taking place in the area of high-voltage and high-current power electronics to enhance control over the power flows and nodal voltages in the high-voltage side of the network during both steady-state and transient regimes. Owing to the many economic and technical benefits that the FACTS technology has delivered, it continues to receive the

[1]Department of Electronic Engineering, Instituto Tecnológico de Morelia, México
[2]Faculty of Electrical Engineering, University of Michoacan, México

backing of the major manufacturers of electrical equipment and utilities worldwide [6,7], as evidenced by the many FACTS and HVDC installations around the globe, as shown in Chapter 2.

5.2 Inherent limitations of conventional transmission systems

In an electric system with no power flow control, the power flows from sources to loads inversely to the relative impedances of the transmission paths. Low-impedance transmission paths carry the largest portion of the flow, which minimizes losses if the *X/R* ratio of the paths is similar. However, this advantage disappears if fixed series capacitors are added to the network or different voltage levels are presented in the power flow paths. Since all transmission lines carry power flow in an interconnected network, utilities operating in a deregulated market might face significant challenges, as their operating state can be adversely affected by power transactions in which they have no direct involvement.

In a typical extra-high-voltage transmission line, the series reactance X is much larger than the resistance R. Hence, the simplified active and reactive power flow equations (5.1) and (5.2) may be derived for the transmission line representation shown in Figure 5.1.

$$P_{km} = \frac{V_k V_m \sin(\theta_k - \theta_m)}{X} \tag{5.1}$$

$$Q_{km} = \frac{V_k^2 - V_k V_m \cos(\theta_k - \theta_m)}{X} \tag{5.2}$$

From (5.1), it is clear that the direction of active power flow is only determined by the voltage angle difference. If V_k leads V_m, the active power direction is from the sending to the receiving node. The active power has the opposite direction if V_k lags V_m. In theory, the maximum transfer of power is given when the relative difference between both voltage angles is 90°.

The voltage magnitude at both nodes determines the reactive power direction. If V_k is greater than V_m, the reactive power direction is from the sending to the receiving node. If V_k is less than V_m, the reactive power has the opposite direction.

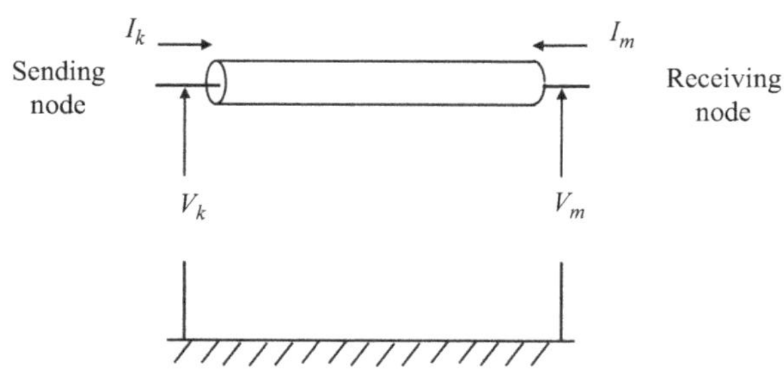

Figure 5.1 Overhead transmission line

For a given operating condition, the maximum electric power transfer in a transmission system is limited by steady-state and transient stability limits [4,5]. The former is determined by one of the following criteria:

1. Angular stability limits.
2. Thermal limits.
3. Voltage limits.

These limits define the maximum electric power that can be transmitted safely without damaging transmission lines or utility and customer equipment.

The angular stability limit is determined by (5.1). However, this limit can be modified by altering the transmission line's natural inductive reactance using series compensation or adjusting the relative phase angle difference at the transmission line terminals using a phase-shifting transformer.

When thermal limits are reached, excessive current flow overheats the transmission line and other electric equipment to the point of permanent damage. The addition of ancillary equipment and reconfiguration of the network topology can normally bring current flows back to safe limits. Conventional solutions include the use of series reactors, series capacitors, and phase shifters.

The voltage magnitudes can also be a limiting factor in power flow transfer. Changes in network configuration caused by equipment outages can give rise to unacceptably high or low voltage conditions, and thermal limits can be exceeded. In this context, the system voltage regulation is affected by the reactive power flowing in the network, as seen in (5.3), which is directly derived from (5.2) by assuming that the angular difference is small.

$$\Delta V_{km} = V_k - V_m = \frac{XQ_k}{V_k} \tag{5.3}$$

It is clear from (5.3) that the voltage drop ΔV_{km} across the transmission element is mainly a function of the reactance and the reactive power flowing through the transmission element. However, if the reactive power demanded by a load is supplied locally by connecting a shunt compensator at the load bus, then the voltage drop across the line reduces.

The nodal voltage variations caused by changes in the system's operating status can be regulated by providing reactive power compensation, governed by (5.4), where V_k is regulated by the injected reactive power Q_k of the shunt compensator.

$$V_k = \frac{-V_m \pm \sqrt{V_m^2 + 4XQ_k}}{2} \tag{5.4}$$

It follows that to correct an instance of low voltage, the remedy is to supply reactive power to improve the load power factor and to reduce reactive power losses in transmission lines and transformers. Nowadays, dynamic voltage control in high voltage nodes is carried out using Static Var Compensators

(SVC) and STATic synchronous COMpensators (STATCOMs), as outlined in Chapter 2. In the past, this task was carried out using rotating synchronous condensers.

Regarding high voltage issues, produced mainly by lightly loading conditions, the shunt capacitors are disconnected, and shunt reactors are brought into service. SVCs and STATCOMs can be set to absorb reactive power dynamically. Also, synchronous generators with automatic voltage regulators and synchronous condensers can be used to absorb reactive power dynamically.

5.3 FACTS controllers

FACTS controllers have been designed to narrow the gap between the non-controlled and the controlled power system mode of operation by providing additional degrees of freedom to control power flow over desired transmission routes, to increase transmission capacity, allowing secure loading of the transmission lines up to their thermal limits, to provide effective utilization of available generation, and to containing outages from spreading to a wider area [2,3].

From the operational point of view, the FACTS technology is concerned with the ability to control, in an adaptive fashion, the path of the power flows throughout the network. Within this context, FACTS devices are designed to control one or more of the following power system parameters: the line impedance and the nodal voltage magnitudes and angles at both the sending and receiving ends of transmission lines. FACTS devices perform beneficial roles during both steady-state and transient conditions; however, some FACTS devices are specifically designed to control one or more of those key network parameters, as illustrated in Figure 5.2. FACTS devices provide fast voltage support and effective control of power flow in key power transmission corridors, significantly enhancing the network's transmission capacity and system security. The specifics of how FACTS devices control active and reactive power flow are described below.

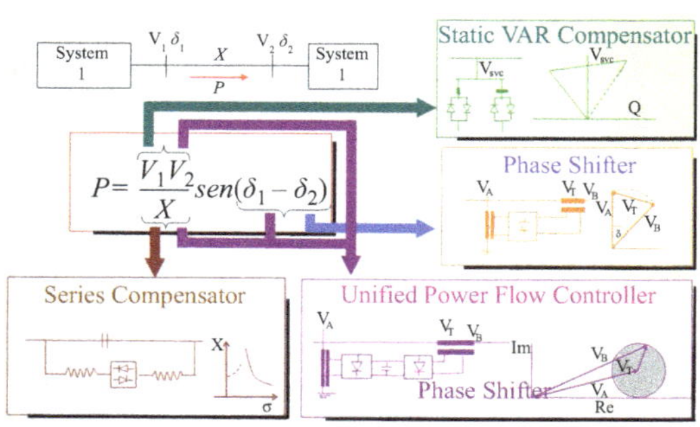

Figure 5.2 FACTS controllers. Adapted from [8]

5.3.1 Thyristor-controlled series compensator

The Advanced Series Compensator (ASC) [9], which belongs to the family of FACTS devices termed Thyristor-Controlled Series Compensator (TCSC), comprises several TCSC modules connected in tandem. A TCSC exerts rapid and flexible control on the series reactance of the compensated transmission line. The basic operating principles and structures of the TCSC are explained in Chapter 2 (Section 2.2.2.1). For instance, the ASC physical structure is shown schematically in Chapter 2 (Figure 2.15). Equivalently, the ASC may be represented by a single TCSC kernel, having the circuit representation, in the inductive and capacitive operating regions, shown in Figure 5.3(a) and (b), respectively. Notice that in principle at least, the TCSC modules making up the ASC can be operated in different modes and have different capacitive and inductive parameters. However, the ASC installed at the Slatt substation—the only one in existence—has six identical TCSC modules. This makes sense from the manufacturing and installation vantages.

The equivalent TCSC module shown in Figure 5.4, as part of the TCSC-compensated transmission line scheme, has the fundamental frequency impedance given by (5.5). This is a function of the thyristor's firing angle α, as reflected in the equivalent reactance vs. firing angle characteristic shown in Figure 5.5, corresponding to the ASC installed at the Slatt substation. Hence, the TCSC can increase the electrical length of the line by supplying positive reactance, thereby reducing the line's ability to transfer power. When the TCSC is operated in the capacitive region, it produces the opposite effect. Notice that at least from the steady-state operation point-of-view, the interest is to reduce the electrical length of the compensated transmission line, hence, operating the TCSC in its capacity region but not

Figure 5.3 Advanced series compensator: (a) ASC in inductive operative region and (b) ASC in capacitive operative region

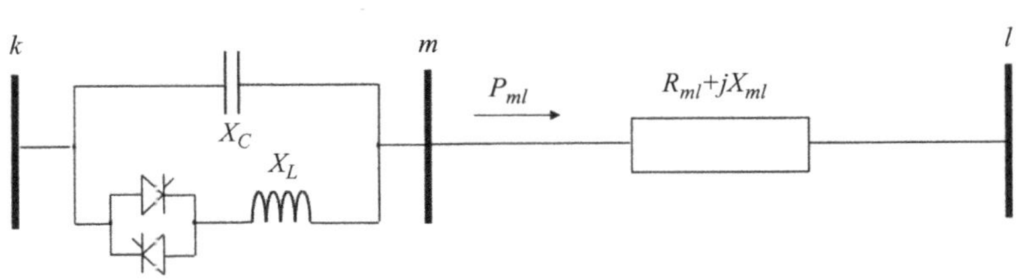

Figure 5.4 Compensated transmission line

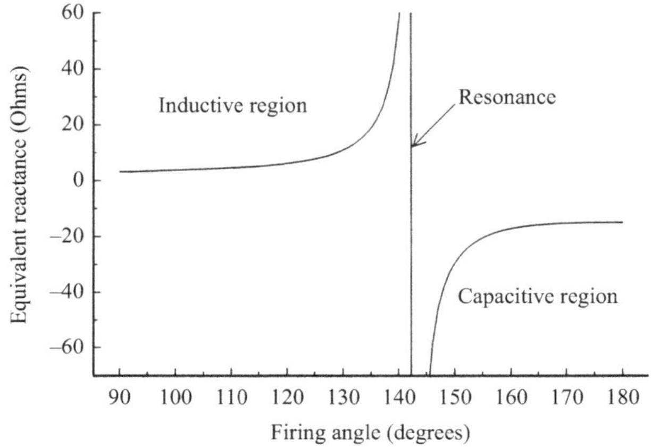

Figure 5.5 TCSC fundamental impedance

too close to the resonant region.

$$
\begin{aligned}
X_{TCSC(1)} &= -X_C + (X_C + X_{LC}) \left(\frac{2(\pi - \alpha) + \sin(2(\pi - \alpha))}{\pi} \right) \\
&- \frac{4\, X_{LC}^2 \cos^2(\pi - \alpha)}{X_L} \left(\frac{k \tan(k(\pi - \alpha)) - \tan(\pi - \alpha)}{\pi} \right)
\end{aligned}
\tag{5.5}
$$

The active and reactive powers flowing through the TCSC module in Figure 5.4 are given by (5.6) and (5.7). The relationship of these equations with (5.5) and Figure 5.5 shows that the power flow control is performed through the TCSC's firing angle, which determines its reactance value.

$$
P_{km} = \frac{V_k\, V_m}{X_{TCSC(1)}} \sin(\theta_k - \theta_m) \tag{5.6}
$$

$$
Q_{km} = \frac{V_k^2}{X_{TCSC(1)}} - \frac{V_k\, V_m}{X_{TCSC(1)}} \cos(\theta_k - \theta_m) \tag{5.7}
$$

Even though the amount of active power controlled by the TCSC depends on the level of compensation the controller provides, it will also depend on the inductive series reactance of the compensated transmission line. Figure 5.6 compares the increase of active power flow as a function of the compensation level for three different transmission lines. This is done in connection with a small test network case reported in [10]. The transmission lines are Lake-Main, North-Lake, and South-Lake with inductive series reactance of 0.3 pu, 0.18 pu, and 0.23 pu, respectively. For a specified level of series compensation, the amount of active power increased by the TCSC will be larger in transmission lines with a high value of inductive series reactance. Moreover, as the compensation increases, the power flow through the compensated transmission line also increases.

The active power control capacity of the TCSC is now being explored in a utility-size power system model suitable for network expansion studies, which includes these

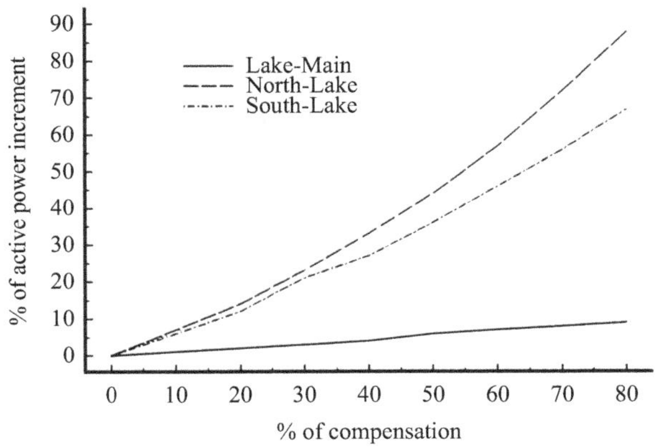

Figure 5.6 Increments of active power flow as a function of series compensation

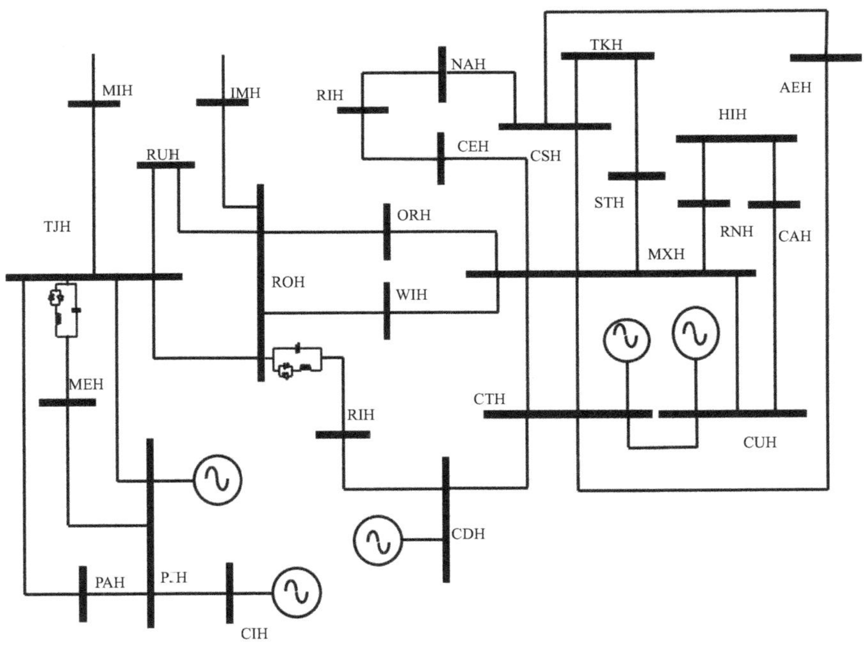

Figure 5.7 A relevant portion of the 166-node with TCSCs

power electronics controllers. It comprises 166 nodes, 108 transmission lines, and 128 transformers [6]. The relevant portion of the power system is shown in Figure 5.7. In real life, this network is part of a much larger interconnected system comprising 160 power plants: 76 hydro and 84 thermal, with 579 generating units, 2,172 nodes, 2,294 transmission lines, and 768 transformers. For the 166-node power system, the active power generation and demand are 2,096 MW and 2,076 MW, respectively. When the system operates under a maximum demand, an active power flow of 60.42 MW is injected into node ROH through line RIH-ROH, and 87.22 MW is injected into node TJH through line MEH-TJH. The relevant part of the upgraded network, with two TCSCs, is shown in Figure 5.7 [10]. The TCSCs have been set to maintain active power flow injected into nodes ROH and TJH at 90 MW and 80 MW, respectively. The TCSCs achieved the required control by adjusting their firing angles at 151.037 and 150.15, respectively, which resulted in the corresponding TCSCs' reactance values of −0493 pu and −0.0053 pu. Even though the TCSCs were used to increase the active power flow through lines RIH-ROH and MEH-TJH, the overall system's performance was improved by moderately reducing production cost from 293.643 to 293.36 \$/h and active power losses from 20.117 to 19.993 MW [10,11].

5.3.2 Unified power flow controller

The unified power flow controller (UPFC) is an advanced power systems device capable of simultaneously controlling voltage magnitude and active and reactive power flows, all in an adaptive fashion. Owing to its almost instantaneous speed of response and unrivaled functionality, it is well-placed to solve most issues relating to active and reactive power flow control while enhancing considerably transient and dynamic stability [5,12].

The basic principles of operation of the UPFC are presented in Chapter 2. This Section is concerned with a UPFC model suitable for steady-state solutions of large-scale power networks, obtained from the schematic representation of the UPFC shown in Figure 5.8 [13]. This controller consists of two back-to-back, self-commutated voltage source converters, sharing a shunt capacitor on the DC side. One converter is coupled to the AC system via a series transformer and the other is coupled to the AC system via a shunt transformer.

The output voltage of the series converter is added to the AC terminal voltage $\boldsymbol{V_o}$ via the series-connected coupling transformer. The injected voltage, $\boldsymbol{V_{cR}}$, acts as an AC series voltage source, changing the effective sending-end voltage as seen from node m. The active and reactive power exchanged between the series converter and the AC system is determined by the product of the transmission line current, $\boldsymbol{I_m}$, and the series voltage source, $\boldsymbol{V_{cR}}$.

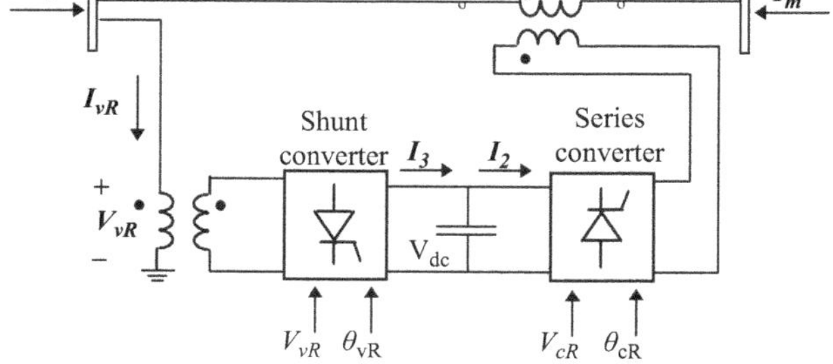

Figure 5.8 UPFC schematic diagram

The shunt converter supplies the active power demanded by the series converter from the AC power system via the common DC link. It can generate or absorb controllable reactive power in both operating modes, rectifier, and inverter. The independently controlled shunt reactive compensation can be used to maintain the shunt converter terminal AC voltage magnitude at a specified value.

Based on the above, the UPFC equivalent circuit for steady-state studies is depicted in Figure 5.9, with the two controllable voltage sources representing the fundamental components of the Fourier series associated with the voltage waveforms at the converters' AC terminals. From a mathematical perspective, these controllable sources are defined as described by (5.8).

$$V_{source} = V_{source}(\cos\,\theta_{source} + j\sin\,\theta_{source})\ source = se, sh \tag{5.8}$$

where *se* and *sh* denote series and shunt, respectively. The voltage phase angles can be controlled between 0 and 2π: $0 \le \theta_{source} \le 2\pi$, $\theta_{source} = \theta_{se}, \theta_{sh}$; while the voltage magnitudes vary from zero to a predefined limit value that is determined by the nominal VA (volt-ampere) of the corresponding converter: $V_{source}^{\min} \le V_{source} \le V_{source}^{\max}$, $V_{source} = V_{se}, V_{sh}$.

Based on the UPFC equivalent circuit, the active and reactive power flow are given as described in (5.9)–(5.17) [13]:

At node k:

$$\begin{aligned} P_{km} &= \frac{V_k V_m}{X_{se}}\sin(\theta_k-\theta_m) + \frac{V_k V_{se}}{X_{se}}\sin(\theta_k-\theta_{se}) \\ &+ \frac{V_k V_{sh}}{X_{sh}}\sin(\theta_k-\theta_{sh}) \end{aligned} \tag{5.9}$$

$$\begin{aligned} Q_{km} &= V_k^2\left(\frac{1}{X_{se}}+\frac{1}{X_{sh}}\right) - \frac{V_k V_m}{X_{se}}\cos(\theta_k-\theta_m) - \frac{V_k V_{se}}{X_{se}}\cos(\theta_k-\theta_{se}) \\ &- \frac{V_k V_{sh}}{X_{sh}}\cos(\theta_k-\theta_{sh}) \end{aligned} \tag{5.10}$$

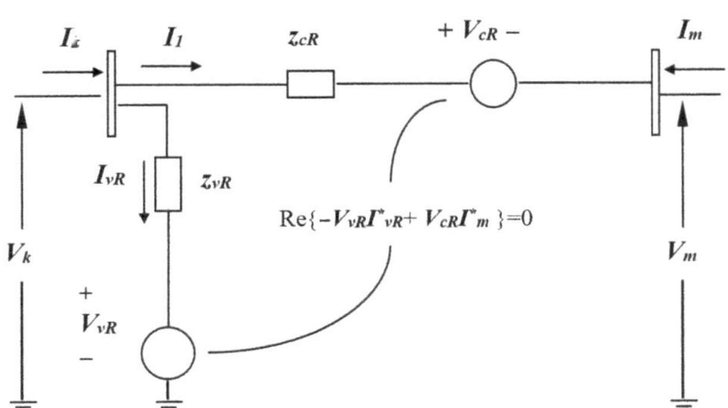

Figure 5.9 UPFC equivalent circuit

At node m:

$$P_{mk} = \frac{V_m V_k}{X_{se}} \sin(\theta_m - \theta_k) - \frac{V_m V_{se}}{X_{se}} \sin(\theta_m - \theta_{se}) \tag{5.11}$$

$$Q_{mk} = \frac{V_m^2}{X_{se}} - \frac{V_m V_k}{X_{se}} \cos(\theta_m - \theta_k) + \frac{V_m V_{se}}{X_{se}} \cos(\theta_m - \theta_{se}) \tag{5.12}$$

Series converter:

$$P_{se} = \frac{V_{se} V_k}{X_{se}} \sin(\theta_{se} - \theta_k) - \frac{V_{se} V_m}{X_{se}} \sin(\theta_{se} - \theta_m) \tag{5.13}$$

$$Q_{se} = \frac{V_{se}^2}{X_{se}} - \frac{V_{se} V_k}{X_{se}} \cos(\theta_{se} - \theta_k) + \frac{V_{se} V_m}{X_{se}} \cos(\theta_{se} - \theta_m) \tag{5.14}$$

Shunt converter:

$$P_{sh} = \frac{V_{sh} V_k}{X_{sh}} \sin(\theta_{sh} - \theta_k) \tag{5.15}$$

$$Q_{sh} = \frac{V_{sh}^2}{X_{sh}} - \frac{V_{sh} V_k}{X_{sh}} \cos(\theta_{sh} - \theta_k) \tag{5.16}$$

The equality constraint described in (5.17) mathematically models the free loss operation of converters.

$$P_{se} = P_{sh} \tag{5.17}$$

The previous equations encapsulate the interaction between the transmission network's nodal voltages and the controllable voltages associated with the UPFC, which allows for the definition of each converter's control modes of operation.

The shunt converter makes it possible to transmit the active power demanded by the series converter from the electrical system. This power supply is performed through the common DC link, which is evident from the last term of (5.9), (5.15) and (5.17). Additionally, this converter can either absorb or inject reactive power from or to the AC system, as demonstrated by the last term on the right side of (5.10) and (5.16). Within this context, there are two control modes of operation. The first involves injecting or absorbing a predetermined reactive power value irrespective of the system's operating conditions. The second control mode consists of dynamically injecting or absorbing reactive power to maintain the voltage magnitude at the system node within a preset range or a specified value.

Figure 5.10 shows the relevant part of the 166-node test system described in Section 5.3.1, where three UPFCs have been used. The UPFCs have been set to increase the active power transfer through each transmission line ROH-RUM, AEH-CSH, and PAH-TJH by 10% relative to the original active power flows, while the reactive power is maintained at the original value.

Table 5.1 gives the UPFC parameters corresponding to an OPF (optimal power flow) solution, while Figure 5.11 shows how the cost of the total active power

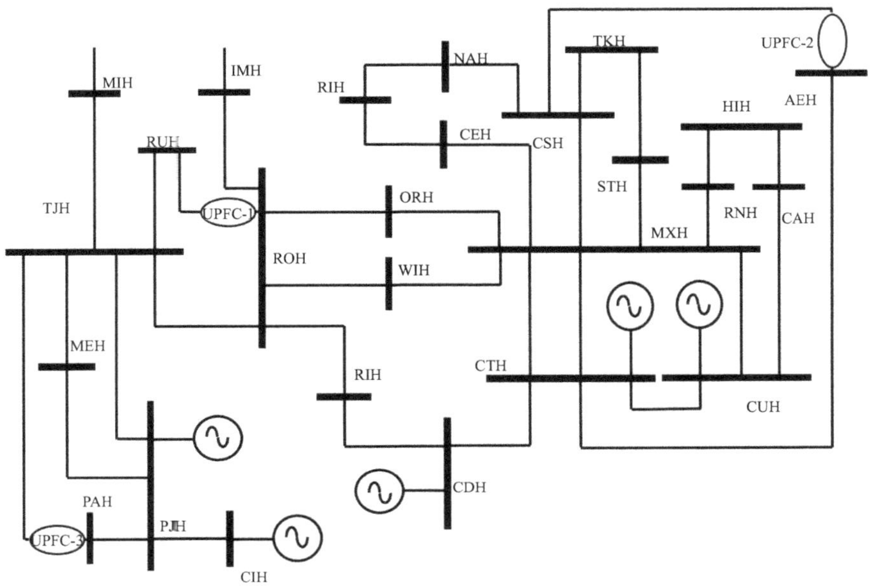

Figure 5.10 A relevant portion of the 166-node test system upgraded with three UPFCs

Table 5.1 UPFCs parameters for the 166-node test system

Device	**Series source**		**Shunt source**	
	V_{cR} **(pu)**	θ_{cR} **(degrees)**	V_{vR} **(pu)**	θ_{vR} **(degrees)**
UPFC-1	0.180	−126.709	1.008	−3.060
UPFC-2	0.132	−111.578	1.012	−1.944
UPFC-3	0.076	−124.137	1.011	−3.212

generation dispatch varies from one iteration to the next during the OPF solution [9]. The economic dispatch solution is slightly more expensive when the UPFCs have been used because the UPFC restrains some variables, which the OPF solution can no longer use in pursuit of minimizing costs further. Nonetheless, the use of UPFCs has improved the network's operating conditions: the amount of power transfer through each selected transmission line increased by 10%, while the total power economic dispatch only increased by 1.4% $/h.

5.4 CSC-high-voltage direct current links

High-Voltage Direct Current (HVDC) transmission is the pioneering application of power electronics technology into electrical power networks, as detailed in

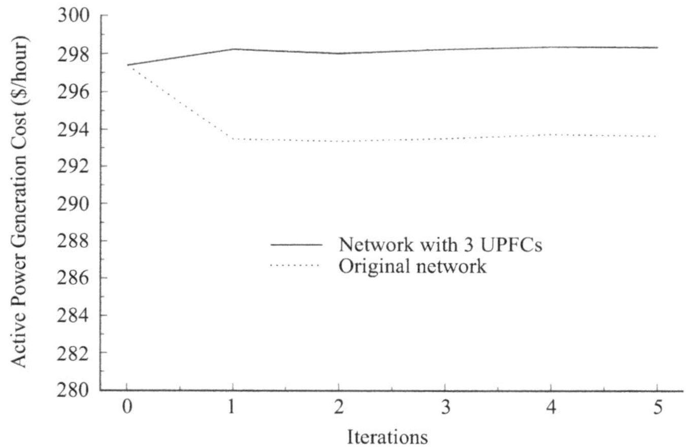

Figure 5.11 Active power generation cost profiles corresponding to the 166-node test system and the upgraded 166-node test system with three UPFCs

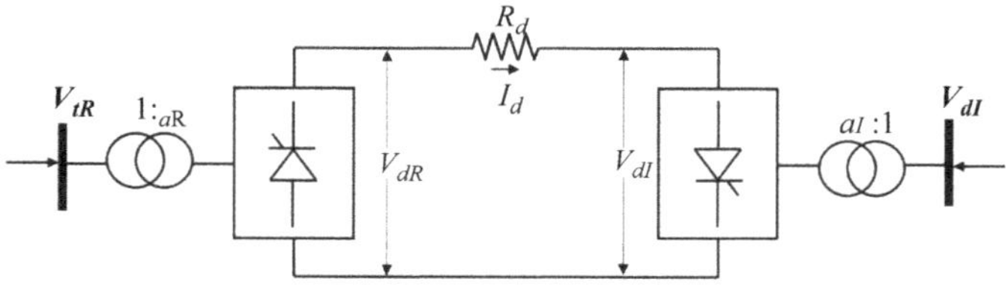

Figure 5.12 Monopole, point-to-point HVDC transmission link

Chapter 2. CSC-HVDC enables fast power control through the link and without contributing short-circuit current to the neighboring AC system. When transmitting large blocks of energy, particularly in undersea transmission applications, the cable capacitance limits the maximum possible transmission distance using the AC option. This limitation does not exist when using DC cable power systems. One key advantage of the HVDC technology is its ability to significantly reduce overall power losses. An HVDC system boasts lower transmission losses than an AC system of comparable power capacity, even when the converter losses are factored in. Another significant advantage of the HVDC technology is its ability to connect systems with different operational frequencies. HVDC links are useful for interconnecting networks that would be quite challenging and costly to link otherwise due to stability issues and frequency oscillation excursions. A schematic representation of a point-to-point, monopole HVDC link is shown in Figure 5.12, which consists of two converter stations and a DC transmission line. The equations describing this HVDC link are summarized by (5.18)–(5.26), derived from basic power electronics theory [14,15],

$$V_{dR} = \frac{3\sqrt{2}}{\pi} np_R \, a_R V_{tR} \cos \alpha_R \; - \frac{3}{\pi} X_{CR} \, np_R I_d \tag{5.18}$$

$$V_{dI} = \frac{3\sqrt{2}}{\pi} np_I \, a_I V_{tI} \cos \alpha_I \; - \frac{3}{\pi} X_{CI} \, np_I I_d \tag{5.19}$$

$$P_{dR} = V_{dR} I_d \tag{5.20}$$

$$P_{dI} = V_{dI} I_d \tag{5.21}$$

$$S_{dR} = k \frac{3\sqrt{2}}{\pi} a_R \, V_{tR} \, np_R I_d \tag{5.22}$$

$$S_{dI} = k \frac{3\sqrt{2}}{\pi} a_I V_{tI} n np_I I_d \tag{5.23}$$

$$Q_{dR} = \sqrt{S_{dR}^2 - P_{dR}^2} \tag{5.24}$$

$$Q_{dI} = \sqrt{S_{dI}^2 - P_{dR}^2} \tag{5.25}$$

The adjusting parameter k is given by (5.26). It aims to correct the apparent powers taking account of the commutation (overlap) period. It is valid when the firing angle of the conducting outgoing and incoming thyristors does not exceed 60° [14].

$$k = \frac{\sqrt{[\cos 2\alpha - \cos 2(\alpha + \mu)]^2 + [2\mu + \sin 2\alpha - \sin 2(\alpha + \mu)]^2}}{4\,[\cos \alpha - \cos(\alpha + \mu)]} \tag{5.26}$$

In the set of (5.18)–(5.26), V_{dR} and V_{dI} denote the DC voltages at the terminals of the rectifier and inverter stations, respectively. Furthermore, np_R and np_I are the number of series connected bridges in the rectifier and inverter stations, respectively. The control angles in the rectifier and inverter are denoted by α_R and α_I, respectively, while X_{CR} and X_{CI} are the commutation resistances in the rectifier and inverter, respectively. V_{tR} and V_{tI} are the effective voltage magnitudes (line-to-line) at the AC terminals of the rectifier and inverter, respectively. I_d is the DC current. Notice that α can be either α_R or α_I in (5.26). If the commutation overlap is neglected then $\mu = 0$ and $k = 1$.

An increase in α_R results in a decrease in the DC voltage reflected in the term $\cos\alpha_R$. It also leads to a higher reactive power consumption by the converter since the current will lag the voltage. This concept is illustrated in Figures 5.13(a) and 5.14(a), where an idealized DC current I_d is assumed. The fundamental current phasor I_{a1} of I_d is shown in Figures 5.13(b) and 5.14(b), together with the voltage phasor E_a representing the line-neutral voltage.

When the fundamental current I_{a1} is in phase with the voltage E_a, $\phi = \alpha_R = 0$, the active power drawn by the rectifier plant is maximum, and the reactive power is null. An increase in α_R produces a decrease in active power and an increase in

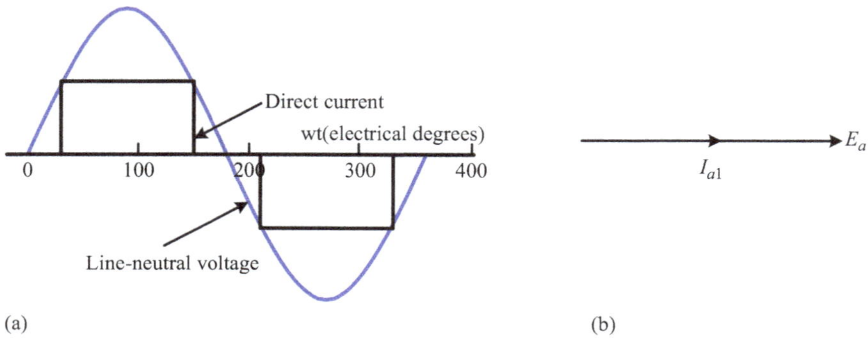

Figure 5.13 Relationship between firing angle (α_R = 0°) and phase displacements: (a) Relationship between current and voltage waveforms and (b) relationship between current and voltage phasors

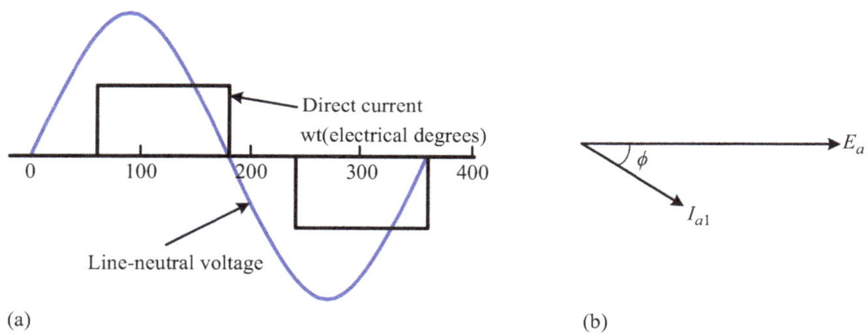

Figure 5.14 Relationship between firing angle (α_R = 30°) and phase displacements: (a) Relationship between current and voltage waveforms and (b) relationship between current and voltage phasors

reactive power consumption. Typically, the reactive power requirement of a converter station is rated at 50% of the active power nominal capacity of the link [16].

During the rectification process, the phase currents are not instantly transferred from the outgoing phase into the incoming phase; it requires a finite time, known as commutation or overlap time. This current transfer yields a self-clearing short circuit of small duration and magnitude between the two commuting thyristors, causing a temporary voltage reduction ΔV_d. This effect does not explicitly appear in the set of equations presented above, but it is mathematically represented by (5.27) for the rectifier operation.

$$\Delta V_d = \frac{3}{\pi} X_{CR}\, np_R I_d \tag{5.27}$$

A similar effect and representation exist for the inverter station.

One additional equation completes the formulation of the DC system configuration. For the point-to-point, monopolar DC-link shown in Figure 5.12, the rectifier voltage equation is (5.28),

$$V_{dR} = V_{dI} + R_d I_d \tag{5.28}$$

where R_d is the DC cable resistance.

The 166-node network used in previous Sections is modified to incorporate three CSC-HVDC links, as shown in Figure 5.15. The transmission lines connecting nodes RIH and ROH and MEH and TJH have been replaced by the HVDC links HVDC-1 and HVDC-2, respectively. A third HVDC-3 link was connected in series through the node AEHDC, with the transmission line connecting nodes AEH and CSH. The parameters of the three links are equal and given in Tables 5.2 and 5.3. In this study, the DC links have not been set to perform any control.

Table 5.4 displays the operational status of the HVDCs' state variables at the optimal solution. Table 5.5 provides information on the active and reactive powers and the DC currents in each HVDC link. Table 5.6 summarizes the total power generated at the optimal operating state, including generation dispatch costs and AC system losses. The total losses in the DC transmission links are 0.230 MWs.

In a second test case, DC current control is activated on the HVDC-1 and HVDC-2 links to increase the current flow by 20% compared to the values reported

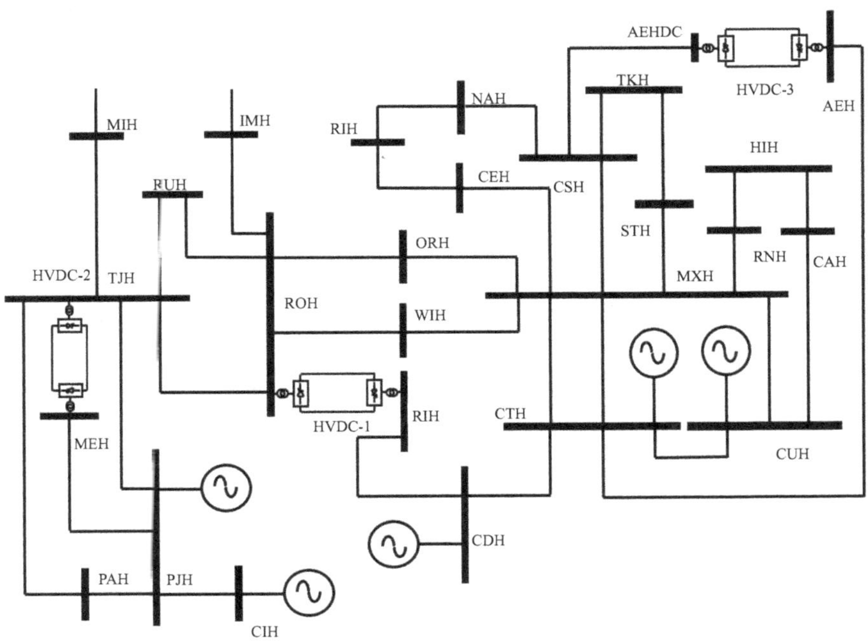

Figure 5.15 A relevant portion of the 166-node test system upgraded with three HVDC links

Table 5.2 DC-link parameters

	Rectifier	**Inverter**
Commutation reactance	0.01260	0.00728
DC-link resistance	0.00334	

Table 5.3 Operational limits of the DC-link

DC variable	**Limits**	
	Min	**Max**
V_{dR} (pu)	1.000	1.500
a_R	0.900	1.100
α_R (degrees)	10.000	20.000
V_{dI} (pu)	1.000	1.500
a_I	0.900	1.100
α_I (degrees)	15.000	25.000

Table 5.4 DC-link control parameters for the 166-node system

HVDC links	V_{dR} **(p.u.)**	a_R **(p.u.)**	α_R **(degrees)**	V_{dI} **(p.u.)**	a_I **(p.u.)**	α_I **(degrees)**
HVDC-1	1.330	1.002	13.800	1.328	1.008	17.103
HVDC-2	1.316	0.998	14.818	1.314	1.007	17.963
HVDC-3	1.309	0.971	14.902	1.309	0.997	17.600

Table 5.5 Power at the converters

HVDC links	P_{dR} **(MW)**	Q_{dR} **(MVAR)**	P_{dI} **(MW)**	Q_{dR} **(MVAR)**	I_d **(p.u.)**
HVDC-1	83.168	20.652	83.038	25.030	0.625
HVDC-2	68.762	18.122	68.670	21.731	0.522
HVDC-3	20.930	5.264	20.927	6.326	0.160

Table 5.6 Summary of power flow solution for the 166-node system

Results	**Optimal solution**
Active power generation Cost ($/h)	293.585
Active power losses (MW)	20.199
Active power generation (MW)	2,096.000
Reactive power generation (MVAR)	608.288

Table 5.7 Power at the converters (controlled case)

HVDC links	P_{dR} (MW)	Q_{dR} (MVAR)	P_{dI} (MW)	Q_{dR} (MVAR)	I_d (p.u.)
HVDC-1	103.000	24.681	102.611	30.912	0.750
HVDC-2	82.708	21.970	82.576	26.245	0.626
HVDC-3	15.654	3.930	15.650	4.761	0.120

Table 5.8 Summary of power flow solution for the 166-node system (controlled case)

Results	Optimal solution
Active power generation Cost ($/h)	293.605
Active power losses (MW)	20.279
Active power generation (MW)	2,096.080
Reactive power generation (MVAR)	625.293

in Table 5.4 of the base case, where the HVDC links do not perform any control action. A summary of the results for this operating scenario is given in Tables 5.7 and 5.8.

The comparison of results from both case studies indicates that an increase in active power flow through the converters leads to a higher absorption of reactive power by the inverters. This is also evident in the total reactive power generated by the generators. Furthermore, the active power generation cost was slightly increased from 293.585 $/h to 293.605 $/h when specific values were set for the DC currents at the HVDC links. Notice that the electric power losses in the non-controlled OPF case are 20.199 MW plus 0.230 MW in the DC transmission links, whereas in the controlled case, these losses amounted to 20.279 MW, with an additional 0.351 MW in the HVDC links.

5.5 VSC-HVDC links

The Voltage Source Converter-based High Voltage Direct Current (VSC-HVDC) technology is one of the most powerful solutions available today to integrate large amounts of renewable energy into AC power grids while ensuring the operational security of the electric power system. In this context, the direct current transmission links are evolving speedily in the direction of multiterminal VSC-HVDC (VSC-MTDC) systems, with the North Sea Super DC Grid being a prime example of this kind of development.

Unlike the CSC technology, the VSC technology enhances operation flexibility through rapid and independent control of active and reactive powers. Moreover, it reverses the power flow direction without altering the DC line voltage's polarity.

This unique capability is a key reason why VSC-MTDC systems can interconnect bidirectional, asynchronous AC transmission systems, enabling new services, such as power balancing between otherwise independent power systems.

5.5.1 VSC power control

Regardless of the topological configuration—whether it is a two-level, a multilevel, or a Modular Multilevel Converter (MMC)—a three-phase VSC serves as a component that connects a three-phase AC voltage with a DC voltage. This component can adjust the ratio between the magnitudes of the AC and DC voltages and determine when each sinusoidal signal begins. These capabilities are referred to as modulation index and phase angle, respectively.

Due to the commutative nature of the VSC semiconductor devices, a smoothing filter on the AC side of the VSC is essential to reduce the harmonic content. Additionally, a reactor is crucial for independent control of the active and reactive powers. A complete VSC-HVDC station always includes a transformer, a key component that ensures the voltage levels between the converter and the AC grid are appropriately adjusted. Figure 5.16 presents all the above elements to illustrate the control modes of operation.

Two controlling variables are indicated inside the VSC block: the modulation index m_a and the phase angle θ_{ac}. In this context, the modulation index, associated with the PWM technique, is a voltage magnitude relation between the *ac* and *dc* sides of the VSC, indicated as "*ac*" and "*dc*" nodes in Figure 5.16, respectively. On the other hand, the phase angle θ_{ac} determines the zero-crossing instant of the VSC AC voltage fundamental component, monitored at the "*ac*" node connected to one side of the reactor. This implies that the converter can control the voltage angle difference $(\theta_{ac} - \theta_{vi})$ in the reactor's terminals.

Based on the classic two-port network π model, the reactor's power flow equations are derived from the *ac* node to the *vi* node as noted in (5.29) and (5.30).

$$P_{acvi} = -G_r V_{ac}^2 + V_{ac} V_{vi}(G_r \cos(\theta_{ac} - \theta_{vi}) + B_r \sin(\theta_{ac} - \theta_{vi})), \tag{5.29}$$

$$Q_{acvi} = B_r V_{ac}^2 + V_{ac} V_{vi}(G_r \sin(\theta_{ac} - \theta_{vi}) - B_r \cos(\theta_{ac} - \theta_{vi})). \tag{5.30}$$

where V_{ac} and V_{vi} are the voltage magnitudes at the *ac* and *vi* nodes, respectively; θ_{ac} and θ_{vi} are the relative voltage phase angles, while G_r and B_r are the reactor conductance and susceptance, respectively.

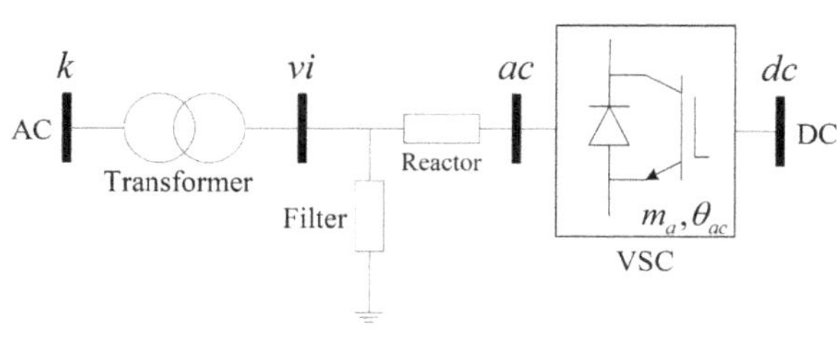

Figure 5.16 VSC HVDC station

Considering that the reactor is highly inductive, $B_r >> 0$ and $G_r \approx 0$, the conductance can be neglected, reducing the power flow equations to (5.31) and (5.32).

$$P_{acvi} = B_r V_{ac} V_{vi} \sin(\theta_{ac} - \theta_{vi}), \tag{5.31}$$

$$Q_{acvi} = B_r V_{ac}(V_{ac} - V_{vi} \cos(\theta_{ac} - \theta_{vi})). \tag{5.32}$$

Analyzing the active power equation P_{acvi}, the power will be positive, making the VSC work as an inverter if the phase angle difference is greater than 0° and less than 180°. In contrast, it will be negative, where the VSC operates as a rectifier, if this difference is greater than −180° and less than 0°. In this case, the active power is controlled between its minimum value $-B_r V_{ac} V_{vi}$ and maximum value $+B_r V_{ac} V_{vi}$ by controlling the angle difference $\theta_{ac} - \theta_{vi}$ between −90° and 90°. In this context, the VSC measures the phase angle θ_{vi} at the *vi* node and builds the voltage at the *ac* node with a leading or lagging phase angle θ_{ac} to work as a rectifier or an inverter, respectively. When no active power is available or desired, the phase angle θ_{ac} is synchronized with the phase angle θ_{vi}, having the same zero-crossing instant in these two nodes, such that $\theta_{ac} - \theta_{vi} = 0$ and $P_{acvi} = 0$.

On the other hand, when analyzing the reactive power equation Q_{acvi}, since the phase angle difference $\theta_{ac} - \theta_{vi}$ will be between −90° and 90°, the value of $\cos(\theta_{ac} - \theta_{vi})$ will always be positive in the range from 0 to 1. In this context, the voltage magnitudes V_{ac} and V_{vi} define the reactive power's magnitude and direction. When $V_{ac} > V_{vi}$, the reactive power Q_{acvi} will be positive regardless of the phase angle difference. In order to have a $Q_{acvi} < 0$, the voltage magnitude V_{ac} would have to be less than the voltage magnitude V_{vi}, and this difference has to be increased as the absolute phase angle difference $|\theta_{ac} - \theta_{vi}|$ is increased.

5.5.2 VSC voltage control

A typical VSC-HVDC station, like the one presented in Figure 5.16, has a node "*k*" as the point of connection to an AC system, whereas the node "*dc*" is the point of connection to a DC system. In this context, the AC system could be as complex as a network of thousands of nodes or as simple as a wind generation coupling point. Conversely, the DC system could be as complex as a meshed MTDC grid or as simple as a set of capacitors or a battery.

The voltage magnitudes on the AC system can be controlled by reactive power, whereas the voltage magnitudes on the DC system can be controlled by active power. In this case, by sending reactive power to the AC node "*k*", its voltage magnitude will be increased, whereas it will be decreased by extracting reactive power from the AC system. Similarly, the voltage magnitude at the "*dc*" node will increase or decrease when the VSC operates as a rectifier or inverter.

The correlation between the AC voltage magnitudes and the reactive power is related to the typical load being inductive. Therefore, it "absorbs" reactive power, which means that the current is getting delayed with respect to the voltage signal, creating a lagging power factor causing a drop in the voltage. To solve this problem, the transmission system can "generate" reactive power by installing capacitors that do the opposite by delaying the voltage signal with respect to the current, increasing the power factor, and recovering the voltage magnitude.

In the case of a VSC, reactive power injection into the AC system is achieved by creating a voltage signal that is delayed to the current to act like a capacitor without actually being one. When reactive power needs to be absorbed from the AC system, the VSC can act like an inductor by advancing the voltage signal with respect to the current signal.

Figure 5.17 shows how the VSC can create a voltage signal to act like a capacitor, a resistance, or an inductor by delaying the voltage, synchronizing the voltage, or advancing the voltage according to the current waveform phase, respectively. Acting as a capacitor injects reactive power; when it behaves as a resistance, no reactive power is created; and acting as an inductor, it absorbs reactive power.

A critical point to note is that although Figure 5.17 shows a difference in the voltage phase angle with respect to the current phase angle, the VSC phase angle θ_{ac} is not the manipulated variable to control the reactive power because θ_{ac} is adjusted with respect to the voltage phase angle θ_{vi} and not with respect to the current angle. In this case, and because of the reactor, the reactive power is controlled by changing the voltage magnitude in the "*ac*" node V_{ac} relative to the voltage magnitude in the "*vi*" node V_{vi}, on the other side of the reactor, utilizing the modulation index m_a as indicated in (5.32).

The relationship between active power and DC voltage magnitude is explained as follows. When only capacitors/batteries are installed on the "*dc*" node, they receive electric charge and increase their voltage when current is injected into their positive side. In this case, the converter functions as a rectifier, transferring energy from the "*ac*" to the "*dc*" node, as shown in Figure 5.18(a). When the VSC acts as an inverter, current is extracted from the positive side of the capacitors/batteries, causing energy transportation from the "*dc*" to the "*ac*" node, discharging these devices and reducing the DC voltage magnitude. The inverter operation is illustrated in Figure 5.18(b). To achieve a current flow from or to the "*dc*" node, the converter creates a voltage difference between the "*ac*" and "*dc*" nodes such that the current will flow from the node with higher voltage to the node with lower voltage.

If a more complex DC system is connected to the converter, the "*dc*" node is connected to one side of one or more DC transmission lines with their corresponding resistance values. Sending current from the "*dc*" node to the other side of

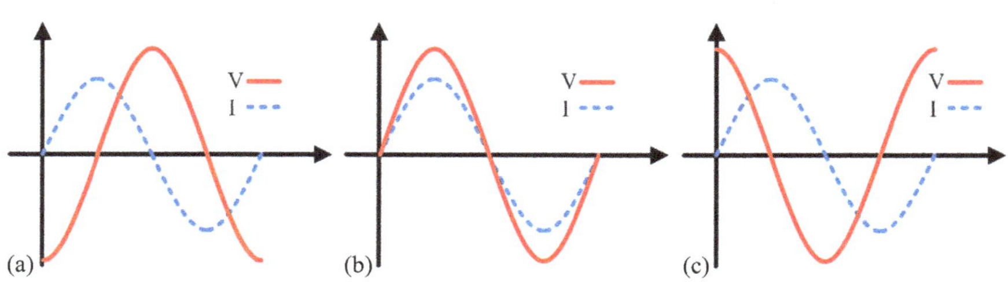

Figure 5.17 VSC acting as a: (a) capacitor, (b) resistance or (c) inductor

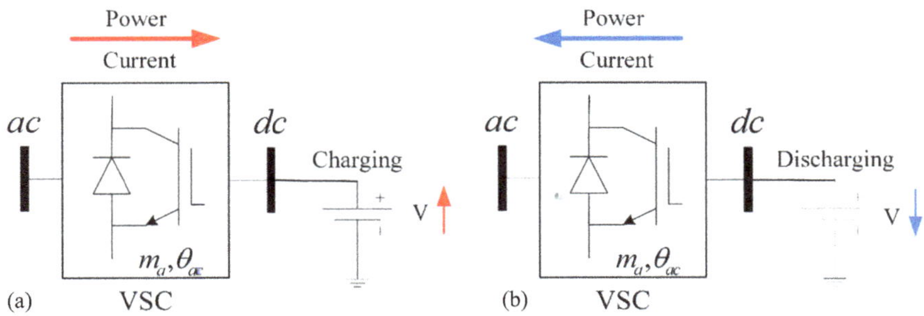

Figure 5.18 VSC acting as a: (a) rectifier or (b) inverter

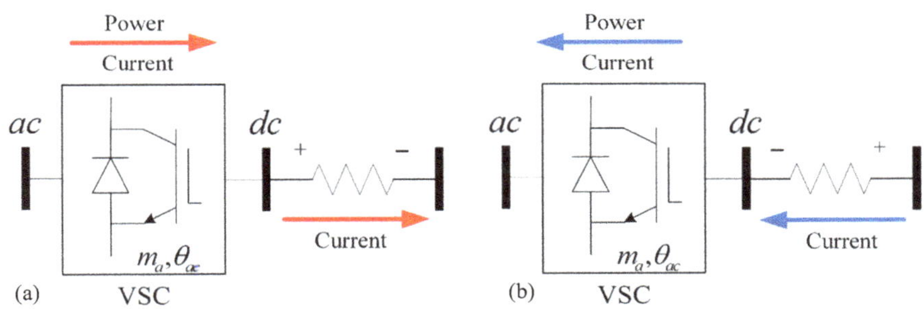

Figure 5.19 VSC DC voltage control as a: (a) rectifier or (b) inverter

a DC line will create a higher voltage on the "*dc*" node than in the receiving node, in proportion to the current as per Ohm's law. The opposite will occur when the VSC withdraws current from the DC line. These concepts are illustrated in Figure 5.19, where the VSC acts as a rectifier or inverter, causing the voltage at the "*dc*" node to be greater or lesser than the voltage on the opposite side of the line, respectively. In both cases, the "*dc*" node voltage is controlled by the VSC's phase angle θ_{ac}, which has a direct impact on the active power P_{acvi}, as indicated in (5.31). Summarizing the abovementioned, the VSC has an outer control that selects between reactive power and AC voltage magnitude control using the modulation index m_a. Independently of this control, the VSC also selects between active power and DC voltage level to be controlled through the phase angle θ_{ac}.

5.5.3 VSC current control

The current control scheme is used to prevent overcurrent operating conditions. To achieve this control, the VSC can be equipped with voltage and current sensors installed on the DC side of the VSC and the *vi* side of the reactor, as well as a Phase Loop Lock (PLL) installed on the phase "*a*", to monitor the operation of the VSC-HVDC station. These measurement spots are illustrated in Figure 5.20.

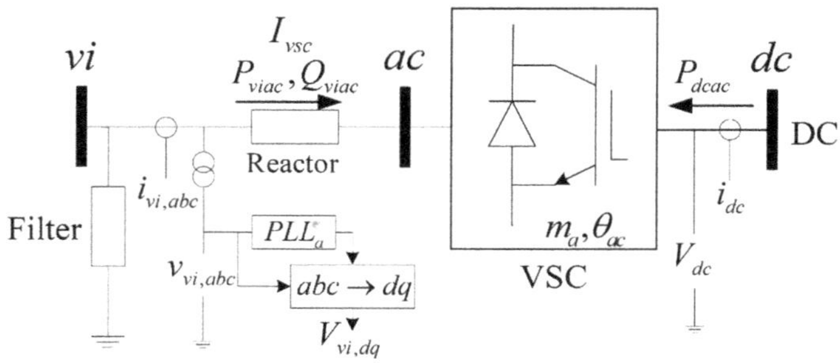

Figure 5.20 VSC-HVDC station measurements

By measuring the "*dc*" node voltage V_{dc} and the direct current i_{dc}, entering the converter on the DC side, the computation of the power P_{dcac} is attainable. On the other hand, with measurements of the three-phase reactor current measurement i_{vi} and the three-phase "*vi*" node voltage v_{vi}, the active and reactive powers P_{viac}, Q_{viac} flowing through the reactor can be computed. Moreover, with a PLL detecting the zero-crossing of phase "a," the transformation from the *abc* to the *dq* reference frame can be made to obtain the voltage of the "*vi*" node $V_{vi,dq}$.

Considering a voltage invariant transformation using a direct axis alignment to the phase "*a*" and a balanced three-phase system, the active power P_{viac} is related to the direct component of the "*vi*" voltage $V_{vi,d}$ and the direct component of the VSC current $I_{vsc,d}$ as described by (5.33):

$$P_{viac} = 1.5 V_{vi,d} I_{vsc,d}. \tag{5.33}$$

On the other hand, the reactive power Q_{viac} is related to the direct component of the "*vi*" voltage $V_{vi,d}$ and the quadrature component of the VSC current $I_{vsc,q}$ as expressed by (5.34):

$$Q_{viac} = -1.5 V_{vi,d} I_{vsc,q}. \tag{5.34}$$

According to these equations, the direct component of the VSC current $I_{vsc,d}$ is directly associated with the active power P_{viac}, whereas the quadrature component of the VSC current $I_{vsc,q}$ is related to the reactive power Q_{viac}. These relationships directly relate the outer power/voltage control setpoints to the values of the VSC direct and quadrature currents, $I^{des}_{vsc,d}$ and $I^{des}_{vsc,q}$, that flow through the converter to accomplish the specified control targets. The feasibility of having these "desired" values of VSC's currents is checked in the current limiter scheme. This protects the VSC from overcurrents with the potential to damage the converters' switches. Following the current limiter validation, the VSC current reference values $I^{ref}_{vsc,d}$ and $I^{ref}_{vsc,q}$ are obtained to be used as setpoints by the inner current control to define the VSC phase angle and modulation index, as shown in Figure 5.21.

The current limiter validation essentially checks the current magnitude, where the desired *dq* current components are treated as real and imaginary components of a current vector. As a result, the desired current magnitude I^{des}_{vsc} is given by (5.35).

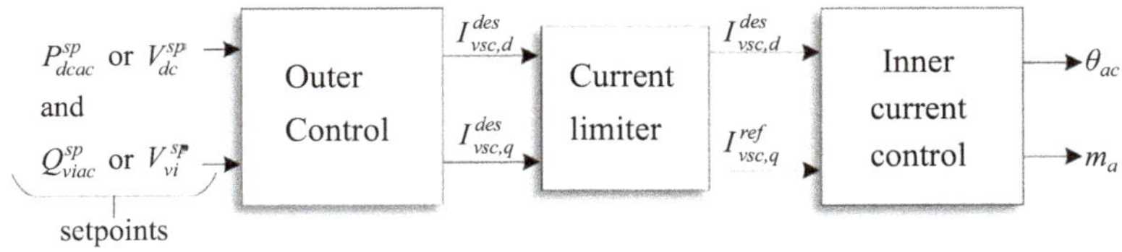

Figure 5.21 Current control scheme

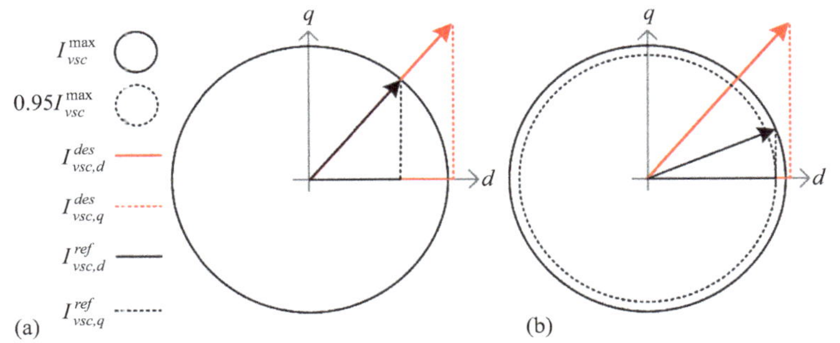

Figure 5.22 Current limiter schemes: (a) vector limiter and (b) active power priority

This magnitude is compared with the maximum permissible VSC current $I_{vsc}^{\max}$ to avoid overcurrent operation. If the desired current magnitude is less or equal to the maximum current, i.e., $I_{vsc}^{des} \leq I_{vsc}^{\max}$, there is no risk of overcurrent operation, and the dq components of the current references are equal to the desired dq components as given in (5.36). On the other hand, if the desired current magnitude exceeds the maximum current, $I_{vsc}^{des} > I_{vsc}^{\max}$, different limiting decisions could be taken based on two different limiter schemes: the vector limiter scheme and the active power priority scheme shown in Figure 5.22(a) and (b), respectively.

$$I_{vsc}^{des} = \sqrt{\left(I_{vsc,d}^{des}\right)^2 + \left(I_{vsc,q}^{des}\right)^2} \tag{5.35}$$

$$I_{vsc,d}^{ref} = I_{vsc,d}^{des}, I_{vsc,q}^{ref} = I_{vsc,q}^{des} \tag{5.36}$$

In the vector limiter scheme, the direct and quadrature components are scaled down in the same proportion. This scale proportion is defined by the ratio between the desired current magnitude I_{vsc}^{des} and the VSC maximum operating current $I_{vsc}^{\max}$ so that the VSC current reference values are in agreement with (5.37).

$$I_{vsc,d}^{ref} = \frac{I_{vsc,d}^{des}}{I_{vsc}^{des} / I_{vsc}^{\max}}, \; I_{vsc,q}^{ref} = \frac{I_{vsc,q}^{des}}{I_{vsc}^{des} / I_{vsc}^{\max}}. \tag{5.37}$$

In the active power priority scheme, the direct component associated with the active power has the priority to use up to a defined percentage of the maximum current, e.g., 95% of the $I_{vsc}^{\max}$. After assigning the direct component reference, the quadrature component will use the remaining current capability. Figure 5.26(b) shows the case where the direct component reference is 95% of the current capacity: $I_{vsc,d}^{ref} = 0.95 I_{vsc}^{\max}$, and the quadrature reference is $I_{vsc,q}^{ref} = 0.31225 I_{vsc}^{\max}$.

In cases where the desired value of the direct component is less than 95% of the current capacity: $I_{vsc,d}^{des} < 0.95 I_{vsc}^{\max}$, the direct component reference will be set to the desired value, $I_{vsc,d}^{ref} = I_{vsc,d}^{des}$, and then the quadrature component reference will be calculated using the remaining current capacity based on the Pythagorean theorem as indicated in (5.38).

$$I_{vsc,q}^{ref} = \sqrt{\left(I_{vsc}^{\max}\right)^2 - \left(I_{vsc,d}^{ref}\right)^2}. \tag{5.38}$$

After obtaining the dq current references, a closed-loop PI control, shown in Figure 5.21 as inner current control, can be applied to achieve these reference values by adjusting the voltage magnitude V_{ac}, through the modulation index m_a, and phase angle θ_{ac} at the "*ac*" node. These two state variables are obtained from the linear plant or VSC model.

First, the controller must have the current's true values to compare them to the current reference values. As the latter have been obtained as described above, the former are obtained as follows. Based on the measurements shown in Figure 5.20 at node "*vi*", the active and reactive powers given in (5.33) and (5.34) are directly computed. From the measurements of $V_{vi,abc}$, the voltages $V_{vi,d}$ and $V_{vi,q}$ are also computed through the PLL by using the transformation from the *abc* to the *dq* reference frame. All these computations allow us to obtain the true values of $I_{vsc,d}$ and $I_{vsc,q}$ using (5.33) and (5.34). Hence, the true and reference values of currents flowing through the VSC are known to adjust the voltage magnitude and phase angle at the "*ac*" node through a PI controller.

To obtain the required modulation index m_a and voltage phase angle θ_{ac}, the true voltage at the "*ac*" node must be first computed from the known values of $V_{vi,d}$,$V_{vi,q}$, $I_{vsc,d}$ and $I_{vsc,q}$ by applying Ohm's law at the VSC reactor's terminals.

Considering that the reactor impedance is composed of a resistance R_r and an inductance L_r, the direct components of voltages at nodes "*vi*" and "*ac*": $V_{vi,d}$ and $V_{ac,d}$, are related to the reactor impedance and the VSC current dq components as described by (5.39) [17]:

$$V_{vi,d} - V_{ac,d} = R_r I_{vsc,d} + L_r \frac{dI_{vsc,d}}{dt} - \omega L_r I_{vsc,q}, \tag{5.39}$$

where ω is the angular frequency. Similarly, the quadrature components of these nodal voltages are obtained from (5.40) [17]:

$$V_{vi,q} - V_{ac,q} = R_r I_{vsc,q} + L_r \frac{dI_{vsc,q}}{dt} + \omega L_r I_{vsc,d}. \tag{5.40}$$

Hence, the required values of $V_{ac,d}$ and $V_{ac,q}$ are directly obtained from (5.39) and (5.40), respectively, yielding the expressions in (5.41) and (5.42):

$$V_{ac,d} = V_{vi,d} - R_r I_{vsc,d} - L_r \frac{dI_{vsc,d}}{dt} + \omega L_r I_{vsc,q}, \tag{5.41}$$

$$V_{ac,q} = V_{vi\,q} - R_r I_{vsc,q} - L_r \frac{dI_{vsc,q}}{dt} - \omega L_r I_{vsc,d}. \tag{5.42}$$

By applying a dq to positive sequence transformation to $V_{ac,d}$ and $V_{ac,q}$, the voltage magnitude V_{ac} is obtained. Finally, the necessary VSC modulation index m_a to achieve the computed V_{ac} is obtained by considering the measured voltage V_{dc} as described by (5.43) [17]:

$$m_a = V_{ac} / \left(V_{dc} \sqrt{3/2} \right), \tag{5.43}$$

where the value $\sqrt{3/2}$ represents the relationship between V_{dc} and the line-to-line root mean squared (RMS) voltage V_{ac} for a unitary modulation index, i.e., $m_a = 1$.

Lastly, the VSC phase angle θ_{ac} is solved relative to the voltage phase angle θ_{vi} measured by the PLL, yielding (5.44) [17]:

$$\theta_{ac} = \theta_{vi} + \arctan\left(\frac{V_{ac,q}}{V_{ac,d}} \right). \tag{5.44}$$

The computed values of m_a and θ_{ac} are adjusted by the PI control loop based on the error between the true and reference values of VSC's currents [17]. The modulation index and the phase angle are two of the three most important parameters to define the switching control for the semiconductor devices inside the VSC, where the third parameter is the frequency. In this context, the frequency could be defined as the VSC following the AC grid frequency or the VSC imposing a certain frequency on the AC grid. The former is usually for robust AC grids, while the latter is applied for weak AC grids.

5.5.4 Droop control

The VSC uses the phase angle θ_{ac} to control the amount of active power and the direction of the active power flow, with specific impacts on the DC and AC sides, respectively. On the DC side, the injection of more active power will increase the DC nodal voltages. In contrast, the active power injection to the AC grid will be considered as a generation such that the frequency of such a grid will increase.

The active power P_{acdc} extracted from the AC grid and injected into the DC grid is opposite in sign but equal in magnitude to the active power P_{acvi} indicated in (5.29), i.e. $P_{acdc} = -P_{acvi}$. When the converter uses the phase angle θ_{ac} to achieve a specified DC voltage V_{dc}^{sp}, the control strategy relation between the DC voltage and the VSC active power is the one illustrated in Figure 5.23. In this case, the specified voltage value will be achieved as long as the VSC operates within its active power capability, defined by the current limiter control. In an MTDC grid, only one VSC can operate with this type of control, which will become the DC slack of the MTDC to balance the active power of the grid. In this case, this converter requires a control coordination communication link with the rest of the VSCs to avoid a grid collapse.

When the converter phase angle θ_{ac} is used to achieve a specified value of active power P_{acdc}^{sp}, this value will depend on the current limiter and the minimum and maximum permissible V_{dc} values, as illustrated in Figure 5.24. In this context,

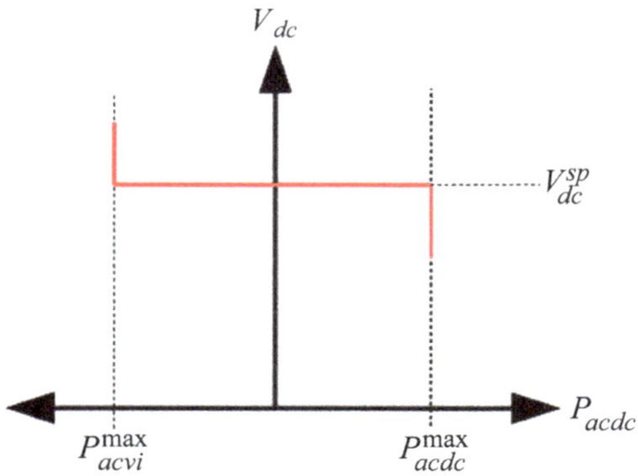

Figure 5.23 Constant DC voltage control

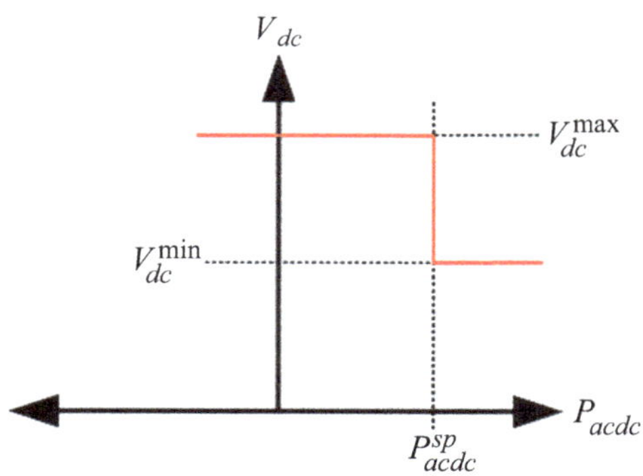

Figure 5.24 Constant P active power control

if the maximum permissible DC voltage value V_{dc}^{max} is sensed, the VSC will reduce its active power P_{acdc} to alleviate this problem, the opposite will occur, where the active power P_{acdc} will be increased under a minimum DC voltage value V_{dc}^{min} scenario.

Considering that the VSC-HVDC stations could be installed several miles/kilometers apart, different control coordination methods that do not require communication links between stations have been proposed. The droop control strategy is the main proposal to coordinate multiple VSC-HVDC stations without a dedicated communication link. The main idea behind the droop control strategy is to modify the amount of active power based on the VSC's DC voltage measurement V_{dc}. This control strategy can be based on a linear relation between the voltage V_{dc} and the active power P_{acdc}, as illustrated in Figure 5.25, which is frequently known as *V–P* droop control. In this case, three values are indicated: the reference of the active power value P_{acdc}^{ref} to be accomplished when the measured DC voltage V_{dc} is equal to the V_{dc}^{ref} value; the DC voltage reference value V_{dc}^{ref}; and the slope defined

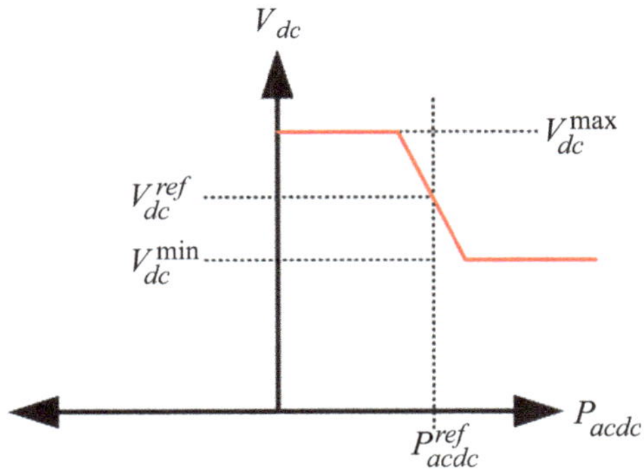

Figure 5.25 V–P droop control

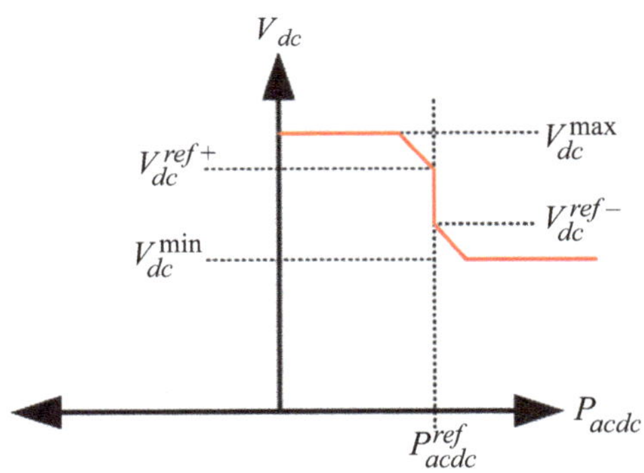

Figure 5.26 V–P droop control with a dead band

with the desired decrement in the active power $-\Delta P_{acdc}$ per increment in the DC voltage ΔV_{dc}. Hence, the VSC will define the active power setpoint according to equation (5.45):

$$P_{acdc}^{sp} = P_{acdc}^{ref} - \frac{\Delta P_{acdc}}{\Delta V_{dc}}\left(V_{dc} - V_{dc}^{ref}\right). \tag{5.45}$$

A variation of the *V–P* droop control is the *V–P* droop control with a dead band. In this control strategy, the specified active power remains constant for a defined DC voltage range, i.e., when $V_{dc}^{ref-} \leq V_{dc} \leq V_{dc}^{ref+}$ holds true, then $P_{acdc}^{sp} = P_{acdc}^{ref}$. Outside this dead band, the active power is adjusted linearly based on a defined slope, as shown in Figure 5.26.

Other similar droop control strategies can be implemented, like the *V–I* (voltage–current) droop control, with or without a dead band, where the slope represents a linear relation between the VSC current I_{vsc} and the DC voltage V_{dc}. In this context, the slope is defined as the desired decrement in the converter current $-\Delta I_{vsc}$ per DC voltage increment ΔV_{dc}. This slope will act as a negative conductance $-G_a$ or as an inverse of a negative resistance $-1/R_a$. The converter current that the current limiter will receive as a desired value is described by (5.46),

$$I_{vsc}^{des} = I_{vsc}^{ref} - \frac{\Delta I_{vsc}}{\Delta V_{dc}}\left(V_{dc} - V_{dc}^{ref}\right). \tag{5.46}$$

The *V–I* droop control is graphically represented in Figure 5.27.

5.5.5 VSC steady-state model

Any VSC-based equipment can be included in the power flow formulation problem by considering the VSC-HVDC station steady-state model represented as a two-port model. The power flow equations flowing between the *dc* and *ac* nodes and between the *ac* and *vi* nodes are necessary to solve the control variables of the VSC, i.e., modulation index m_a and phase angle θ_{ac}.

In this context, the VSC-HVDC station presented in Figure 5.28 is modeled by the positive sequence equivalent circuit, shown in Figure 5.29, using the concept of an ideal phase-shifting transformer [18–21].

The power flow equations (5.47)–(5.54) are derived from this positive sequence model, as detailed in [21]. The value $K = \sqrt{3/2}$ represents the relationship between the DC voltage and the line-to-line RMS voltage at the VSC's AC terminals for a unit modulation index. To model the converter power losses, the series resistance R_1 represents the ohmic losses and the conductanceG_{sw}represents switching losses [21]. On the other hand, the values G_{pr} and B_{pr} are the phase reactor conductance and susceptance, respectively, whereas G_{sf}^{pr} and B_{sf}^{pr} are the real and imaginary parts of $\bar{\mathbf{Y}}_{sf}+\bar{\mathbf{Y}}_{pr}$, respectively.

$$P_{dcac}(\boldsymbol{x}) = V_{dc}{}^2\left(G_{sw} + K^2 m_a{}^2 R_1^{-1}\right) - K m_a V_{dc} V_{ac} R_1^{-1} \cos(\phi - \theta_{ac}) \tag{5.47}$$

$$Q_{dcac}(\boldsymbol{x}) = -K m_a V_{dc} V_{ac} R_1^{-1} \sin(\phi - \theta_{ac}) \tag{5.48}$$

$$P_{acdc}(\boldsymbol{x}) = R_1^{-1} V_{ac}^2 - K m_a V_{dc} V_{ac} R_1^{-1} \cos(\theta_{ac} - \phi) \tag{5.49}$$

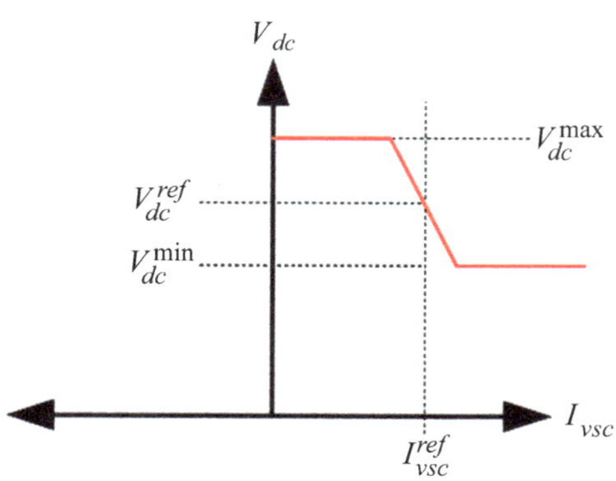

Figure 5.27 V–I droop control

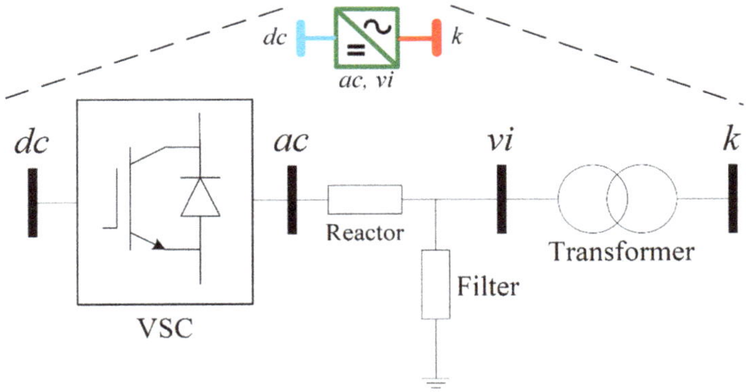

Figure 5.28 Detail of VSC-HVDC station

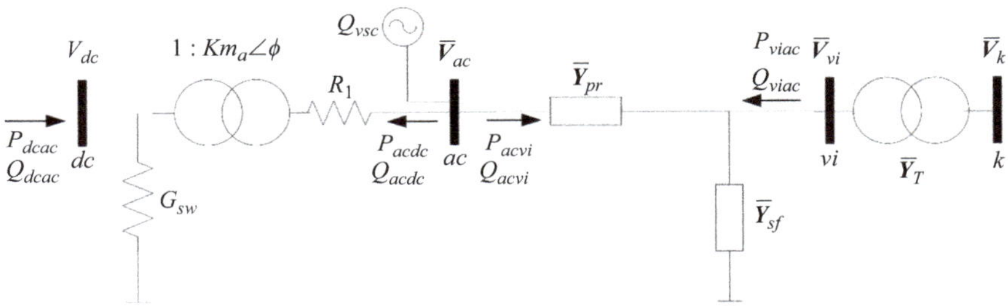

Figure 5.29 Positive sequence equivalent circuit model of VSC-HVDC station

$$Q_{acdc}(\boldsymbol{x}) = -Km_a V_{dc} V_{ac} R_1^{-1} \sin(\theta_{ac} - \phi) \tag{5.50}$$

$$P_{viac}(\boldsymbol{x}) = G_{sf}^{pr} {V_{vi}}^2 + V_{vi} V_{ac}\big(G_{pr} \cos(\theta_{vi} - \theta_{ac}) + B_{pr} \sin(\theta_{vi} - \theta_{ac})\big) \tag{5.51}$$

$$Q_{viac}(\boldsymbol{x}) = -B_{sf}^{pr} {V_{vi}}^2 + V_{vi} V_{ac}\big(G_{pr} \sin(\theta_{vi} - \theta_{ac}) - B_{pr} \cos(\theta_{vi} - \theta_{ac})\big) \tag{5.52}$$

$$P_{acvi}(\boldsymbol{x}) = -G_{pr} {V_{ac}}^2 + V_{ac} V_{vi}\big(G_{pr} \cos(\theta_{ac} - \theta_{vi}) + B_{pr} \sin(\theta_{ac} - \theta_{vi})\big) \tag{5.53}$$

$$Q_{acvi}(\boldsymbol{x}) = B_{pr} {V_{ac}}^2 + V_{ac} V_{vi}\big(G_{pr} \sin(\theta_{ac} - \theta_{vi}) - B_{pr} \cos(\theta_{ac} - \theta_{vi})\big) \tag{5.54}$$

Since the VSC's phase angle ϕ is equal to the *ac* node voltage angle θ_{ac}, i.e., $\phi = \theta_{ac}$, (5.47)–(5.50) are reduced to

$$P_{dcac}(\boldsymbol{x}) = {V_{dc}}^2\big(G_{sw} + K^2 {m_a}^2 R_1^{-1}\big) - Km_a V_{dc} V_{ac} R_1^{-1} \tag{5.55}$$

$$Q_{dcac}(\boldsymbol{x}) = 0 \tag{5.56}$$

$$P_{acdc}(\boldsymbol{x}) = R_1^{-1} V_{ac}^2 - K m_a V_{dc} V_{ac} R_1^{-1} \tag{5.57}$$

$$Q_{acdc}(\boldsymbol{x}) = 0 \tag{5.58}$$

By using this approach, the DC grid is treated as an AC grid working at a unity power factor with null reactive power, voltage angles, and line reactances. For this reason, the reactive power source Q_{vsc} connected to the *ac* node is considered to represent the reactive power capabilities of the VSC.

The phase reactor and the shunt filter are almost pure reactances with very low resistances such that the conductances G_{pr} and G_{sf}^{pr} are almost null. This allows the converter to control the active power with the *ac* voltage angle θ_{ac} and control the reactive power with the *ac* voltage magnitude V_{ac}, which is controlled with the modulation index m_a.

5.5.6 Multiterminal HVDC system

Multiterminal HVDC (MT-HVDC) systems transmit energy with higher efficiency and lower electrical losses over long distances [17]. The reduction in power losses is derived from the DC energy transmission, which has no reactive power associated with it.

An MT-HVDC grid comprises several VSC-HVDC stations connected on their DC side through DC lines. For power flow studies, where the steady state is analyzed, the DC lines only contain series and shunt conductance, only active power exists, and the nodal voltages are reduced to only a magnitude.

With this kind of grid, different AC systems operating at different frequencies can share active power. This adaptability ensures that a strong system can deliver active power to a weak system for energy support, providing reassurance in the face of changing energy needs. A good example of this kind of development is the North Sea Super DC Grid [22]. Moreover, HVDC links represent an ideal solution to incorporate big amounts of renewable energy into the AC power grid, particularly offshore wind power [23].

For power flow studies, the CIGRE test system, presented in Figure 5.30, has become the main standard because it contains different kinds of AC and DC systems; a complete mathematical description of the power flow equations to solve this kind of system can be found in [21]. The CIGRE test system is composed as follows: (i) two onshore AC networks of two and four nodes, respectively, operating at 380 kV; (ii) four offshore AC systems operating at 145 kV represent a connection with offshore wind farms; and (iii) three VSC-HVDC systems corresponding to one point-to-point link at $\pm$200 kV, another monopolar radial network with five nodes at $\pm$200 kV, and lastly, a bipolar 8-node meshed grid at $\pm$400 kV. There are 11 VSC-based AC-DC interfaces, six of which have a monopolar configuration, while the rest have a bipolar arrangement. Additionally, two DC-DC converters supply specific power, and one enables power exchange from the meshed bipolar grid to the radial monopolar network. Finally, the total number of nodes corresponds to 33 in the AC system and 15 in the DC systems, with four

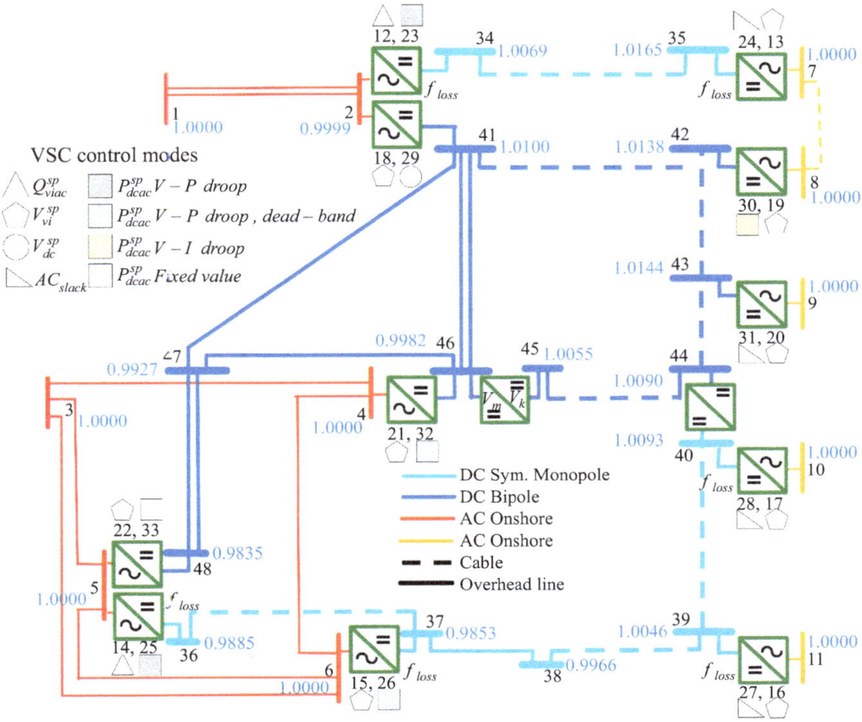

Figure 5.30 CIGRE test system

having no connections to the AC system. The system parameters are given in [24]. Those corresponding to the transformers, filters, and reactors of the VSC-based AC-DC station model detailed in Figure 5.29 are $\bar{Y}_T = -j10000$, $\bar{Y}_{sf} = j0.05$, and $\bar{Y}_{pr} = -j1000$, respectively.

The power flow study of the CIGRE test system is performed by considering the current capacity detailed in [24] for all converter stations. The particular control modes of each VSC-HVDC are indicated in Figure 5.30, while the two DC-DC converter stations operate using active power control mode. The VSCs connected to *dc* nodes 34 and 36 operate under the *V–P* droop control with reference voltages at 1 p.u. and 0.99 p.u., respectively. The active power is controlled by these converters at 350 MW and −120 MW, respectively. The VSCs connected to *dc* nodes 37 and 46 are operating using a *V–P* droop control with a dead band from 0.98 to 1 p.u., reference voltages of 0.99 p.u., and active power setpoints of 800 MW and 1,500 MW, respectively. Furthermore, the VSC connected to *dc* node 42 uses a *V–I* droop with a reference voltage of 1.01 p.u. and a reference current of −5.9406 p.u. All the droop controls consider a slope of $1/10P_{nom}$. Lastly, the power losses of six VSCs are given by $f_{loss} = a + bI_{vsc} + cI^2_{vsc}$, where a is a constant power loss obtained as the 1% of the converter nominal power P_{nom}, b represents the current

Table 5.9 Results of CIGRE test system case 4

Nodes				Case 4	
dc	ac	vi	m_a	P_{dcac} (MW)	Q_{viac} (Mvar)
34	23	12	0.8147	405.46	0
35	24	13	0.7988	−410.34	31.56
36	25	14	0.8245	−132.37	0
37	26	15	0.8342	800	−18.25
39	27	16	0.8076	−487.37	−0.25
40	28	17	0.8130	102.80	−0.01
41	29	18	0.8029	−2019.7	−141.18
42	30	19	0.7998	−571.80	−39.63
43	31	20	0.7999	−992.09	−1
46	32	21	0.8233	1,500	−22.25
48	33	22	0.8362	1,700	−19.31

dependent power loss obtained as the 0.3% of the converter nominal voltage V_{nom}, and $c = 0.01\ V_{nom}^2/P_{nom}$, representing the internal conduction resistance of the VSC's that will dissipate power depending on the current squared value. These VSCs are identified in Figure 5.30 with the symbol f_{loss}. The active and reactive powers exchanged between the *dc* and *ac* nodes of the 11 VSCs are reported in columns 5 and 6 of Table 5.9, respectively.

In Figure 5.30 of the CIGRE test system, next to the node number, the nodal voltage magnitude is indicated in blue color using the format x.xxxx. In the case of DC lines, it is easy to observe the power direction with the nodal voltages, since the power will always flow from the higher voltage to the lower voltage, e.g., the DC line between nodes 34 and 35 has power flowing from the node 35, with a nodal voltage magnitude of 1.0165 p.u., to the node 34, with a nodal voltage magnitude of 1.0069 p.u. Moreover, from Table 5.9, since the power flowing from 35 to 24 is −410.34 MW, the power flowing from 35 to 34 will be near 410.34 MW due to shunt power losses. The power received in the VSC at node 34 is 405.45 MW, as can be seen in Table 5.9, representing a power transmission efficiency of almost 99%.

5.6 Conclusions

As electric utilities transition into more competitive generation supply regimes, the optimization of existing transmission corridors for power transfer becomes a matter of paramount importance. With limited scope to expand transmission facilities, the role of FACTS and HVDC technologies, which enhance system operation flexibility, becomes increasingly significant.

This chapter has focused on presenting suitable steady-state models of FACTS devices and HVDC links, demonstrating their practical benefits and influence on the overall performance of real-life electric power systems. The effectiveness of these technologies in achieving control of power flows and nodal voltage magnitudes has been addressed, together with the economic and operational benefits afforded by this new form of power electronic control.

The VSC-HVDC equipment allows great flexibility in the operation of electrical power systems due to the rapid and independent controllability of active and reactive powers. Owing to the unrivaled functionality of these control modes and the voltage-sourced nature of the VSC-HVDC systems, it would be possible to build advanced multiterminal DC networks of arbitrary configuration, with utmost controllability and reliability. This is particularly important when the requirement is to tap into large amounts of renewable power generation. In this context, a comprehensive and flexible mathematical model of a VSC-based AC–DC converter station for power flow studies has been put forward in this chapter.

References

[1] Hingorani N.G.: 'High Power Electronics and Flexible AC Transmission System', *IEEE Power Engineering Review*, vol. 8, no. 7, 1988, pp. 3–4.

[2] Hingorani N.H.: 'Flexible AC Transmission Systems', *IEEE Spectrum*, vol 30, no. 4, 1993, pp. 40–45.

[3] Ledu A., Tontini G. and Winfield M.: 'Which FACTS Equipment for Which Need?', *International Conference on Large High Voltage Electric Systems (CIGRÉ)*, paper 14/37/38-08, Paris, September 1992.

[4] Gomis-Bellmunt O, Sau-Bassols J, Prieto Araujo E, and Cheah-Mane M: 'Flexible Converters for Meshed HVDC grids: From Flexible AC Transmission Systems (FACTS) to Flexible DC Grids', *IEEE Transactions on Power Delivery*, vol. 35, no. 1, 2020, pp. 2–15.

[5] IEEE/CIGRE: 'FACTS Overview', Special Issue, 95TP108, IEEE Service Center, Piscataway, NJ, 1995.

[6] Aboytes F, Arroyo G, and Villa G. 'Application of Static Var Compensators in Longitudinal Power Systems', *IEEE Transactions on Power Apparatus and Systems*, vol. Pas-102, no. 10, 1983, pp. 3460–3466.

[7] Bailey A, Faulkner B, Chou H, McVey M, Parker D, and Tatsumi T, Okuyama R. 'Installation and Commissioning of Mitsubishi Electric's MMC STATCOM (SVC-Diamond®) at Dominion Energy's Clover & Rawlings Substations' In *Proceedings of the 2022 Grid of the Future, CIGRE US National Committee*, Chicago, IL, USA, Nov. 2022, pp. 1–5.

[8] Song S.H., Lim J.U., and Moon S. 'Installation and operation of FACTS devices for enhancing steady-state security', *Electric Power Systems Research*, vol. 70, no. 1, 2004, pp. 7–15.

[9] Larsen E., Bowler C., Damsky B. and Nilsson S.: 'Benefits of Thyristor Controlled Series Compensation', *International Conference on Large High Voltage Electric Systems (CIGRÉ)*, paper 14/37/38-04, Paris, 1992.

[10] Ambriz Pérez H., "Flexible AC Transmission Systems Modelling in Optimal Power Flows Using Newton's Method," *Ph.D. dissertation*, Department of Electronics & Electrical Engineering, University of Glasgow, Glasgow, Scotland, UK, 1998.

[11] Ambriz-Pérez H., Acha E., and Fuerte-Esquivel C.R. "TCSC-Firing Angle Model for Optimal Power Flow Solutions Using Newton's Method", *International Journal of Electrical Power and Energy Systems*, vol. 28, no. 2, 2006, pp. 77–85.

[12] Gyugyi L., Schauder C.D., Williams S.L. Rietman T.R., Torgerson D.R. and Edris A.: 'The Unified Power Flow Controller: A New Approach to Power Transmission Control', *IEEE Transactions on Power Delivery*, vol. 10, no. 2, 1995, pp. 1085–1097.

[13] Fuerte-Esquivel C.R., and Acha E. y Ambriz-Perez H.: "A Comprehensive Newton-Rapshon UPFC for the Quadratic Power Flow Solution of Practical Power Networks", *IEEE Transactions on Power Systems*, vol. 15, no. 1, 2000, pp. 102–109.

[14] Kimbark E.: *Direct Current Transmission*, Vol. I, Wiley-Interscience, New York, 1971.

[15] Ambriz-Peréz H., and Acha E. y Fuerte-Esquivel C.R.: "High Voltage Direct Current Modelling in Optimal Power Flows", *International Journal of Electrical Power & Energy Systems*, vol. 30, no. 2, 2008, pp. 157–168.

[16] Kundur P.: *Power System Stability and Control*, McGraw-Hill, New York, 1994.

[17] A. Yazdani and R. Iravani, *Voltage-Sourced Converters in Power Systems: Modeling, Control, and Applications*, Wiley, New York, 2010.

[18] E. Acha, B. Kazemtabrizi, and L. M. Castro, "A New VSC-HVDC Model for Power Flows Using the Newton-Raphson Method", *IEEE Transactions on Power Systems*, vol. 28, no. 3, ago. 2013, pp. 2602–2612, doi 10.1109/TPWRS.2012.2236109

[19] E. Acha y L. M. Castro, "A Generalized Frame of Reference for the Incorporation of Multi-Terminal VSC-HVDC Systems in Power Flow Solutions", *Electric Power System Research*, vol. 136, 2016, pp. 415–424, doi:10.1016/j.epsr.2016.03.009.

[20] E. Acha y B. Kazemtabrizi, "A New STATCOM Model for Power Flows Using the Newton–Raphson Method", *IEEE Trans. Power Syst.*, vol. 28, no. 3, 2013, pp. 2455–2465, doi:10.1109/TPWRS.2012.2237186.

[21] R. Martínez-Parrales, C. R. Fuerte-Esquivel, B. A. Alcaide-Moreno, and E. Acha, "A VSC-based Model for Power Flow Assessment of Multi-terminal VSC-HVDC Transmission Systems", *Journal of Modern Power Systems and Clean Energy*, vol. 9, no. 6, 2021, pp. 1363–1374, doi:10.35833/MPCE.2021.000104.

[22] T. K. Vrana, R. E. Torres-Olguin, B. Liu, and T. M. Haileselassie, "The North Sea Super Grid – A Technical Perspective", *9th IET International Conference on AC and DC Power Transmission (ACDC 2010)*, 2010, pp. 65–65, doi: 10.1049/cp.2010.0992.

[23] B. Mitra, B. Chowdhury, and M. Manjrekar, "HVDC Transmission for Access to Off-Shore Renewable Energy: A Review of Technology and Fault Detection Techniques", *IET Renewable Power Generation*, vol. 12, no. 13, 2018, pp. 1563–1571, doi:10.1049/iet-rpg.2018.5274.

[24] T. Vrana, S. Dennetiere, Y. Yang, J. A. Jardini, D. Jovcic, and H. Saad, "The Cigré B4 DC Grid Test System", *CIGRE Electra*, vol. 270, 2013.

Chapter 6

Power electronics in electrical power distribution systems

Héctor Francisco Ruiz Paredes[1] *and Enrique Acha*[2]

6.1 Introduction

Traditionally, electrical power has been supplied to industrial, commercial, and domestic end-users employing AC voltages and currents using urban and rural electrical power distribution networks. In broad terms, such networks comprise medium and low-voltage overhead transmission lines, underground power cables, medium and low-power capacity transformers, and ancillary protective gear to safeguard the integrity of utility personnel and main equipment [1].

The classic AC power distribution system, derived from a high-voltage power transmission system, comprises a primary (medium-voltage) and secondary (low-voltage) power distribution circuit. The former exhibits a mesh or a radial structure, whereas the latter exhibits a radial or a ring structure, even though these ring structures are operated in a radial fashion using open points circuit breakers for simplicity of operation and protection. Besides, the classic distribution system is said to be passive – it contains no in-situ power generators – and voltage regulation relies on electro-mechanical actuators (devices) with electronic control, but the distribution system itself lacks power electronic-based voltage regulators. In the 1990s, breakthroughs in low-capacity power sources of various kinds and the emergence of power electronics valves, bridges, and new equipment started challenging tradition in the electricity distribution industry. Nowadays, it may be affordable and reliable to embed electrical power generators at practically any point of the power distribution network, and off-the-shelve power electronics-based controllers are available for active and reactive power control, network reconfiguration and the enhancement of power quality and increased reliability of operation. Network reconfigurations are carried out if needed using SCADA systems comprising a master computer, remote terminal units, measuring devices and Information & Communication Technology (ICT) packs, and automatic schemes to locate and clear electrical faults in primary circuits [2]. All these components are collectively termed Distribution Management System (DMS).

[1]Programa de Graduados e Investigación en Ingeniería Eléctrica, Instituto Tecnológico de Morelia, México
[2]Electrical Energy Consultant, Power Systems Laboratory, Bristol, United Kingdom

In particular, the widespread inception of low-capacity power generators of various kinds – wind power, solar photovoltaic, biomass, and diesel, to name a few – has turned the otherwise passive power distribution network into an active one. The local generators meet a significant portion of the local power demand, resulting in a reduced power flow import from the power transmission grid to satisfy the total local power demand [3]. The large penetration of distributed generation, however, may come accompanied by some undesirable side-effects, such as reversible power flows, high steady-state voltages, low power factors, voltage and current harmonics, transient over-currents, etc. Hence, copious studies are carried out to assess the maximum distributed power generation that can be accommodated (i.e., hosted) in each medium voltage feeder before any kind of undesirable effect surfaces and exceeds the limits established by applicable grid codes. The analyses are complemented by considering annual load growth forecasts, together with estimates of the growth of distributed power generation, with reference to their types, sizes and most probable locations.

Indeed, during periods of low power demand and high winds or solar radiation, the direction of power flows might be from the low voltage end towards the high voltage end of the power network – giving rise to the so-called reverse power flow phenomenon [4]. This has changed the long-standing protection practice of power distribution networks based on simplicity, becoming more elaborate and expensive owing to the need to use bidirectional relays instead of inexpensive overcurrent relays. Furthermore, the inception of distributed power generators in the low-voltage distribution network, such as solar PV generators, paved the way for the widespread adoption of voltage source inverters and DC/DC converters in the power distribution system. The PV panels use the solar radiation resource to produce voltage and current in DC form and to enable this power resource to be injected into the AC power grid, it needs to be converted from its natural DC-generated form into an equivalent AC form – the DC/AC inverter enables such a coupling between the DC power generator and the three-phase AC power grid to take place [5].

Over the past three decades, the nature of the industrial, commercial, and domestic loads has evolved beyond recognition with the availability of affordable power electronics. Today's industrial, commercial, and domestic loads are highly efficient in energy consumption. However, the drawback of the ubiquitous use of power electronics circuits is the pervasive generation of harmonic distortion in the power distribution network, making the power grid more difficult to plan and operate [6]. To compound matters in this score, the widespread use of electric cars is just around the corner. This calls for a staggering pace of development to expand and to adapt the electrical power grid to incorporate the many infeed points that the EV recharging stations require globally. In addition to the new power generation capacity, revamped substations, and new feeders, to face the new challenge would require adding further flexibility and resilience to the electrical power distribution grid by incorporating innovative power electronics-based solutions. This is beginning to happen as the power engineering community realises about the great many challenges and opportunities that lie ahead. For one, the so-called switching centres are a novel concept based on leading-edge power electronics equipment and methods that would merge seamlessly with existing conventional distribution feeders [7,8]. The structure and role of switching centres are at an incipient stage. It would enable the flexible incorporation of some types of embedded generation, and new loads, such as EV charging hubs and battery stations. It should be remarked that the power electronics structure of a switching centre corresponds to that of a back-to-back, multiterminal, VSC-HVDC link and that when embedded within the AC power distribution feeders, together they give rise to a hybrid AC/DC power distribution grid. These issues are addressed in this chapter.

The AC/DC power distribution grid may be developed beyond the incorporation of switching centres, by simply using the existing AC power circuit infrastructure to incorporate DC power circuits to run parallel to the AC feeders. As discussed in the chapter, several AC/DC schemes can be implemented using existing AC cables and newly installed VSC converters. This opens the doors for directly connecting DC power sources and DC loads into the power distribution grid. Furthermore, a new business opportunity may be brewing within the active power distribution system paradigm with the widespread embedding of VSCs and DC/DC converters [8]. These power electronics-based equipment, provided they are fitted with suitable control functions, may be made to exert control actions beyond their most immediate roles, e.g., exerting nodal voltage control for the case of a D-STATCOM, and extended to provide power factor correction, active harmonic filtering functions, voltage balancing, etc. Concerning the battery energy storage systems (BESS), whose primary role is to provide active power smoothing, may be made to provide inertial response and to take a grid-forming (GFM) role, by embedding suitable control functions in the inverter. Such functionality may be made to act on the power distribution network following a third-party business practice and obeying technical/economic signals – following similar procedures to those done in high-voltage power transmission systems using generic *ancillary services* [9].

Further developments may turn the operation of parts of these active power distribution systems, into autonomous microgrids operating in a cooperative and supporting role for the rest of the power distribution system from which they have sprouted [10]. Moreover, the large-scale adoption of Distributed Energy Resources (DERs), ICT, sensors, internet-driven functions, and AI-distributed controls, will see these microgrids developing into low-voltage Smart Grids [11].

6.2 The classic power distribution system [1]

The power distribution system is that part of the power delivery infrastructure that takes electrical energy from the high-voltage power grid and delivers it to customers at sufficiently low voltage levels to ensure its safe usage in factories, offices, and homes. In power distribution systems, this process may involve the use of two or more transformers to bring the voltage down from those employed at the transmission level, e.g., 500 kV, to those employed at the supply points, e.g., 11 kV, 240 V or 13.8 kV, 220V, depending on specific countries. This is illustrated schematically in Figure 6.1 indicating the various subsystems that make up a utility-level electrical power system. This encompasses power generation employing mostly large

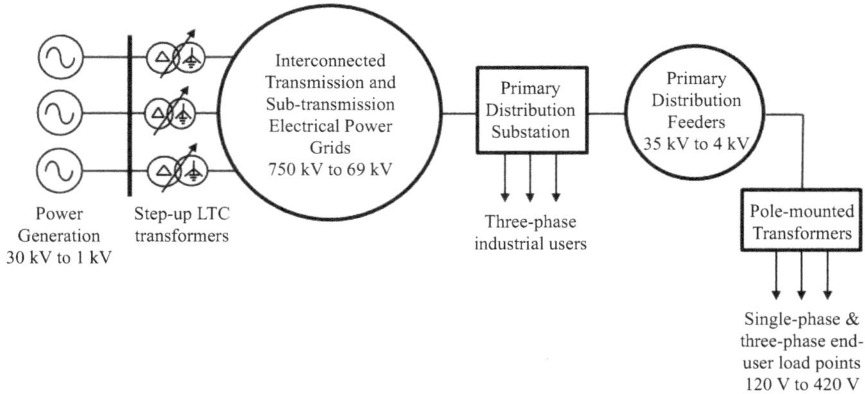

Figure 6.1 Generic structure of a utility electrical power system

synchronous generators, stepping the voltage up to transmission levels using LTC power transformers. Primary distribution substations sit between sub-transmission power transmission points and primary distribution feeders, reducing the voltage level to safe values for feeding large industrial customers and feeding in numerous primary distribution feeders, which may be either low-voltage, overhead transmission lines or underground cables.

The structure of classic power distribution systems is characterised by being radial or forming a ring, and the service to the end user being either three-phase or single-phase, depending on the service contracted by the customer, to satisfy its load requirements. Classic power distribution systems do not have power generators embedded within. Hence, the power flows are unidirectional, from the primary distribution substation towards the end-user loads. The protection system employs simple overcurrent relays, circuit breakers, and fuses. Also, power electronics-based equipment is absent from the power distribution system: neither fast reactive power control nor fast voltage regulation. The equipment used to achieve these operational functions is switchable banks of capacitors and LTC transformers, respectively, with their slow-acting servo-mechanical actuators. There is no energy storage in the classic power distribution system.

At a primary distribution substation, a power transformer takes the incoming sub-transmission voltage (115 to 69 kV) and steps it down to several distribution primary circuits, which branch out of the substation. Primary distribution lines are medium voltage circuits ranging from 4 kV to 35 kV. Close to end-users, a distribution transformer takes the primary distribution voltage and steps it down to a low-voltage secondary circuit. From the distribution transformer (e.g., a pole-mounted transformer), the secondary distribution circuits connect to the end user using a three-phase (e.g., 460, 440, 420, 230, 220, 210 V, depending on specific country) or a single-phase supply (240 V or 120 V).

Power utility companies have achieved a great deal of standardisation in their layouts, transformer capacities, relaying protection schemes, SCADA, and monitoring systems. Nonetheless, there is a limit to what can be done in that respect owing to the wide range of customers they serve, which may be scattered over large geographical areas, following rather irregular patterns. For instance, the layout and rating of an urban distribution substation differ from that required to supply rural customers or the electricity supply of a densely populated area such as the Square Mile in the City of London, Shanghai, or Manhattan. For example, Figure 6.2 shows typical layouts of rural and urban power distribution substations.

Rural substations normally have small ratings and only a few feeders spanning out of the substation. However, some rural feeders may have long lengths. In some countries, in remote areas, feeders of 20 km or more are not uncommon; a condition that causes poor voltage regulation and the use of voltage booster transformers becomes necessary. Alternatively, the loads served by such long feeders may benefit from installing small capacity, renewable power generators at the load end. Urban substations have much larger ratings, and several inputs feed into the substation to meet the energy requirements of the larger number of end-users served by the substation. During peak times, this may result in low voltages in the substation and capacitive reactive power compensation may be needed to regulate voltage.

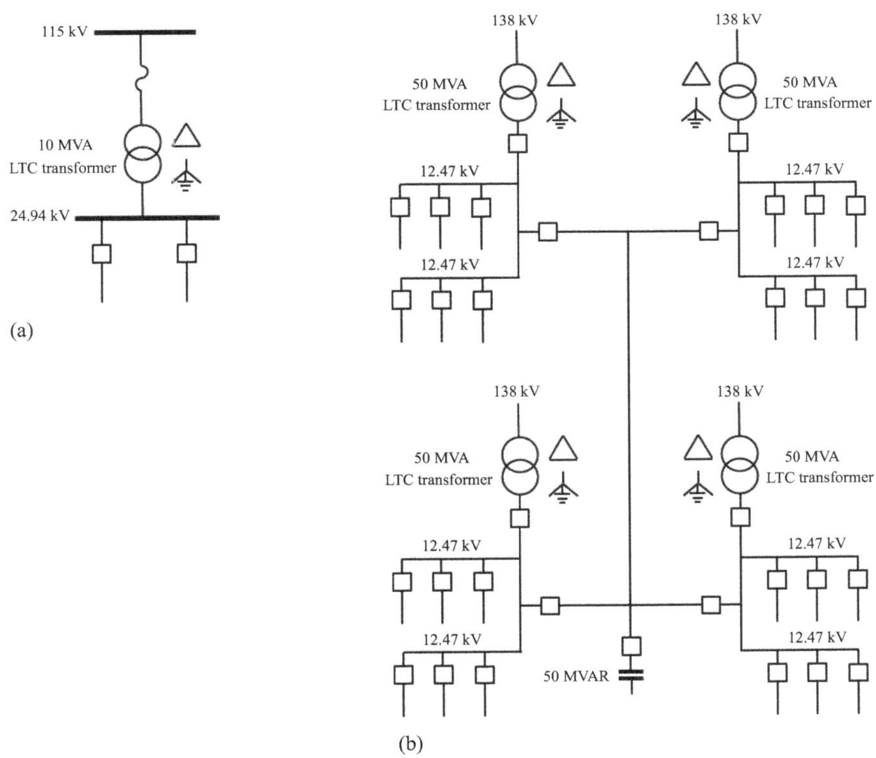

Figure 6.2 Typical layouts of power distribution substations: (a) rural and (b) urban

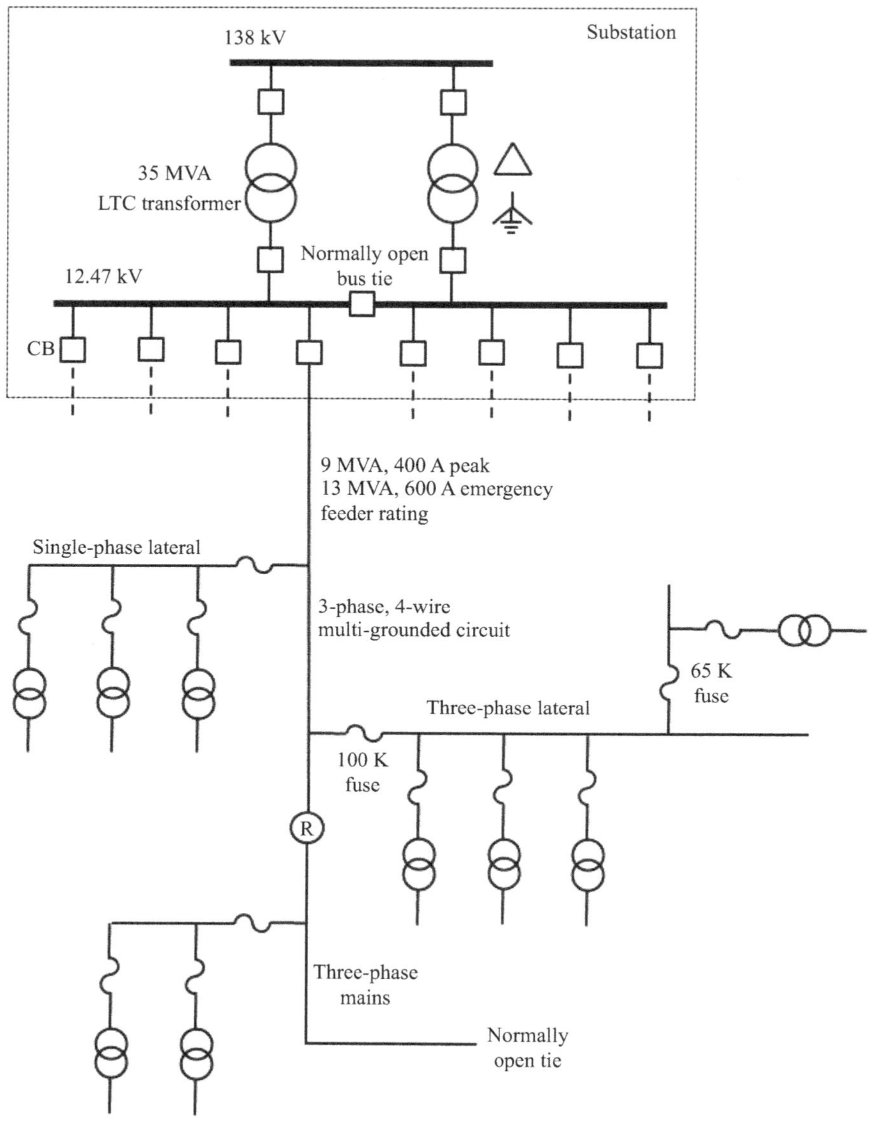

Figure 6.3 Primary power distribution substation and main feeder

The substation layout in Figure 6.3 gives more detail for one of the feeders. The substation is fed by two 138 kV lines feeding two 138/12.47 kV step-down transformers. Four feeders of 9 MVA peak capacity span out of each bus. Only one of the main feeders is drawn in certain detail. It is a three-phase feeder rated 400 A and 600 A during emergencies. According to the requirements, single-phase, two-phase, and/or three-phase lateral branches may exist along the main feeder. Note the use of fuses between the main feeder and the lateral feeders to provide isolation in cases of faults developing in the lateral feeders.

The main feeders are normally three-wire plus a multi-grounded neutral. Many single-phase loads are served by pole-mounted transformers, connected between one phase and the neutral. The single-phase loads along the feeder are alternatively connected to each one of the phases (and the neutral), aiming to balance the overall "three-phase load" connected to the feeder as much as possible. The neutral acts as a return feeder and as an equipment safety ground. Most distribution systems have radial structures, and even those with a ring structure are operated radially for simplicity of operation. A radial system uses simple protection methods, has lower fault currents, simple voltage regulation mechanisms, lower installation and maintenance costs, etc. Typically, power distribution systems, which represent up to 40% of the total investment in an electrical power system, is the interface of an electricity company to industrial and domestic consumers. Great many steps have been taken recently to increase the efficiency and quality of service.

Primary and secondary distribution systems are designed, built, and upgraded to fulfil long-term, forecasted load requirements. Distribution systems reinforcements and expansions tend to be gradual since these works are programmed and executed in response to long-term, load-growth forecasting. Much larger infrastructure upgrades would take place if the power distribution structure had to host a large power generation station or many power generation sources of small and medium size ratings. This would push a classic power distribution system into being an active power distribution system, involving major structural and operational changes. For one, active power may flow from the low-voltage power grid towards the sub-transmission power grid, i.e., reverse power flows. Ensuring the integrity of the power distribution system may require a change of the protection philosophy; the long-enduring, simple overcurrent protection method may not be adequate any longer.

6.3 Active electrical power distribution systems

Classic power distribution structures have served the electricity supply industry and its customers well for almost one century. However, starting in the 1980s, the power system distribution load shifted gradually but relentlessly towards more versatile and energy-efficient industrial equipment and domestic appliances using power electronics circuits and control. Within a relatively short period, solid-state electronics became ubiquitous in the electrical power distribution systems' load composition, introducing many great improvements and bringing with it many teething problems, such as harmonic distortions, reactive power deficits, low power factors, etc. This has introduced new challenges to the operation of the power grid. Paradoxically, to combat these problems effectively, utility-level, power electronics-based equipment has been developed to restore flexibility and resilience to the electrical power distribution grid [12].

In the 1990s, low-capacity power generators (i.e., microturbines) started becoming economically viable, and wind power turbines, fuel cells, and solar PV power panels for grid-connected applications became accessible. These electrical power renewable technologies had humble beginnings in terms of sizes, operational reliability, and costs. Still, over the last three decades, all three technologies have made superlative progress on all three counts. To a greater or lesser extent, low-capacity power generators have been connected, following a distributed fashion, into power distribution systems since the early 1990s [3]. This is exemplified in schematic form in Figure 6.4.

Recently, electric cars have become competitive with petrol cars, and the power grids' capacities would need to grow swiftly to provide the many power grid infeed points that the EV recharging stations would require globally. Due to the ongoing mushrooming of EVs worldwide, time and economic resources are in short supply to develop the required infrastructure [13]. One way to proceed apace is to merge the existing AC primary distribution feeders with strategic DC infrastructure in the form of so-called Switching Centres, which house extended DC buses to recharge EVs, embedded PV power generators, fuel cells, and BESS [7].

In a nutshell, electrical power distribution systems are undergoing a deep transformation, spearheaded by the so-called Distributed Energy Resources (DER) and ICTs [14]. Most notably, distributed power generation, EVs, and energy storage are changing for good the way electrical power distribution systems are planned, operated, and run. It should be noted that internally, DERs are made up mainly of solid-state circuits and work with DC powers. Currently, DERs are connected to the AC power grid using Voltage Source Inverters (VSIs) – one VSI per DER. Incidentally, this is where a large portion of the costs of installing DERs lie. The alternative would be to connect all DERs to DC feeders or, even better, a DC power grid. This would require far fewer VSIs to interconnect many DERs to the AC power grid [8]. This

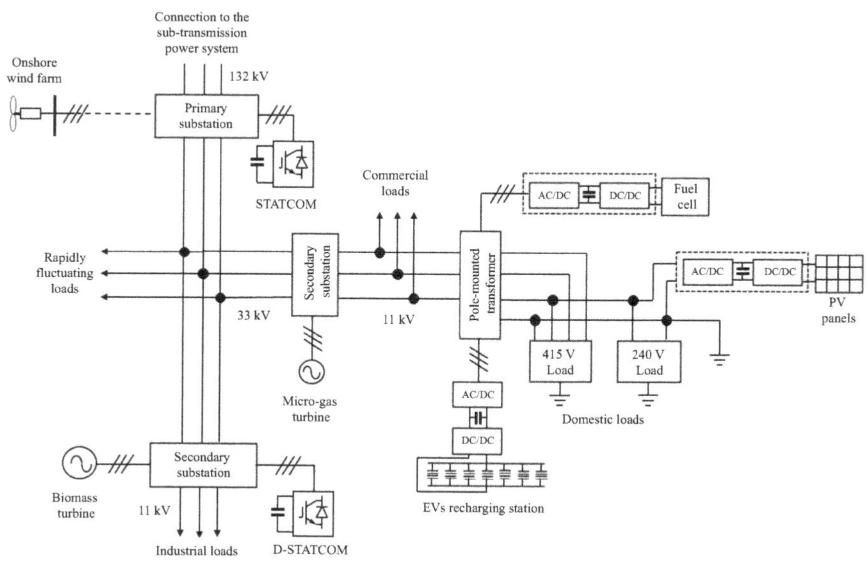

Figure 6.4 Schematic representation of an active power distribution system

would give rise to a hybrid AC/DC power distribution system. This concept has not yet been implemented in practice anywhere, but it deserves full study and consideration as we move towards the future power distribution system.

The drivers for such a swift transformation are very pressing indeed. Forecasts from the International Energy Agency indicate that the current fleet of 2 million EVs will rise to 56 million by 2030 [13]. Concerning grid-connected PV power generation, the International Renewable Energy Agency (IRENA) predicts a global increase from the current 233 TWh up to 1.104 TWh by 2030 [15]. The third pillar of DERs, BESS, is expected to become one more in the array of household appliances to smooth out the electrical energy usage within the home environment, and, at the utility level, BESS will be employed strategically to combat network congestion. Using BESS instead of a percentage of new transmission line construction gives rise to the Virtual Transmission idea. IRENA predicts an increase of BESS installed power from 1 GW to 250 GW by 2030 [16].

The simplest upgrade of an AC power distribution system to take advantage of current power electronics technology would be installing a multiterminal, back-to-back, VSC-HVDC link at the far end of existing AC power distribution feeders [5]. Figure 6.5 illustrates this idea schematically.

An alternative, fuller approach to developing hybrid AC/DC power distribution systems is shown in Figure 6.6 [8]. It involves using the existing AC feeder infrastructure to lay out brand-new DC conductors. For instance, a point-to-point VSC-HVDC link may be built to run in parallel with an existing three-phase power distribution feeder, as illustrated in Figure 6.6(a). The parallel DC link may be an underground cable under the overhead conductors if the existing AC feeder is an overhead transmission line. The DC link may be either a monopole or a bipole scheme. On the other hand, if the existing AC feeder is an underground cable installation, the parallel DC link may use the existing trenches to lay out the new DC cables. This hybrid AC/DC configuration supplies AC loads as usual, while DC loads and DERs are plugged into the DC network using DC/DC converters. This point is

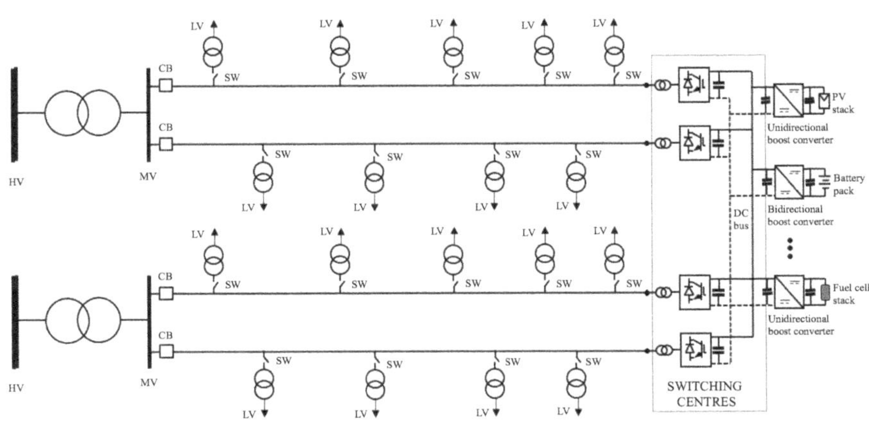

Figure 6.5 Power distribution feeders with a switching centre at the far end

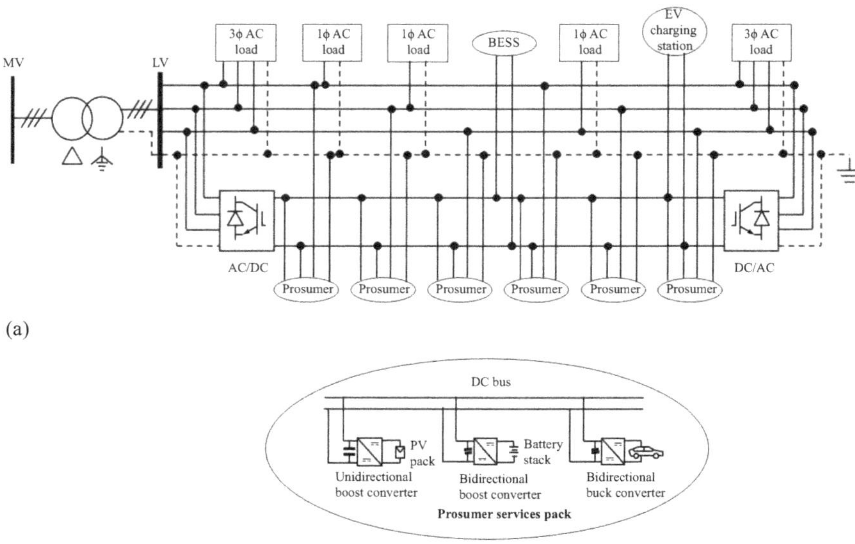

(a)

(b)

Figure 6.6 (a) Hybrid 3ϕ AC/DC power distribution feeder with Prosumers and DERs and (b) example of Prosumer service pack, adapted from [8]

illustrated in Figure 6.6(b) in connection with the recently emerging figure of *Prosumer*. This involves a customer having the dual role of producer and consumer, following online economic signals. The Prosumer may produce electrical power using solar PV static generators, fuel cells, wind power generators, microturbines, and even diesel power generators. The energy stored in BESS and EVs may also inject unused energy into the utility power distribution system if economically advantageous to the Prosumer. On the other hand, the Prosumer buys electrical power from the utility company when its power generation is insufficient to meet its own demand.

The many power electronics converters, AC/DC and DC/DC enable the bidirectional flow of active power and the tight voltage and power factor regulation. This enables optimum power flow exchanges between the AC and DC feeders, with minimum power losses and good power quality indices.

A further option is to carry out a partial reconfiguration of the existing three-phase AC feeder, using the four conductors to make up a single-phase AC feeder in parallel with a DC link, as illustrated in Figure 6.7 [8]. Note that the VSC rectifier is three-phase, whereas the invertor is single-phase in this scheme. This hybrid AC/DC feeder system enables the supply of single-phase AC loads and DC loads and the embedding of DERs. Large three-phase AC loads may be fed from the DC link using a dedicated VSI.

Both hybrid AC/DC schemes presented in Figures 6.6 and 6.7 yield higher electrical energy to flow within the distribution feeder, with a high percentage generated by the local power generators. These would supply the local loads and recharge BEES and EVs. In the scheme presented in Figure 6.6, there is a marked increase in

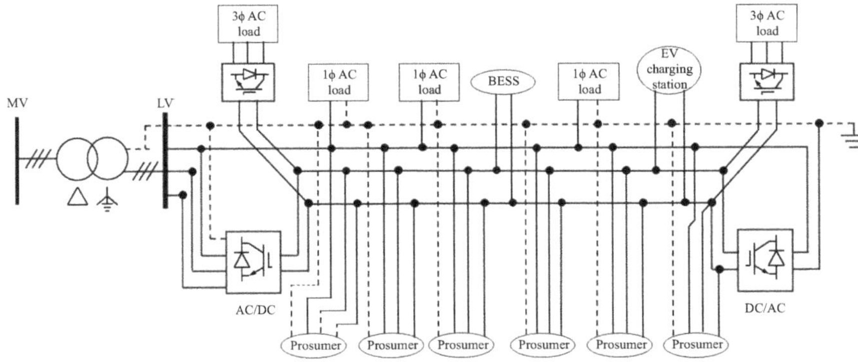

Figure 6.7 Hybrid 1ϕ AC/DC power distribution feeder with Prosumers and DERs

the loadability of the feeder because of the deployment of the DC link. To make this upgrade effective, the MVA capacity of the MV/LV transformer in the secondary substation should be upgraded accordingly. In the hybrid AC/DC scheme shown in Figure 6.7, the increase in loadability would be more modest than in the previous case. The main reason is that no additional conductors are being added to the power distribution circuit. It is true that a DC circuit carries more power than a comparable three-phase AC circuit, but the parallel AC feeder is only a single-phase one. In the latter scheme, the single-phase AC loads and DERs are fed from the AC and DC feeders, respectively. Both hybrid AC/DC feeder schemes have advantages and disadvantages that require a detailed technical/economic assessment at the planning stage, with the size and characteristics of the load playing a decisive major role.

6.3.1 Numerical example [17]

Figure 6.8 shows an active power distribution system comprising 16 three-phase AC nodes (i.e., n1 ... n16) and seven DC nodes (i.e., n17 ... n23). This primary distribution system forms an underground ring configuration with six main nodes (i.e., n2, n3, n5, n8, n11, n14) and operates at 132 kV.

The steady-state operating conditions of the active power distribution system in Figure 6.8 will be assessed using a power flow algorithm comprising the basic AC and DC equipment models described in Chapter 5 but expressed in its three-phase form. In this test case, the power flow is a full three-phase solution. With reference to the power flow theory and the test network (Figure 6.8), node 1 is taken to be the slack bus with a per unit value voltage magnitude of 1.05 and the three-phase voltage phase angles set at 0°, 240°, and 120°, respectively. The voltage magnitudes of all other AC buses are set to 1.0 p.u. and 2.0 p.u. for all DC buses. Note that CB (circuit breaker) between nodes 11 and 14 is taken to be as normally closed. All transformers have grounded star connections on both the primary and secondary windings. The per-phase transformer parameters are given in Table 6.1. Note that all transformers are LTC transformers with fixed taps and that the parameters of the three transformer groups n1-n2, n5-n6, and n8-n9 are given as equivalent parameters.

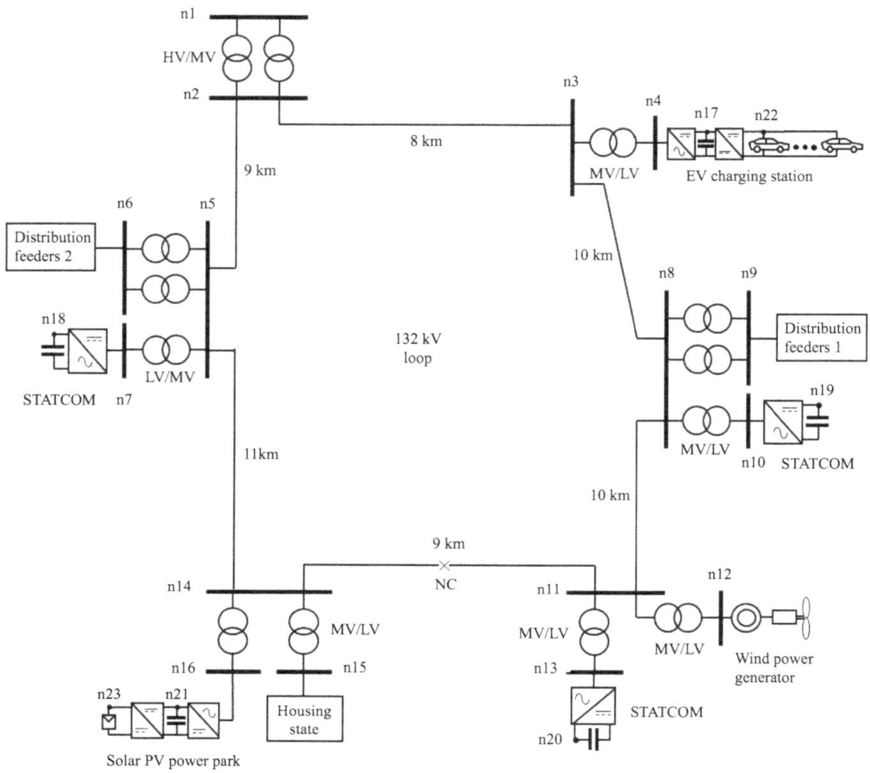

Figure 6.8 An active power, primary distribution system forming a ring configuration

Table 6.1 Transformer parameters

Transformer	Connecting nodes		Resistance (p.u.)	Reactance (p.u.)	Tap position
1	n1	n2	0.012	0.120	0.97
2	n3	n4	0.030	0.300	0.99
3	n5	n6	0.015	0.150	0.95
4	n5	n7	0.030	0.300	1.05
5	n8	n9	0.015	0.150	0.95
6	n8	n10	0.030	0.300	1.05
7	n11	n12	0.030	0.300	0.99
8	n11	n13	0.030	0.300	1.03
9	n14	n15	0.030	0.300	0.93
10	n14	n16	0.030	0.300	1.02

The transmission lines are non-bonded, single-conductor cables placed in buried ducts in a flat configuration. The cable parameters are calculated with a computer program [18] using the cable parameters given in Table 6.2.

The cable's three-phase series impedance and shunt admittances in Ohms/km and Siemens/km are given in the matrices below:

Z_{abc} =	0.1700+j0.1653	0.0945−j0.0127	0.0732−j0.0286	Ω/km
	0.0945−j0.0127	0.1583+j0.1505	0.0945−j0.0127	
	0.0732−j0.0286	0.0945−j0.0127	0.1700+j0.1653	

Y_{abc} =	j71.99	−j3.22	−j1.51	μS/km
	−j3.22	j72.10	−j3.22	
	−j1.51	−j3.22	j71.99	

The following operating conditions are assumed for conducting a full three-phase power flow solution. The wind farm and the solar PV park inject 250 MW and 50 MW, respectively. The EV charging station draws 50 MW. The two distribution feeder systems and the housing estate are non-symmetrical loads. This information is given in Table 6.3. The reactive powers produced/consumed by the wind farm, the VSCs of the solar PV park, and the EV charging station are not known *a priori* but are calculated by the power flow solution. The same applies to the other three STATCOMs in the test case.

The initial conditions to kickstart the power flow iterative solution, corresponding to nodal voltages, are as follows: nodal voltage magnitudes for nodes n2 to n16 and phases *a*, *b*, *c*, are taken to be 1 p.u. For n1, designated to be the slack node, the voltage magnitudes are 1.05 p.u. for the three phases. The voltage phase angles for all 16 nodes are: 0°, −120°, and 120° for phases *a*, *b*, and *c*, respectively. For the five VSCs, their DC voltages are taken to be 2 p.u., and the amplitude modulation ratios and phase-shifting angles are set at 1 and 0, respectively. Note that the VSCs are set to control neither voltage nor power flow. The two DC/DC converters are set to control voltages at the EV battery-charging side and the PV side, at 2 p.u., respectively. The nodal voltages furnished by the three-phase power flow solution are given in Table 6.4.

Table 6.2 Cable parameters

Element	Thickness (mm)	Diameter (mm)
Conductor, Copper, 5 segments	–	48.0
Conductor shield	3.2	54.4
Insulation, XLPE	15.8	86.0
Insulation screen	1.9	89.74
Concentric wires, Copper	2.0	93.8
Jacket, Polyethylene	4.6	103.0

Table 6.3 Sources and sinks of active and reactive powers

Power sources and loads	Active power (MW)				Reactive power (MVAR)			
	a	*b*	*c*	3ϕ	*a*	*b*	*c*	3ϕ
Wind power farm	83.333	83.333	83.333	250	?	?	?	?
PV power park	16.667	16.667	16.667	50	?	?	?	?
EV charging station	−16.667	−16.667	−16.667	−50	?	?	?	?
Housing estate	−31.667	−35.000	−33.333	−100	−15.833	−17.500	−16.667	−50
Distribution feeders 1	−84.407	−76.733	−69.060	−230	−25.263	−22.967	−20.670	−68.9
Distribution feeders 2	−69.060	−76.733	−84.407	−230	−20.670	−22.967	−25.263	−68.9

Table 6.4 Three-phase nodal voltages and DC voltages

	Node	Voltage magnitude (p.u.)			Voltage phase angle (deg)		
		a	*b*	*c*	*a*	*b*	*c*
AC nodes	n1	1.05	1.05	1.05	0	−120	120
	n2	1.1000	1.0971	1.0972	−7.2656	−127.4895	112.5715
	n3	1.0980	1.0097	1.1006	−7.9506	−127.4895	112.5715
	n4	1.1664	1.1661	1.1679	−10.7687	−130.9733	109.0228
	n5	1.0936	1.0926	1.0894	−8.1344	−128.2761	111.4854
	n6	1.1053	1.0982	1.0884	−13.1550	−133.8973	105.2256
	n7	1.0933	1.0923	1.0891	−8.5975	−128.7393	111.0222
	n8	1.0723	1.0746	1.0800	−8.3380	−128.5909	111.3568
	n9	1.0915	1.0929	1.0986	−14.5717	−134.2091	106.3837
	n10	1.0908	1.0985	1.1108	−8.8012	−129.0540	110.8937
	n11	1.0912	1.0926	1.0983	−7.7035	−127.9063	112.1103
	n12	1.1022	1.0995	1.1035	4.1815	−115.9593	123.9680
	n13	1.1124	1.1095	1.1137	−8.1858	−128.3885	111.6280
	n14	1.1255	1.1227	1.1268	−8.0551	−128.2163	111.6380
	n15	1.0986	1.0963	1.0973	−12.5842	−133.2800	106.8423
	n16	1.1187	1.1085	1.1135	−6.4030	−126.5639	113.2903
		Voltage magnitude (p.u.)					
DC nodes	n17	1.9458					
	n18	2.0000					
	n19	1.8758					
	n20	1.8800					
	n21	1.9332					
	n22	2.0					
	n23	2.0					

Figures 6.9 and 6.10 display the three-phase power flows for active and reactive powers, respectively. They are given in per unit values on a 100 MVA basis.

The active power load of this system stands at 610 MW, and the power generated by the wind turbine and the solar PV park is 300 MW. To provide for the deficit of generated active power and power losses, 350 MW are imported from the high-voltage transmission network. The active power grid exports nearly 100 MVAR to the high-voltage transmission network. This reactive power excess is produced mainly by the five VSCs and, to a lesser extent, by the underground cables. Notice that the reactive power requirements of the wind farm, the two distribution feeder systems, and the housing estate are all met by the reactive power generated by the three STATCOMs, connected at nodes n7, n10, and n13, and the two VSCs, connected at nodes n14 and n16. Notice that the nodal voltages in the power grid are on the high side. The reason is that the network is very reactive and that the VSCs characteristics of the load play a decisive major role.

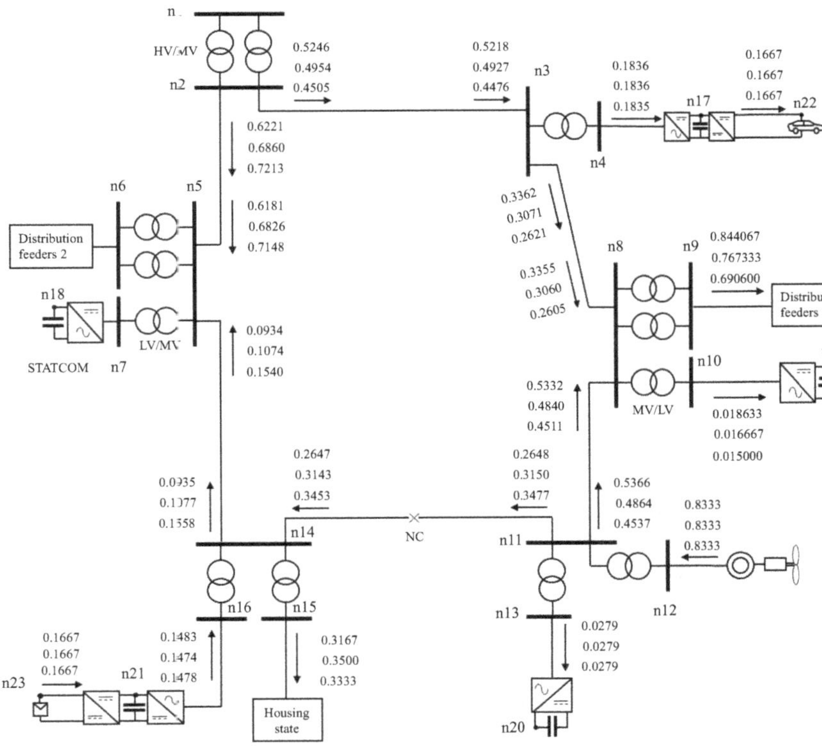

Figure 6.9 Three-phase active power flows

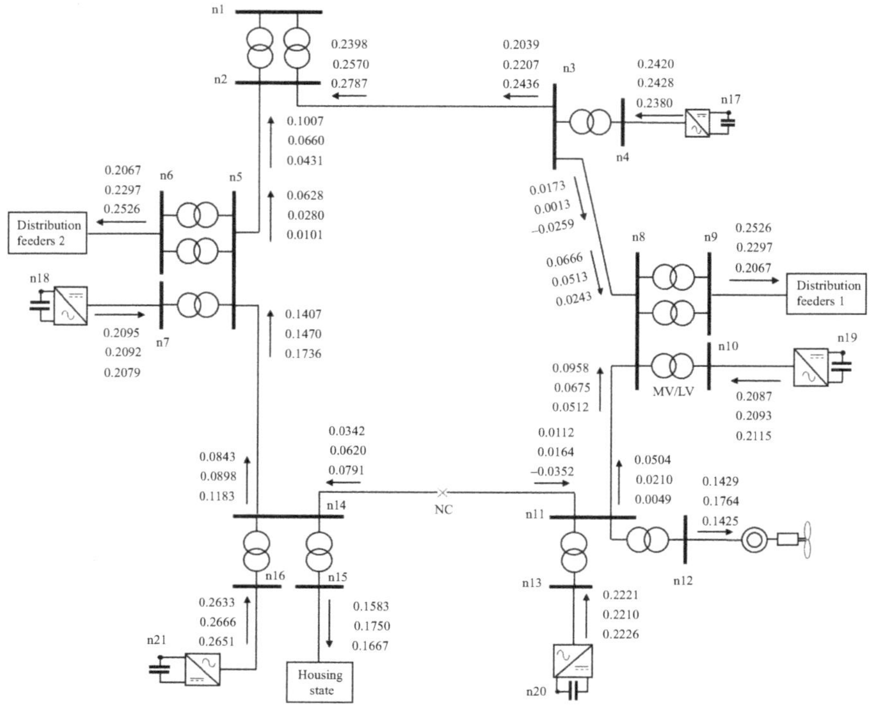

Figure 6.10 Three-phase reactive power flows

6.3.2 Third-party services in active power networks

The many VSCs in a hybrid AC/DC power distribution system may be used to exert fast and precise AC and DC voltage control and active and reactive power flow control. Furthermore, the availability of fast-acting energy conditioning devices, such as BEES and EVs – provided they are connected to the power grid, yields an effective means for counteracting frequency oscillations arising from active power imbalances. Moreover, several other control functions become possible, such as power factor correction, harmonic distortion amelioration, overall power quality indices improvements, and enhanced continuity of energy supply.

If the new power electronics-based equipment belongs to the electrical power utility, all the new control functionality, coordinated from a control centre, would be available to make up for a more flexible and resilient active power grid. However, suppose a utility company receives control services from the power electronics-based equipment, and this is owned by another company. In that case, an economic remuneration or barter operation may occur between the two parties, following an established inter-utility services and commercial framework.

Table 6.5 Third-party services in active power distribution grids

Service	Steady-state regime		Dynamic regime	
	AC side	**DC side**	**AC side**	**DC side**
Voltage magnitude control	STATCOM	DC/DC converter	STATCOM	DC/DC converter
Voltage dip compensation	–	–	DVR	–
Power factor correction	STATCOM	–	STATCOM	–
Harmonic distortion mitigation	STATCOM	DC/DC converter	STATCOM	DC/DC converter
Voltage unbalance reduction	STATCOM	–	STATCOM	–
Power smoothing	BESS	–	BESS	–
Inertial response	BESS	–	BESS	–
Power frequency response	BESS	–	BESS	–
Continuity of supply	SSTS	–	SSTS	–

To a larger or lesser extent, this practice would follow what are common commercial/operational procedures in today's high-voltage power transmission systems, the so-called *ancillary services* [9]. Even though such a practice does not seem to have been implemented in distribution power systems in the general sense [8], there have been successful instances of third-party special services provided by electrical power utility companies to large customers. A case in point is the Dynamic Voltage Restorer (DVR) installed in 1997 at the Caledonian papermill's infeed substation, owned by Scottish Power Ltd, and designed to provide premium power quality supply to the papermill [19]. This service comes under the umbrella title of *Custom Power* [20]. Further details are given for this installation in Section 6.3.

In the fullness of time, as more experience accumulates with the widespread usage of these devices in a network-wide context, additional third-party services may emerge in the power distribution system. Some of these potential services are listed in Table 6.5. This table includes as entries power electronics-based equipment currently available, such as STATCOM, BESS, DVR, Solid-State Transfer Switch (SSTS), and DC/DC converter, but new equipment may yet be developed to provide additional services that may be required in the future. It should be noted that to fulfil some of the tasks assigned to STATCOM in Table 6.5, specific control functions should be programmed within its control system. This table has been inspired by the work listed in [8].

6.4 D-FACTS for power quality amelioration

6.4.1 Background

Robots, adjustable-speed drives, microprocessor-based controllers, and other modern electronic control devices have replaced mechanical controls in all industries throughout the last century. However, adopting such technological advancements has made the modern industrial process more vulnerable. Notwithstanding the efficiency and precision of modern electronic controllers, we must contend with the stark reality of voltage sags, interruptions, surges, and other power quality issues in the waveforms of the electricity supply, which are highly damaging to electronic control devices [20].

Large electricity customers seek to purchase bulk power from reliable electricity suppliers able to deliver green electricity with high Power Quality (PQ) content at competitive prices. To enable the high PQ content element, the electricity companies have embarked on ambitious maintenance and infrastructure reinforcement programs [21]. In particular, the deployment measures that prevent dynamic disturbances affecting power quality and reliability of supply. Leading-edge technologies using power electronics solutions are available to combat a wide range of power quality issues with the potential to stop assembly lines and production processes in the industry. The Custom Power or D-FACTS array of equipment has proved very effective in improving PQ indices in industries with such equipment [19].

6.4.2 Power quality

Two electrical power systems exist today to supply electrical energy to industrial, commercial, and domestic customers: AC power systems and DC power systems. The former may be three-phase and single-phase, operating at 50 Hz or 60 Hz. In a three-phase AC power system, the voltages and currents are expected to be sinusoidal with constant amplitude and frequency while operating at the unity power factor. Furthermore, the phase angles of two consecutive phases are phase-shifted by $2\pi/3$. In a DC power system, the voltages and currents are expected to be direct lines with constant amplitude. Any significant departure from these idealised waveform conditions may jeopardise the normal operation of the electrical power system, and all efforts are made to ensure that those voltage and current waveforms stay as near as practicable to their idealised conditions all the time [22–24].

In the parlance of electrical power engineers, the quality of voltage and current waveforms has been termed Power Quality. It should be clear that Power Quality is an umbrella term that embraces all aspects associated with the amplitude, phase, and frequency of voltage and current waveforms in electrical power installations.

The installation is said to have good PQ when the voltage and current waveforms are near their ideal waveshapes. The opposite indicates bad PQ. Adverse power quality may result from one or various of the following aspects:

(i) Transient conditions develop in the power circuit due to switching operations, short-circuit faults, or the start of a large induction motor.
(ii) The installation of one or more non-linear loads with relatively large ratings, such as an electric arc furnace installation or a power electronic rectifier.
(iii) The installation comprises grid-connected intermittent generators or generators connected to the power grid using power electronics inverters.
(iv) Large numbers of non-linear loads of low ratings, such as energy-saving lamps, colour TV sets, DVD players, microwave ovens, fridges, personal computers, printers, and arc discharge lamps.

Many years ago, apart from low power factors and flicker concerns in industrial installations, PQ problems were confined almost entirely to regulated public utilities, particularly in distribution power systems. However, in the early 1990s, industrial processes started incorporating sophisticated electronics for precise and continuous control, and the sensitivity of such electronics to adverse PQ environments could jeopardise the industry process's reliability and productivity. By the mid-1990s, large customers worldwide started reporting problems with their electricity supply, adversely impacting their processes. It was reported that every year, electrical equipment problems cost companies billions of pounds in wasted material, downtime, damaged data, and delayed orders [21].

This encouraged many large customers to develop strategic alliances with their electricity suppliers and power electronics equipment manufacturers to form partnerships to develop advanced equipment capable of deterring some of the most adverse PQ phenomena affecting them. The industries included paper mills, large data centres, electronics chips manufacturers, steel power plants, airports, etc. [19]. The PQ phenomenon identified as the most damaging were voltage sags and swells, harmonics, and voltage imbalances. Some PQ phenomena exhibit a slow dynamic, others are stationary, and others are fast transients. At the time, the equipment identified to solve such pressing PQ problems was the D-STATCOM, the DVR, and the SSTS [20]. This equipment will be addressed below.

Table 6.6 shows some of the more common disturbances. They include voltage and current harmonics, voltage sags (dips) and swells, spikes, short-time interruptions, and voltage and current unbalance. Some others are not shown, such as electric noise, notches, and flickers.

Starting in the mid-1980s, the issue of PQ gained widespread interest in the industry and academic circles due to the increasing use of loads sensitive to service quality, e.g., computers, industrial drives, communications, and medical equipment. Several books on power quality and harmonics were published, and, to a greater or lesser extent, the curricula of most electrical engineering courses include the topic of PQ [21–24].

PQ is a complex technical issue because the new loads are not only sensitive to service quality but also responsible for adversely affecting the quality of the power supply. Customer and utility equipment suffers undue stress when operated in an unfriendly PQ operating environment. This has prompted professional bodies to enact recommended practices to limit the severity and number of disturbances to customer equipment. Notably among these is IEEE through its standards committee on Power Quality [25] and the family of standards IEC 61000, which does not use the term power quality but instead uses the overlapping term electromagnetic compatibility. The definition of power quality stated in IEEE standard 1100 is: "Power Quality is the concept of powering and grounding sensitive equipment in a manner that is suitable to the operation of that equipment" [26].

PQ disturbances at the utility level may result in the malfunctioning of remote controls, protective devices, energy metering, and the overheating of cables, transformers, and rotating machinery. PQ disturbances in the customer premises

Table 6.6 Common PQ phenomena: (a) voltage sag, (b) voltage swell, (c) short-term voltage interruption, (d) harmonic voltages, (e) voltage spikes, and (f) voltage imbalance

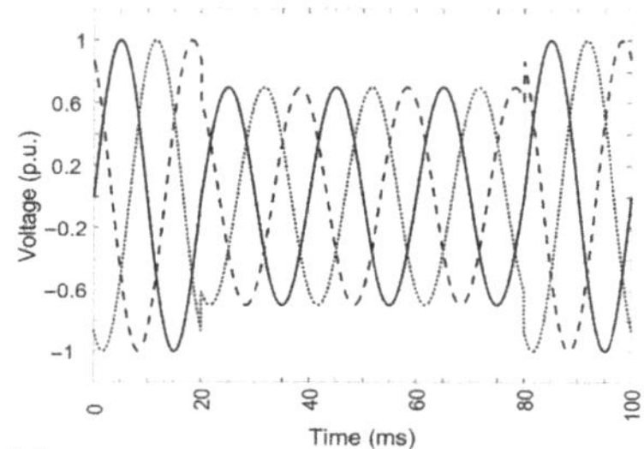

(a) Sag

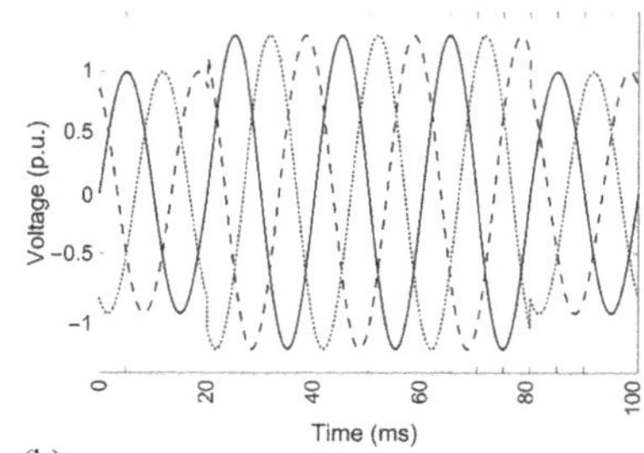

(b) Swell

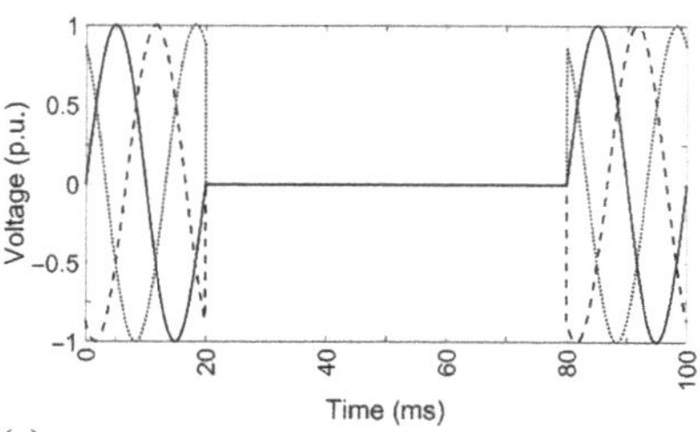

(c) Short-term interruption

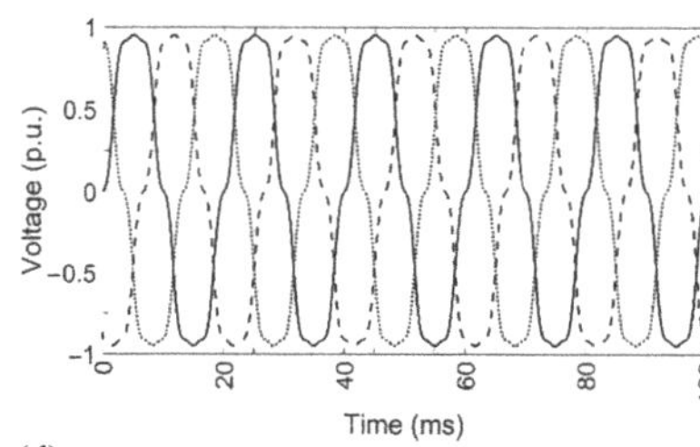

(d) Harmonics

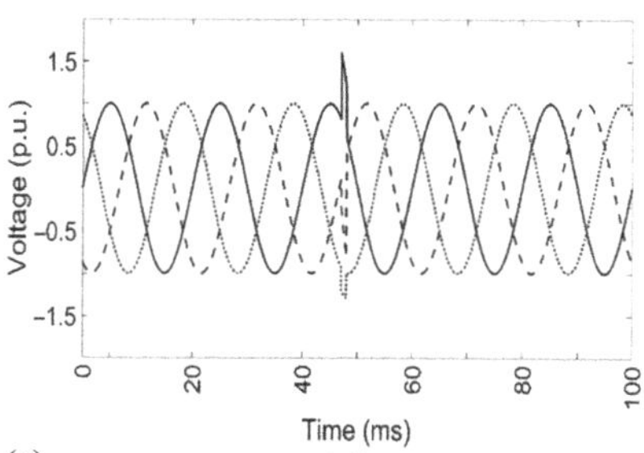

(e) Spikes

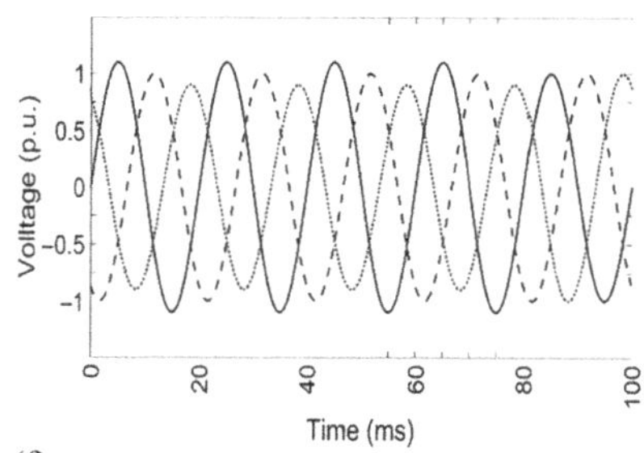

(f) Unbalance

may result in computer data loss due to voltage sags lasting only a few milliseconds. Even shorter voltage sags may cause the tripping of industrial drives [20].

6.4.3 Variations and events [21]

Many PQ disturbances have been reported in the open literature, some transient in nature and others related to periodic, steady-state operations. To assess power quality phenomena, the deviations may be divided into variations and events.

(i) Variations relate to small voltage and current amplitude deviations relative to their nominal values, e.g., frequency, and power factor. Variations are monitored continuously.

(ii) The latter refers to large deviations in voltage or current from their normal rated value. Examples of events are a sudden drop in voltage to zero due to the operation of a switching breaker (voltage event) and an overcurrent due to the switching of an unloaded transformer, i.e., inrush current. Events are monitored using a triggering mechanism.

6.4.3.1 Variations

1. **Voltage and current unbalance:** Unbalance in a three-phase system refers to the three RMS (Root Mean Square) magnitudes not having equal values or the phase angles of two consecutive phases not being phase-shifted by 120 degrees. The main culprit for voltage unbalance is load imbalances, hence, unbalanced current.
2. **Voltage fluctuation:** If the voltage variations are sufficiently large or in a certain critical frequency range, the performance of equipment can be affected because the power flow to that equipment will also vary; however, such cases are rare, except for fluorescent lamps. If the supply voltage to the lamp contains a frequency component in the range of 1–10 Hz, above a threshold value, then the human eye becomes irritable. This phenomenon is termed light flicker, and the fast variation in voltage is termed voltage fluctuation.
3. **Harmonic voltage distortion:** The voltage waveform is seldom a sinusoid of power frequency. Often, the voltage waveform will be non-sinusoidal but periodic. In the language of Fourier, a periodic waveform may be decomposed into a series of sine waveforms comprising a fundamental frequency waveform and several harmonically related waveforms.
4. **Harmonic current distortion:** Harmonic current distortion is the complementary phenomenon of harmonic voltage distortion. Both phenomena are strongly interrelated, with the severity of one determining the existence of the other. Harmonic currents require over-rating plant components such as transformers and cables. The resistance increases with frequency due to the skin effect, with a distorted current causing more loss than a sinusoidal current.
5. **Inter-harmonic current distortion:** Some equipment achieves its main operating state by producing current components with a frequency that is not an integer multiple of the fundamental, referred to as inter-harmonics. Examples of equipment that produce inter-harmonic currents are cyclo converters and certain kinds of heating controllers. Inter-harmonic currents are normally of small enough magnitude to cause any concern, but they can occasionally excite resonances made up by the interaction of transformer inductances and capacitor banks. Sub-harmonics is a special case of inter-harmonics, which can damage synchronous generators and turbines.
6. **Periodic voltage notching:** In three-phase rectifiers, the commutation from one diode or thyristor to another creates an effective short circuit with a duration of less than 1 ms, with an ensuing reduction in the supply voltage. Notching may be considered a periodic phenomenon, with the capacity to yield high-order harmonic voltage generation. IEEE defines a "line voltage notch" as any reduction in voltage lasting less than half a cycle.

6.4.3.2 Events

Power quality events exhibit random behaviour in terms of their severity, duration, and time of occurrence. An interruption of the voltage supply is the clearest example of a PQ event, but they do not occur too often, whereas transient over-voltages occur all the time. PQ events can potentially cause equipment tripping, halt entire industrial production lines in factories, and even lead to wide-area blackouts at the utility company level. Voltage events are of most concern for end-user equipment and are given priority over current events. The following PQ events are well-characterised:

Interruptions: A voltage interruption may be characterised as a condition where the supply voltage drops below 10% of the nominal voltage [IEEE Std. 1159]. IEC terms this phenomenon as a supply interruption. This may be set by short-circuit faults and the ensuing protection triggering.

Other causes may include ill-tripping of protection circuits, broken conductors and transformers, and planned and unplanned operator action. IEC uses the terms long and short-term interruptions. The latter applies if the interruption lasts for more than 3 min. In IEEE documents, the terms momentary, temporary, and sustained interruptions are used concerning the following time spans: ½ cycle to 3 s, 3 s to 1 min, and more than 1 min, respectively.

Undervoltages: Short-duration undervoltages are termed voltage sags by IEEE and voltage dips by IEC. They are mainly due to short-circuit faults and the starting of large induction motors. For IEC, the voltage reduction is 90%–1% of the nominal voltage, whereas for IEEE, it is 90%–10%.

Over-voltages: Over-voltage events receive different names according to their duration and magnitude. Over-voltages of very short duration and high magnitude are termed transient over-voltages or voltage spikes. Over-voltages with a duration between 1 cycle and 1 min are voltage swells. The name over-voltage is kept for longer over-voltage durations. Their sources are lightning strikes, switching operations, sudden load reductions, and asymmetrical short-circuit faults. An instance of ferro-resonance is an over-voltage of a very long duration. This may involve the magnetising reactance of a saturated transformer and a capacitive element like a capacitor bank or an underground cable in the same substation.

6.4.4 Amelioration of PQ disturbances at the utility company level

Even though events like capacitor switching and voltage sags due to motor starting play an important role in equipment tripping, large equipment tripping is due to short-circuit faults in the distribution system. Figure 6.11 gives a picture to indicate where the various mitigation methods would apply: (i) reducing the number of short-circuit faults; (ii) implementing system changes that reflect in less severe events; (iii) installing special equipment between the utility company and users with sensitive equipment; (iv) increasing the immunity of equipment.

Reducing the number of faults. Several actions may be taken that directly raise the system's immunity, but to a greater or lesser extent, these come at a cost.

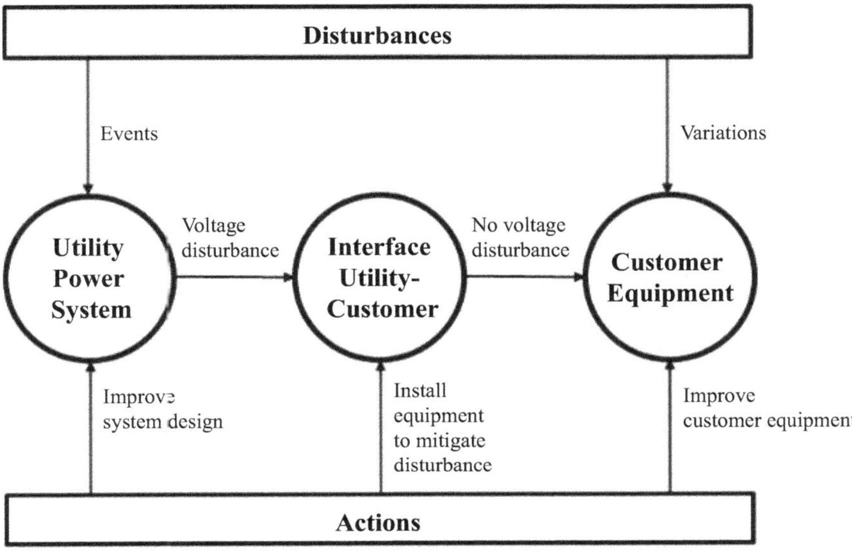

Figure 6.11 Measures to ameliorate adverse PQ disturbances

(i) Replace overhead distribution lines with underground cables. The fault rate in an underground cable is an order of magnitude less than in an overhead transmission line. (ii) Use covered wires (cables) for overhead transmission lines. (iii) Implement a strict policy of tree trimming. (iv) Install additional shielding wires. (v) Increase the insulation level. (vi) Increase maintenance and inspection frequencies.

Reducing the fault clearing time. This does not reduce the number of events but only their severity. Current-limiting fuses can clear a fault within one half-cycle. In some countries, fault location and isolation schemes have been implemented in medium-voltage distribution systems using smart reclosers. These devices are sensitive to changes in the direction of power flows and can detect voltages on both sides of the recloser. This enables fault location, feeder sectionalizing, load reconnection to another feeder, and restoration to the original state once the fault has been cleared.

Improving the power system infrastructure: Install a power generator near the sensitive load, split buses in substations in the supply path, and install current-limiting reactors to increase the electrical distance from areas with high fault incidence to areas with sensitive loads.

Improving equipment immunity. Improving equipment immunity is probably the most effective solution to screen equipment against trips due to voltage sags. By adding capacitance in the DC bus, adjustable-speed drives may be more tolerant to voltage sags due to short-circuit faults.

Installing mitigation equipment. Uninterruptible Power Supplies (UPS) are very popular for protecting computers against power interruptions. Motor-generator sets are very common in industrial installations. Power electronics-based solutions

(D-FACTS) installed at the interface between the utility company and the customer are becoming the preferred option to mitigate voltage sags and interruptions.

6.4.5 FACTS and custom power [20]

The concepts of flexible alternating current transmission systems (FACTS) and Custom Power were put forward in the mid-1980s by Dr N.G. Hingorani while working for EPRI at Palo Alto, CA, USA. Both ground-breaking developments used leading-edge power electronics valves, bridges, control techniques, and methods to solve various operational problems in the electrical power grid. As illustrated in Figure 6.12, FACTS focuses on AC transmission systems and Custom Power, also termed D-FACTS, focuses on power distribution systems.

As outlined in Chapter 1, FACTS addresses issues associated with maximising power throughputs in key AC power transmission corridors of the power grid while enhancing system stability. On the other hand, D-FACTS focuses on the reliability of supply and quality of power flows in low-voltage distribution systems. However, this technology also benefits voltage regulation and balancing, power factor correction, and harmonic cancellation. Custom Power was originally developed in response to reports of poor PQ and supply reliability affecting factories, offices, and homes. The vulnerability of automated equipment and production lines to PQ disturbances on the utility's network was at the centre of such problems. It was correctly surmised that with Custom Power solutions in place, the end-user should see tight voltage regulation, no power interruptions, low harmonic voltages, and acceptance of rapidly fluctuating and non-linear loads in the vicinity.

The most popular Custom Power equipment is the D-STACOM, the Dynamic Voltage Restorer (DVR), and the Solid-State Transfer Switch (SSTS). The

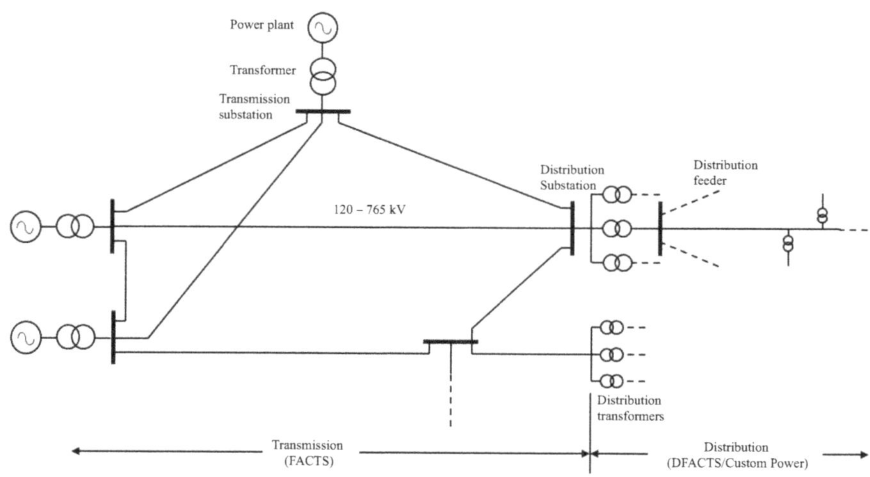

Figure 6.12 Simplified one-line diagram showing the areas of application of FACTS and Custom Power equipment within the electrical power grid, from [20]

D-STATCOM is a two-level VSC with an array of control functions that range from voltage regulation to active filtering. More specialised functions of the D-STATCOM are power factor correction and voltage balancing. The power electronics structure of the two-level STATCOM is addressed in detail in Chapter 4. The DVR is basically a two-level STATCOM connected to the secondary winding of a unity transformer whose primary winding is connected in series with the AC power system. In this respect, the DVR is not different from the Solid-State Series Compensator (SSSC) addressed in Chapter 1, except that the control function of the DVR is very specific: to continuously monitor the incoming voltage waveform at the sending-end node of the DVR. In the event of a voltage amplitude deviation from the declared voltage (increase/decrease), the DVR injects an in-phase voltage with the incoming voltage, restoring the declared voltage at the receiving end of the DVR. In the jargon of electrical power engineers, the main objective of the DVR is to complete the missing voltage cycles. This is illustrated in diagram form in Figure 6.13.

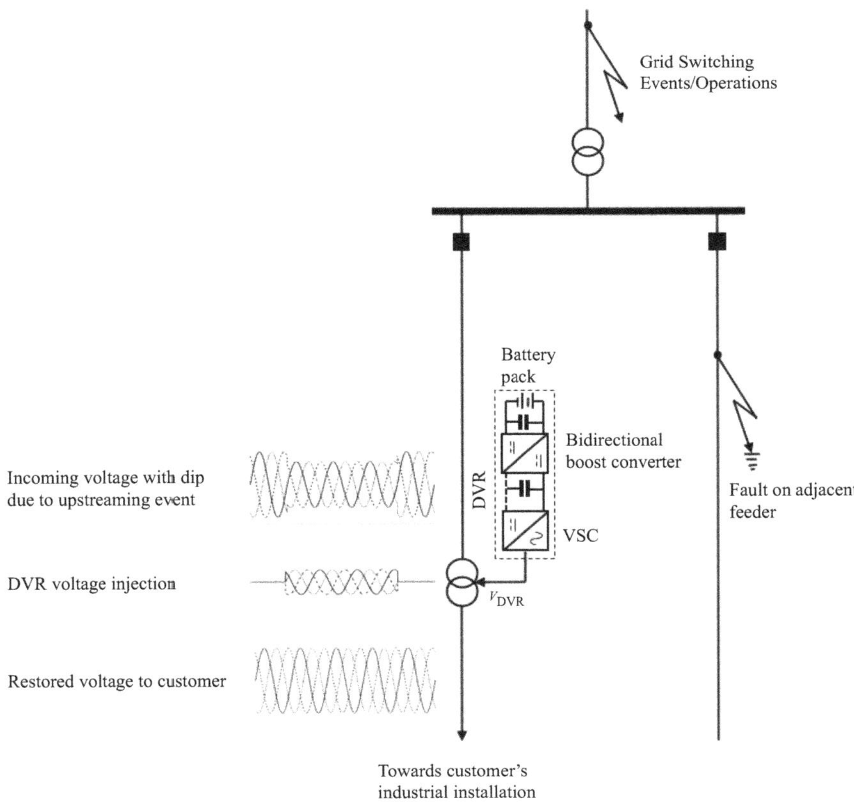

Figure 6.13 DVR contribution to the incoming supply with a voltage sag due to an upstream switching event or a short-circuit fault in a neighbouring feeder

This is exemplified by a DVR designed by Westinghouse (now Siemens) and EPRI in 1997. At the customer end, Scottish Power and Scottish Caledonian paper mill partnered with them. The 4 MW DVR, operated at 50 Hz, was installed at the papermill plant feeding substation [19].

The 4 MVA DVR was designed with a 50% voltage injection to protect the plant drives and control system, improving performance and security against power supply disturbances. Once the system was commissioned, running costs were low, and minimum maintenance was required. Before DVR installation, the mill could not run at full capacity due to potential risks of plant damage caused by voltage dips.

Figure 6.14 shows statistics relating to voltage sags in the Caledonian papermill installation taken over 1 year: 6 months before the DVR was installed and 6 months after it had been commissioned.

The three-dimensional plots give information on dip events at the papermill plant: the number of dip events, their duration in milliseconds, and the retained voltage in percent with respect to the declared voltage. In the figure on top – with no DVR yet installed – 69 dips were registered: 46 with a duration of fewer than

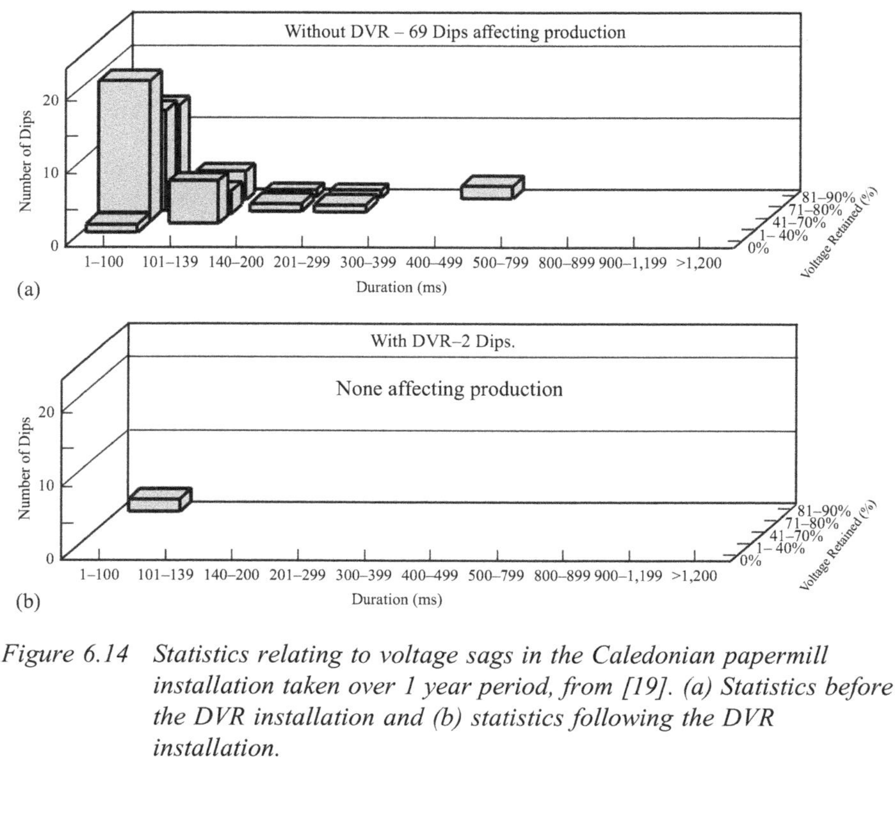

Figure 6.14 Statistics relating to voltage sags in the Caledonian papermill installation taken over 1 year period, from [19]. (a) Statistics before the DVR installation and (b) statistics following the DVR installation.

seven cycles, 12 with a duration of less than ten cycles, four with a duration of less than 15 cycles, four with a duration of less than 20 cycles and three with a duration of 40 cycles. It should be remarked that these voltage dip events were registered 6 months before the DVR equipment was in operation. The retained voltage varied 90%–1%, with the vast majority laying between 90% and 40% and durations of less than ten cycles. Nonetheless, all these events stopped the papermill operations.

The plot at the bottom shows information on dip events after the DVR started satisfactory operation, covering 6 months. Only two dips were registered, lasting less than seven cycles and a retained voltage of more than 80%. More importantly, these events did not have the "power" to trip out the production lines at the papermill plant. It should be remarked that the investment made by Caledonian Papermill towards this cooperative venture, which rounded the £M 3 in 1996–1997, had been recovered by reducing the number of dip events from 69 to 2.

The SSTS can be used effectively to protect sensitive loads against voltage sags, swells and other electrical disturbances. The SSTS ensures continuous, high-quality power supply to sensitive loads by transferring the load from a faulted bus to a healthy one within milliseconds. The basic configuration of this device consists of two three-phase solid-state switches, one for the main feeder and one for the backup feeder. These switches have an arrangement of back-to-back connected thyristors, as illustrated in the schematic diagram of Figure 6.15. Each time a fault condition is detected in the main feeder, the control system swaps the firing signals to the thyristors in both switches, i.e. Switch 1 in the main feeder is deactivated and Switch 2 in the backup feeder is activated. The control system measures the peak value of the voltage waveform at every half cycle and checks whether it is within a prespecified range. If it is outside limits, an abnormal condition is detected, and the firing signals to the thyristors are changed to transfer the load to the healthy feeder [20].

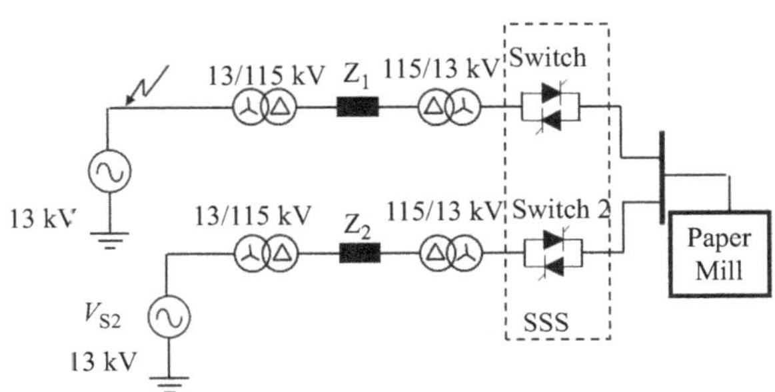

Figure 6.15 Schematic representation of an SSTS as a D-FACTS solution to protect a sensitive load

6.5 Conclusions

The face of the long-serving, classic electrical power distribution system is changing to adapt to a new technological era. One of the drivers is the power systems load, which has undergone great change since the mid-1980s. This encompasses industrial loads, urban loads, and domestic loads. In industry, variable speed drives and microprocessor-based controllers were introduced to modernise assembly lines to shorten production times with utmost quality control. The reliability of these highly automated processes rests on being supplied with a constant electricity supply, i.e., constant voltage amplitude and constant frequency. Unfortunately, this cannot always be warranted, adversely impacting the steadiness of industrial processes, and costing companies billions of dollars in wasted materials, downtime, delayed orders, and damaged electronic control devices. The change in urban loads has been more gradual but relentless: Large call centres and office complexes filled with computers, flat screens, printers, etc., are highly vulnerable to being supplied with electricity with poor power quality, resulting in damaged data, clients' irritations for poor service and missing productivity targets. Power electronics-based equipment has been developed expressly to counteract adverse power quality environments to provide premium protection to sensitive industrial/commercial loads – this was termed Custom Power by its developer. An even deeper makeover of urban loads is associated with the impending mushrooming of EVs and the great many power grids' infeed points that the EV recharging stations would require globally. Besides, the EV load is further transforming the domestic load array, giving rise to a new kind of energy player – prosumer – inserted in low-voltage power distribution systems. This new energy player can buy and sell energy to the utility electricity supply company. EVs, BESS, and solar PV panels may be some energy resources the Prosumer may need to support their business.

Another key driver that is changing the face of the classic electrical power distribution system has been the commercial availability of low-capacity power generators suitable for grid-connected applications, starting in the early 1990s. Collectively, these form part of what has been termed DERs, with wind power turbines, fuel cells, solar PV panels, EVs, and energy storage systems forming part of this array of grid-connected energy resources. This has transformed the classical power distribution grid, which is passive, into an active power distribution grid. The internal circuitry of most DERs is made up mainly of solid-state circuits and works with DC powers. A DER is connected to the AC power grid using a VSI. As DERs proliferate in active power distribution networks, it makes sense to migrate from the use of only AC power distribution grids into hybrid AC/DC power distribution grids, requiring far fewer VSIs needed to interconnect many DERs to the AC power grid in a more economical manner for the owners of DERs. This comes into the realm of future power distribution systems. The many VSCs may exert fast and precise AC and DC voltage control and active and reactive power flow control. Furthermore, fast-acting energy conditioning devices, such as BEES and EVs, are an effective means for counteracting frequency oscillations. Several other control functions become possible, such as power factor correction, harmonic distortion amelioration, overall power quality indices improvement, and enhanced continuity of energy supply. In the fullness of time, as more experience accumulates with the widespread usage of these devices in a network-wide context, additional third-party services may emerge in the power distribution system. Some progress has been made in this direction with the Custom Power initiative. Still, we may yet to witness the full convergence of power grid operational flexibility, resilience, and fully-fledged business opportunities in low-voltage power distribution networks, spearheaded by power electronics technologies.

References

[1] Short TA *Electric Power Distribution Handbook*. Boca Raton, FL: CRC Press, 2014.

[2] Northcote-Green J, and Wilson R. *Control and Automation of Electrical Distribution Systems*. Taylor and Francis, 2007.

[3] Chowdhury S, Chowdhury SP, and Crossley P. *Microgrids and Active Distribution Networks*. Stevenage: IET, 2009.

[4] Acosta Campas CN, Madrigal M, Ruiz Paredes HF. PV Power Curtailment and BESS Management for Distribution Networks: A Practical Approach. *IEEE Latin America Transactions*, vol. 21, no. 1, 2023, pp. 133–141.

[5] Maza-Ortega JM, Acha E, García S, and Gómez-Expósito A. Overview of Power Electronics Technology and Applications in Power Generation Transmission and Distribution. *Journal of Modern Power Systems and Clean Energy*. 2017;5(4), pp. 499–514.

[6] Sallam AA, and Malik OP. Harmonics in Electric Distribution Systems. *In Electric Distribution Systems*. New York: Wiley-IEEE Press, 2019, pp. 379–402.

[7] Romero-Ramos E, Gómez-Expósito A, Marano-Marcolini A, Maza-Ortega JM, and Martínez-Ramos JL. Assessing the Loadability of Active Distribution Networks in the Presence of DC Controllable Links. *IET Generation, Transmission & Distribution*, vol. 5, no. 11, 2011, pp. 1105–1113.

[8] Maza-Ortega JM, Mauricio JM, Barragán-Villarejo M, Demoulias C, and Gómez-Exposito A. Ancillary Services in Hybrid AC/DC Low Voltage Distribution Networks. *In Special Issue on Grid Services in Electric Power Systems published in Energies*. 2020.

[9] Kaushal A, and Van Hertem D. An Overview of Ancillary Services and HVDC Systems in European Context. *In Special Issue on Grid Services in Electric Power Systems published in Energies*. 2020.

[10] Lasseter R, Akhil A, Marnay C, *et al.* White Paper on Integration of Distributed Energy Resources: The CERTS Microgrid Concept. *Consortium for Electric Reliability Technology Solutions*. Prepared for the US Department of Energy, 2002.

[11] Refaat SS, Ellabban OS, Bayhan S, Abu-Rub H, Blaabjerg F, and Begovic MM. *Smart Grid and Enabling Technologies*. New York: Wiley-IEEE Press, 2021.

[12] Chen Z. Power Electronics in Modern Power Systems. *2022 IEEE 9th International Conference on Power Electronics Systems and Applications (PESA)*, Hong Kong, 2022.

[13] Global EV Outlook 2017; *International Energy Agency (IEA)*. Paris, France, 2017.

[14] Acha S. *Modelling Distributed Energy Resources in Energy Service Networks*. Stevenage: IET, 2013.

[15] *REmap: Roadmap for a Renewable Energy Future; International Renewable Energy Agency (IRENA)*. Masdar City, Abu Dhabi, 2016.

[16] *Battery Electricity Storage Costs and Market Outlook to 2030; International Renewable Energy Agency (IRENA)*. Masdar City, Abu Dhabi, 2016.

[17] Three Phase Power Flow Computer Program. Tampere University of Technology, Tampere, Finland, 2015.

[18] De Leon F, Marquez-Asensio ML, and Alvarez-Cordero G. Effects of Conductor Counter-Transposition on the Positive-Sequence Impedance and Losses of Cross-Bonded Cables. *IEEE Transactions on Power Delivery*, vol. 26, no. 3, 2011, pp. 2060–2063.

[19] Burden, A. Caledonian Paper DVR—The Utility Perspective. *IEE Half Day Colloquium on Dynamic voltage Restorers – Replacing Those Missing Cycles*, Glasgow, UK, 1998, pp. 2/1–2/2.

[20] Hingorani NG. Introducing Custom Power. *IEEE Spectrum*. vol. 32, no. 6, 1995, pp. 41–48.

[21] Bollen MH. *Understanding Power Quality Problems: Voltage Sags and Interruptions*. New York: Wiley-IEEE Press, 2000.

[22] Dugan RC, McGranaghan MF, and Beaty HW. *Electrical Power Systems Quality*. New York: McGraw-Hill, 2002.

[23] Schlabbach J, Blume D, and Stephanblome T. *Voltage Quality in Electrical Power Systems*. Stevenage: IET, 2001.

[24] Baggini A (ed). *Handbook of Power Quality*. New York: Wiley, 2008.

[25] IEEE Draft Standard for Harmonic Control in Electric Power Systems. IEEE P519/D5.1.2021. 29 April 2021.

[26] IEEE Recommended Practice for Powering and Grounding Electronic Equipment. IEEE Std 1100-2005 (Revision of IEEE Std 1100-1999). 24 May 2006.

Chapter 7

Power electronics in power generation and storage

Francisco De La Rosa[1]

7.1 Basis and prognosis of renewable generation

7.1.1 Introduction

Variable renewable energy (VRE) is provided as a prodigy of Mother Nature. This power source is harnessed through conversion from physical phenomena – wind and sun – to electricity, which can be coupled to nearby electric power lines. These sources of energy are variable and somehow intertwined. The energy that drives wind is created by the warm air from the uneven heating produced on earth by the sun, creating wind streams. These generation assets must be strategically deployed to maximise output power.

This chapter describes the basis and prognosis of VRE, the regulatory aspects related to the increased integration of renewable energy, the power electronics and control functions for IBRs (Inverter-Based Resources), the inertia loss and synthetic inertia, and the energy storage functionality to support the integration of VRE into the power grid. Note that a full list of the acronyms used in the chapter can be found in the appendix at the end.

Figure 7.1 sketches the structural design of a modern RES (Renewable Energy Source) power generation system with the control functions carried out by the contained power electronics. The power transfer from the renewable source to the power grid is done by precise and adaptable electronic switching in the inverter. This creates a series of power conversion actions to inject the renewable power supply into the grid. Intelligent control functions are needed to ensure that power transfer is performed consistently with the established and expected operation planning specifications, considering the variable nature of renewable energy resources, as stated in [1].

Among the expected inverter control actions, as it transfers the renewable power to the electric grid, the following can be mentioned:

(a) Reliability
(b) Reactive power management

[1]Senior Electric Power Systems Consultant, San Antonio, TX, USA

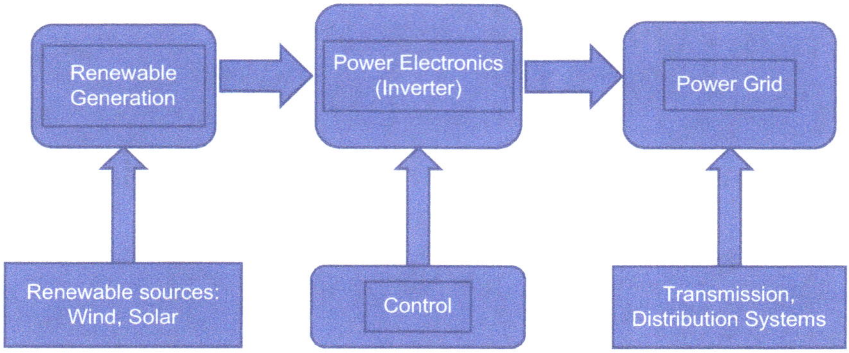

Figure 7.1 Modern RES power generation system

(c) Dynamic support
(d) System monitoring
(e) Communications
(f) Protection

7.1.2 VRE is a growing player in the electric power generation ballpark

Globally, combined variable energy generation from solar and wind was expected to account for two-thirds of the growth in renewable power in 2021. Regarding energy production, the projected amount for the same year was set to reach about 8,300 TWh, marking the largest increase in successive years since the 1970s [2]. China alone should have accounted for almost half of the global increase in renewable electricity in 2021, followed by the United States, the EU, and India [2].

The most significant increase in renewable energy generation is anticipated to come from wind, which in 2021 grew by 275 TWh, or nearly 17 per cent, above 2020 levels. Due to Chinese and American policy deadlines, a record amount of capacity was completed by developers in the fourth quarter of 2020, which resulted in a notable increase in generation as early as the first two months of 2021. Throughout 2021, China was projected to produce 600 TWh, and the US, 400 TWh, or more than half of the world's wind output [2]. Notably, predictions for 2021 were established, expecting a fast recovery in economic production and energy demand across all sectors after having been impacted by Covid 19.

Figure 7.2 depicts the electric energy contribution for the period 2020–2021 [2]. China, the United States, and the EU accounted for the largest growth in wind energy production, with just the wind energy generated in China and the USA exceeding 50 per cent of the global wind energy production.

Figure 7.3 illustrates the global electricity generation by technology for the periods 2015 and 2021 and the projected growth for 2027, with renewables accounting for almost 40 per cent of the global electricity generation [3].

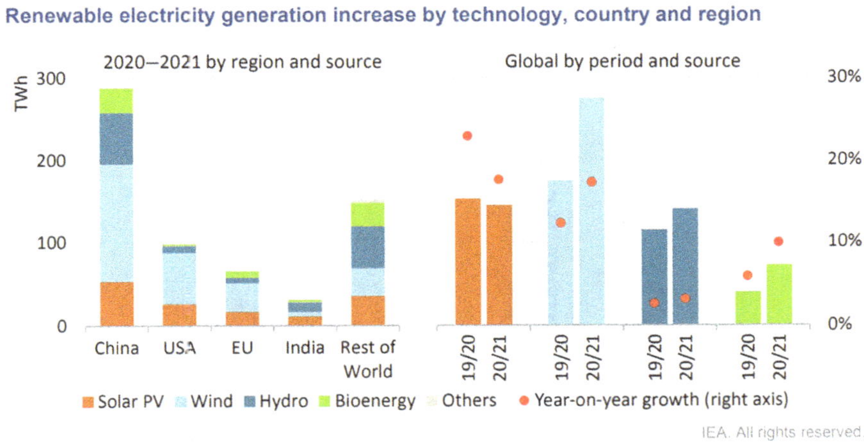

Figure 7.2 Global contribution of renewable energy rise for the period 2020–2021, from [2]

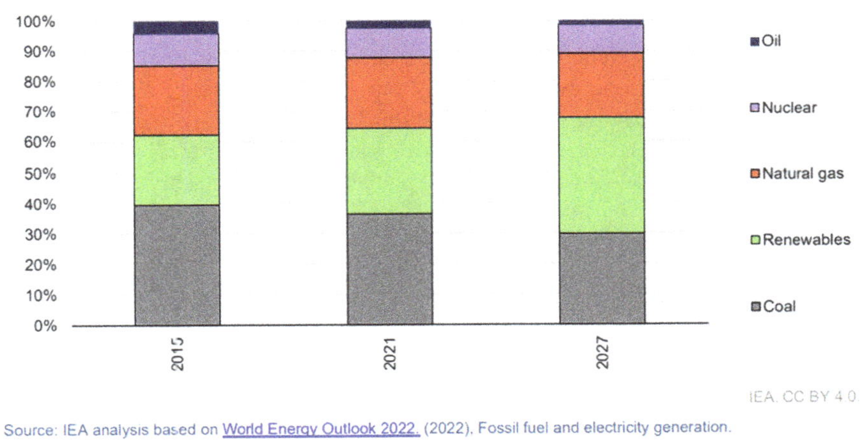

Figure 7.3 Global growth of renewable generation including projections for 2027, from [3]

While China will continue to be the world's largest PV market, expansion in the US will proceed with continued federal and state legislative funding [2]. India's PV market was anticipated to recover quickly in 2021 after seeing a significant decline in new solar PV capacity additions in 2020 due to COVID-related delays. In Brazil and Vietnam, increases in generation are being driven by strong policy support for distributed solar PV applications.

To reach 1,000 TWh in 2021, the world's solar PV electricity generation had to rise by 145 TWh or over 18 per cent [2]. Likewise, annual solar PV expansion was

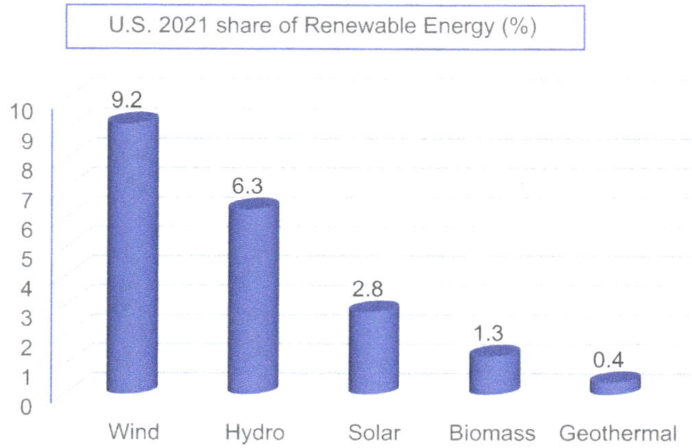

Figure 7.4 2021 share of renewable generation, adapted from [5]

expected to reach 162 GW in 2022, setting new records and accounting for nearly 55 per cent of all renewable energy expansion [4].

The United States is rich in renewable energy resources, ostensibly with around 100 times the electricity needed annually. According to the US Energy Government, in 2021, energy production from renewable energy showed a clear lead by wind power, leaving solar power in third place below hydro generation and biomass and geothermal in fourth and fifth places, as illustrated in Figure 7.4. Renewable power accounted for around 20 per cent of the electricity demand in the USA [5].

The cited US Energy Government source [5] elucidates that 2022 expected a significant increase in generating capacity, mainly from wind (46 per cent) and solar (17 per cent), which would account for over 60 per cent of the total generation production in the USA. This was further endorsed by the International Energy Agency report [6], which estimated that renewable capacity would rise over 80 per cent in 2022 relative to 2021, closely advancing to a 300 GW mark for the first time. Likewise, it mentioned that the projected photovoltaic power would account for 60 per cent of the growth in renewable energy capacity in 2022, a 25 per cent increase over 2021. Mention was made that along the lines of a robust policy environment in China and in the European Union pushing for faster deployments of renewable power, utility-scale projects in the US accounted for almost two-thirds of overall photovoltaic expansion in 2022.

Figure 7.5 depicts the participation of renewable energy in USA electricity production up to 2022 and the projections till 2050, as reported in [7]. The remarkable growth of intermittent generation is significant, expected to approach around 2,000 billion kilowatt-hours for solar and above 1,000 billion kilowatt-hours for wind by 2050.

Due to population and industrial growth, the demand for power electronics has recently increased. This is observed in industrial processes using variable frequency

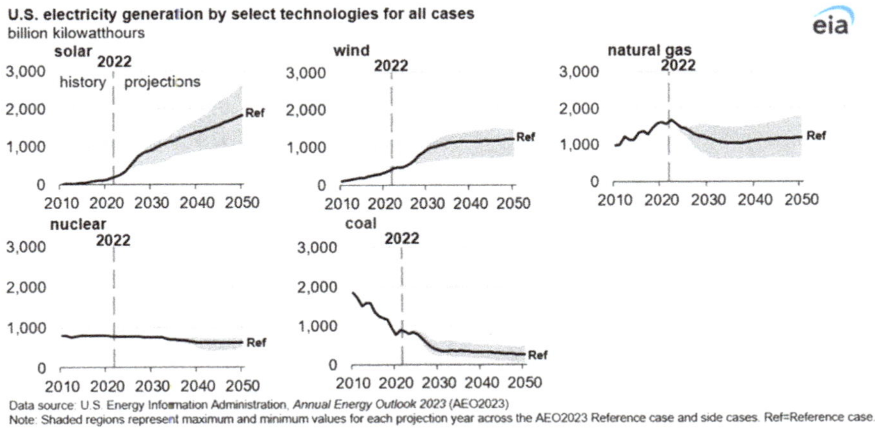

Figure 7.5 US electricity generation, from [7]

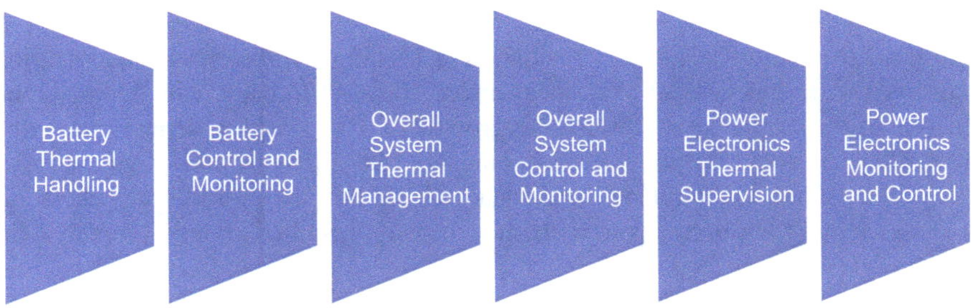

Figure 7.6 Typical battery energy storage system control functions

drives, consumer electronics, electric vehicles, power generation, and particularly from variable renewable resources.

Globally, and with the participation of new onshore wind farms resulting from a 32 per cent year-on-year decline in 2021, new onshore wind installations worldwide are expected to show a slight recovery to extend close to 80 GW. On the other hand, the global increase in offshore wind is anticipated to lessen by about 40 per cent in 2022, doubling up the 2021 additions and remaining solid for 2023 [6].

7.1.3 Participation of energy storage in the integration of VRE

Likewise, utility battery storage also requires power electronics for grid connection and to carry out its duty to monitor and control thermal status and performance as it goes through cycles of charge/discharge and carries out all needed power conversion actions. This is illustrated in Figure 7.6.

The EU solar energy strategy [8] mentions that introducing high shares of solar PV and wind will influence how the electricity grid is managed. As renewable power from PV solar is produced in Direct Current (DC) form, conversion to Alternating Current (AC) to feed into the grid energy and any conversion to store energy leads to energy losses. These losses would increase further since batteries, heat pumps, data centres, electric vehicles, and other applications operate in DC. This would suggest that boosting the use of DC transmission and distribution infrastructure could thus benefit the electricity system. Naturally, this comes up with substantial power electronics deployment.

According to some estimates, rooftop PV could provide almost 25% of the EU's electricity consumption, which is more than the share of natural gas today. The EU-wide European Solar Rooftops Initiative announced in the Commission's REPower-EU Communication that it aims to unlock rooftops' vast, underutilised solar generation potential to achieve clean, more secure, and affordable energy. To achieve this swiftly, immediate action is necessary. EU members are pursuing an expansion of the cross-border electricity infrastructure that can allow efficient integration of solar electricity produced in the EU [8]. The use of power electronics goes hand-in-hand with the growing rooftop PV installations.

Among other countries firmly committed to solar energy deployment, it is significant to mention that the EU is assisting India through technical cooperation and business-to-business interactions under the EU-India Clean Energy and Climate Partnership. The exponential growth of PV markets demonstrates the adaptability of solar technologies in countries like Vietnam and Japan. Similar developments are taken place in Africa, which has the world's richest solar resources and 600 million people without electricity, installed about 5 GWs of solar PV in 2019 [9].

Increasingly large amounts of intermittent electrical energy will be injected into the power grid. When assisted by grid-scale battery storage, delivers enhanced reliability of supply. Using power electronics, inverter-based resources (IBR), including wind, solar, and energy storage, can quickly detect frequency deviations and respond to system imbalances. Opting for electronic-based resources can enable response rates many times faster than the traditional mechanical response from conventional generators, thereby conveniently handling fast-frequency response events and reducing the need for inertia [9]. This is particularly important to guarantee extremely reliable power electronics strategies, especially considering the added control functions utilised in power conversion. Transmission and distribution of power operations and battery storage provisions require reliable incorporation of variable renewable energy into the power system.

7.1.4 Meteorological conditions and their impact on renewable generation

Climate conditions affect the generation of electricity in several ways. Depending on cold or warm seasons, power demand changes as cooling or heating systems start. Rising temperatures due to ongoing climate changes lead to increased power generation demand. On the supply side, wind and solar changes determine the timing and availability of intermittent renewable generation. Likewise, hydro resources are impacted by the variability of rainwater.

However, climate conditions vary worldwide, making it challenging to extrapolate observations from one location to another unless they correspond to areas with similar weather patterns. Therefore, a power system's resilience must be assessed considering the climate variability across the regions involved in the relevant power utility assets and variable renewable generation sites.

Integrated assessment models (IAMs) in [10] used specific search words to review selected papers issued in 2000 or later and published in peer-reviewed journals. These were related to climate changes on the supply side, such as primary resources, power generation technologies, and energy transport. They conducted impact studies to simulate the effects of changing climatology on one or several elements of the supply system, using detailed climate change scenarios. This aimed to determine how aspects of the energy systems were impacted in the different studies.

The findings in [10] revealed that average wind speeds in Europe and North America would likely endure variations ranging from ±15–30% of current values by the end of the century. In contrast, limited studies over China and Southern Africa did not observe a significant change in wind resources.

On the other hand, projections on Solar Climate from these studies revealed that cloud cover will decrease in low-to-mild altitude levels and that the resulting rise in solar irradiation will be offset by rising temperatures, which will impact efficiency. Therefore, changes in solar generation were projected to be less than ±10% by the end of the century.

Forecasting is complicated because of variability and unpredictability. Geography and topography also play important roles, and environmental climate changes must be figured out.

Following the IAM discussion, careful consideration of temporal and spatial resolution and scope is required to represent climate effects in larger energy system models. For instance, a few decades may be selected as the time horizon for energy system optimisation modelling to reduce the effects of changing technology and cost uncertainty.

On the other hand, a longer time horizon may be desired to investigate the bigger implications of climate change or the significance of technology lock-in [11]. Some impact models, such as process-based hydrological models, require high-resolution input data, such as local topography features. Due to computational expense, these are frequently performed at regional scales, making it more difficult to incorporate them with global-scale energy system models. Global models may employ simpler simulations and cover a much greater geographic area [11].

Finally, because global IAMs often represent the world as only 10–30 regions, climate impact data must be aggregated to this scale. Variations in influence within various locations may provide challenges for this aggregation. Regarding impact research, several IAM authors [12] avow that their findings will depend on the climate model employed as the driver. Some studies employ an ensemble mean to decrease the influence of selecting a single climate projection [11,13]. However, there is little discussion of the techniques' benefits and limitations.

7.1.5 Solar and wind resources assessment

An excellent open source for preliminary assessments of solar energy projects is the Global Solar Atlas version 2.0 [14]. It provides resources to carry out site-specific and regional spatial PV evaluations. It is a high-resolution global database handled by models using satellite, atmospheric, and meteorological data. The main database consists of solar and meteorological data series calculated in intervals of 10/50/30 minutes for the periods 1994, 1999, 2007, and up until 2018, depending on the region. Figure 7.7 depicts the global horizontal irradiation map, and Figure 7.8 portrays the Photovoltaic Power Potential map published by the World Bank Group [14].

Figure 7.9(a) depicts an extended mean wind speed map, and a close-up view is shown in Figure 7.9(b) [15], while Figures Figure 7.10 (a) and (b) portray world wind power density potential maps, all showing wind at 150 m above the surface level, for the period 2008–2017 [16]. These are high-resolution maps based on a

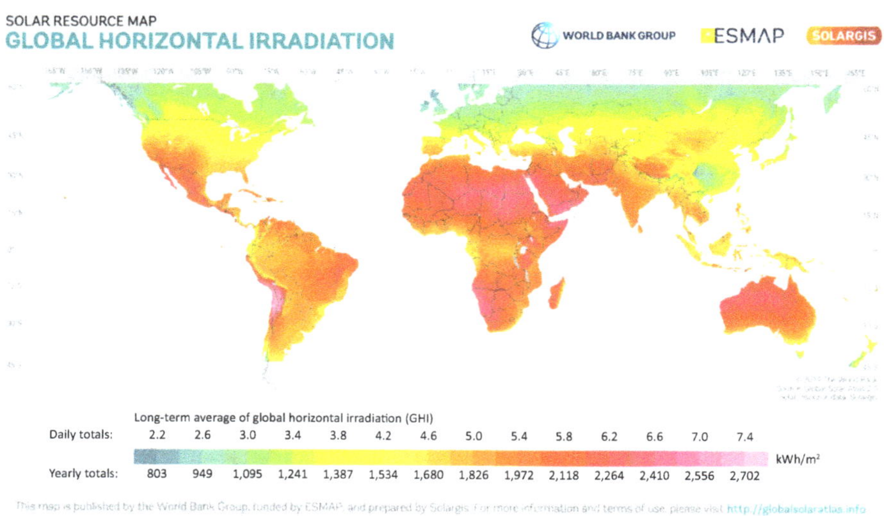

Figure 7.7 Global horizontal irradiation map, from [14]. Data/information/map obtained from the "Global Solar Atlas 2.0, a free, web-based application is developed and operated by the company Solargis s.r.o. on behalf of the World Bank Group, utilising Solargis data, with funding provided by the Energy Sector Management Assistance Program (ESMAP). For additional information: https://globalsolaratlas.info.

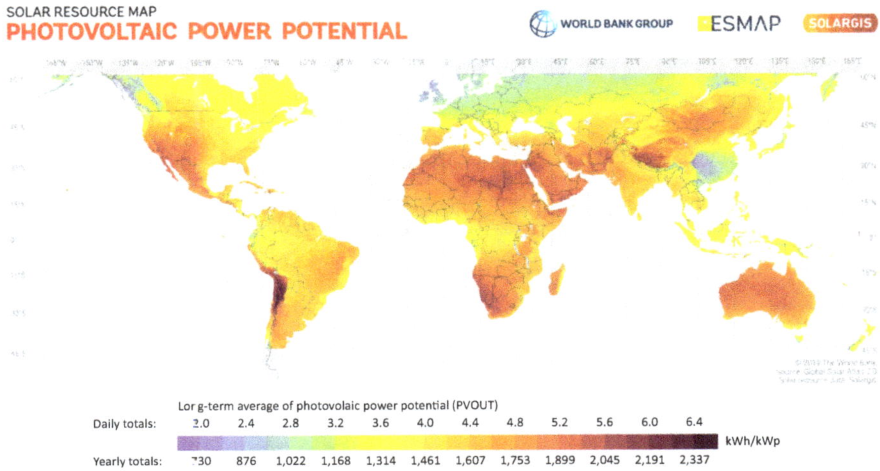

Figure 7.8 Photovoltaic power potential map, from [14]. Data/information/map obtained from the "Global Solar Atlas 2.0, a free, web-based application is developed and operated by the company Solargis s.r.o. on behalf of the World Bank Group, utilising Solargis data, with funding provided by the Energy Sector Management Assistance Program (ESMAP). For additional information: https://globalsolaratlas.info.

group of models, which scale back winds from global models (around 30 km) to mesoscale (km) to microscale (250 m).

Figures 7.11 and 7.12 show the worldwide energetically optimal stand-alone global electric power produced by photovoltaic-wind hybrid renewable energy systems, respectively [17].

Given the stochastic nature of wind and photovoltaic variable generation, its seasonal aspects, spatial and time dependency, and high variability in a short period, PV and wind optimal locations should consider the valuable information contained in these solar and wind maps to account for the different potentials that these renewable resources may have given their geographically dispersed arrangements. This can provide aggregate reliability to the power system.

7.1.6 VRE forecasting

VRE forecasting is a new topic, as is integrating VRE resources. The CAISO was the first to implement it in 2014 [18]. Grid operators and regulators use VRE generation forecasting to address the barriers to integrating VRE into the grid. As the share of VRE in power systems increases, grid integration concerns hinder the scaling up of renewable energy. VRE forecasting is crucial for grid reliability and cost efficiency. Failure to accurately forecast VRE can lead to curtailment,

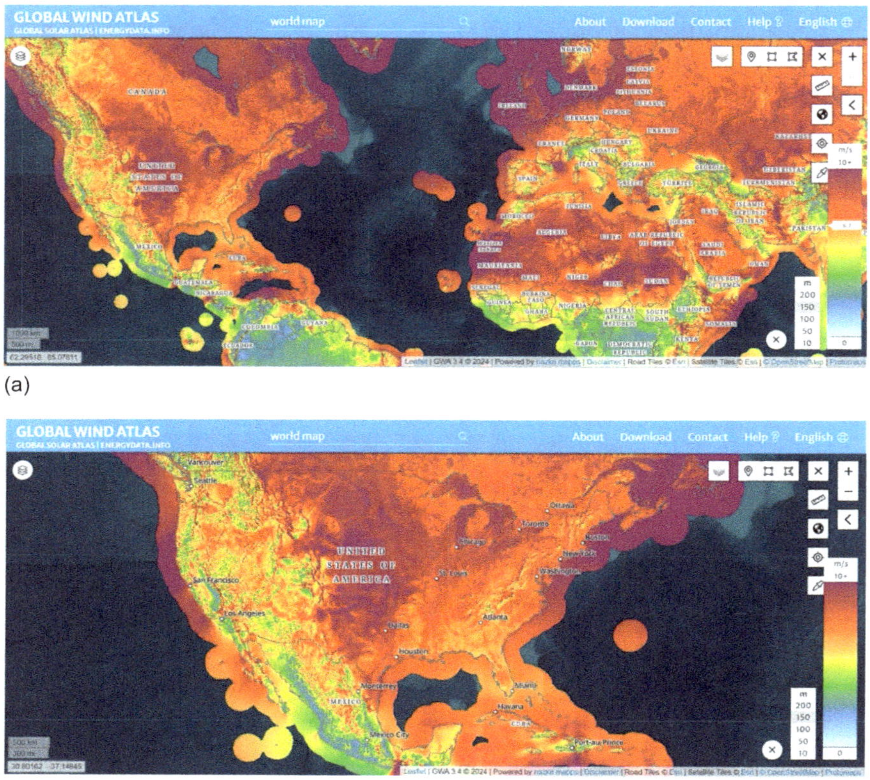

Figure 7.9 A global mean wind speed map, from [15]: (a) extended map and (b) close-up view. Data/information/map obtained from the Global Wind Atlas version 3.3, a free, web-based application developed, owned and operated by the Technical University of Denmark (DTU). The Global Wind Atlas version 3.3 is released in partnership with the World Bank Group, utilising data provided by Vortex, using funding provided by the Energy Sector Management Assistance Program (ESMAP). For additional information: https://globalwindatlas.info.

frequency and voltage disturbances, negative market prices, price volatility, and dropped load.

VRE forecasting predicts future VRE power plant output on week-ahead, day-ahead, or intraday timescales. It supports system operators in balancing generation with load and managing reserves. It reduces the impact of VRE integration on the grid and enhances grid flexibility. VRE forecasting is vital for system operators to make informed decisions about conventional generator capacity factors, reduce curtailments, and increase VRE penetration. VRE forecasting improves unit

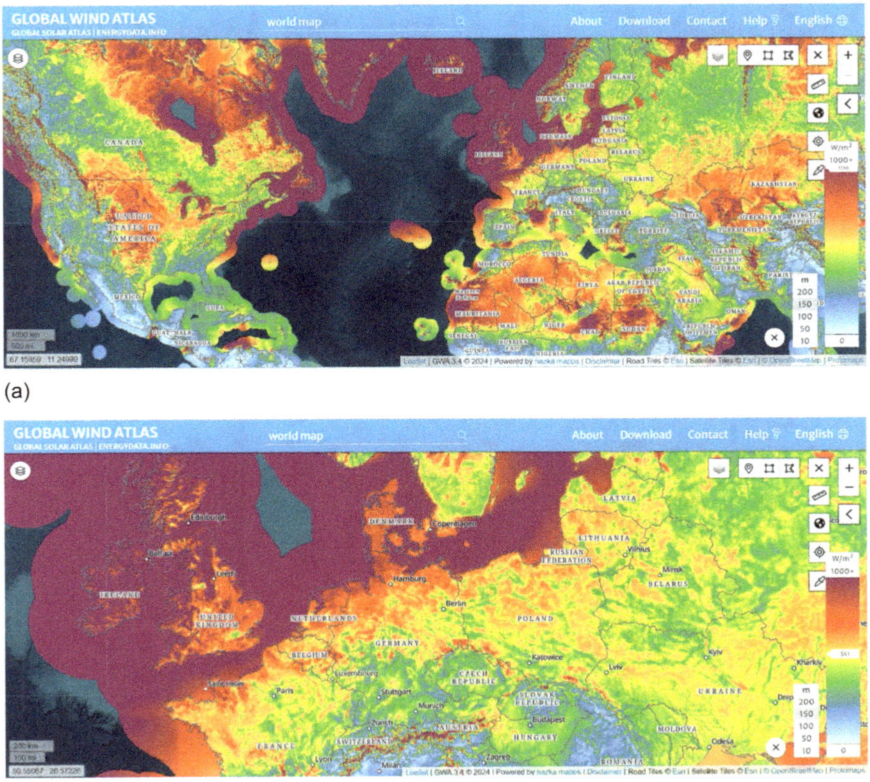

(a)

(b)

Figure 7.10 World wind power density potential map, from [16]: (a) extended map and (b) close-up view. Data/information/map obtained from the Global Wind Atlas version 3.3, a free, web-based application developed, owned and operated by the Technical University of Denmark (DTU). The Global Wind Atlas version 3.3 is released in partnership with the World Bank Group, utilising data provided by Vortex, using funding provided by the Energy Sector Management Assistance Program (ESMAP). For additional information: https://globalwindatlas.info.

commitment, economic dispatch, real-time balancing, and reserves planning in grids that do not operate with real-time power markets.

VRE forecasting affords essential benefits for emerging market grid operators and regulators, as illustrated in Figure 7.13.

7.1.7 Variable renewable generation management

As renewable energy grows, managing variable generation becomes crucial for system resource planning. Unscheduled variability based on resource characteristics

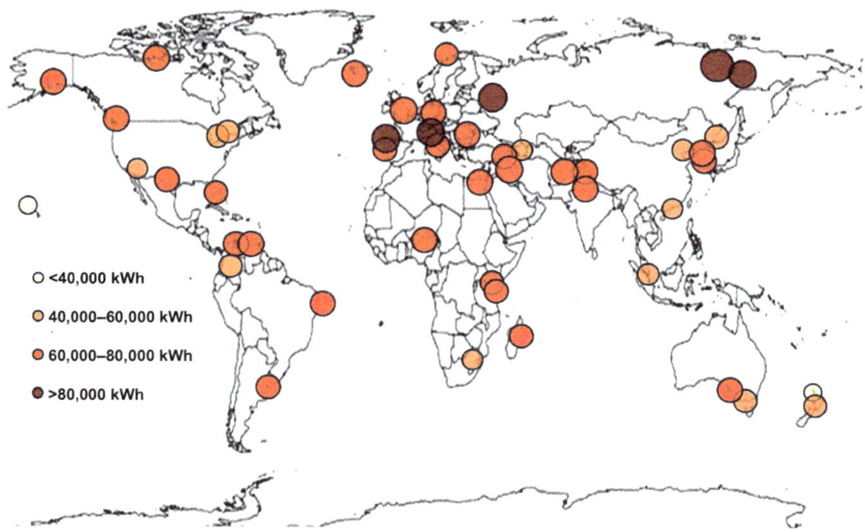

Figure 7.11 PV electric power generated by the energetically optimal stand-alone systems, from [17]

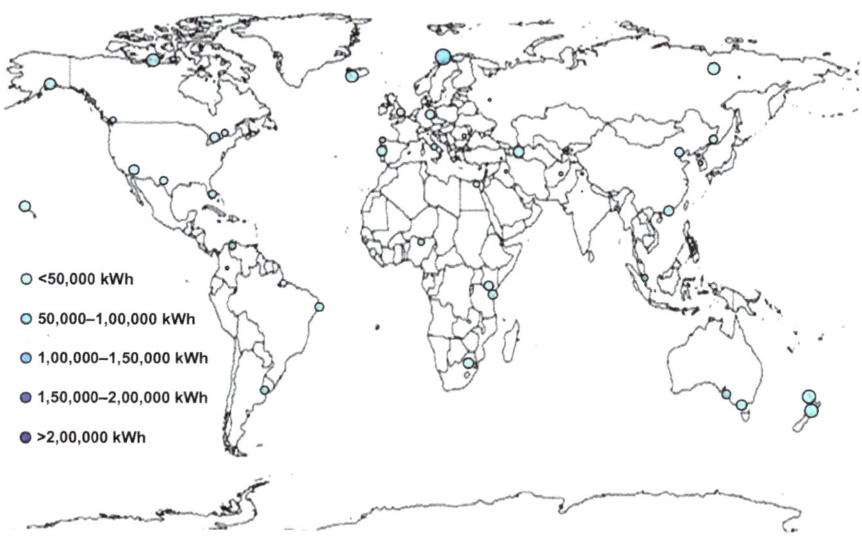

Figure 7.12 Wind energy generated by the energetically optimal stand-alone systems, from [17]

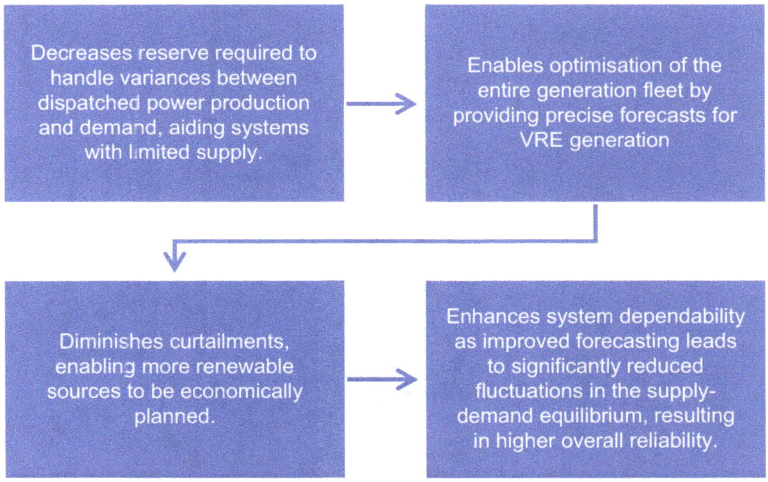

Figure 7.13 VRE forecasting benefits, adapted from [18]

can be coordinated in different manners. One way is limiting penetration caps or curtailment to manageable grid impacts. Another way is to dedicate fast-ramping fossil-fuelled generation to address renewable energy variability beyond grid stability mechanisms. Improved wind forecasts can also leverage slowly ramping resources, as demonstrated in the Canadian Maritime Provinces [19].

Extending renewable generation capacity limits can also be achieved without additional dispatchable generation. This includes expanding electric grid flexibility by adding energy storage, which is technically effective but economically challenging.

Understanding and maximising the effects of resource aggregation and transmission can allow smooth renewable outputs and address oversupply or undersupply. Transmission's efficacy relies on the spatial variability of renewable resources. Analyses should be carried out to identify regions with renewable resources that, when combined, minimise the electric grid's costs.

7.1.8 Developmental challenges and the way forward

Several issues in estimating solar and wind power apply to vertically integrated and deregulated electricity industries. Countries new to VRE or have not yet built a complete VRE forecasting program can apply recommended solutions based on best practices. This is recapped in Figure 7.14, adapted from [19].

Additional advice on the process of VRE forecasting, from the types and accuracy of forecasting models to the VRE forecasting ecosystem, how VRE forecasts are used in scheduling and dispatch decisions, the system architecture of VRE forecasting, and the reasons why VRE forecasting is important for grid integration can be found in [19].

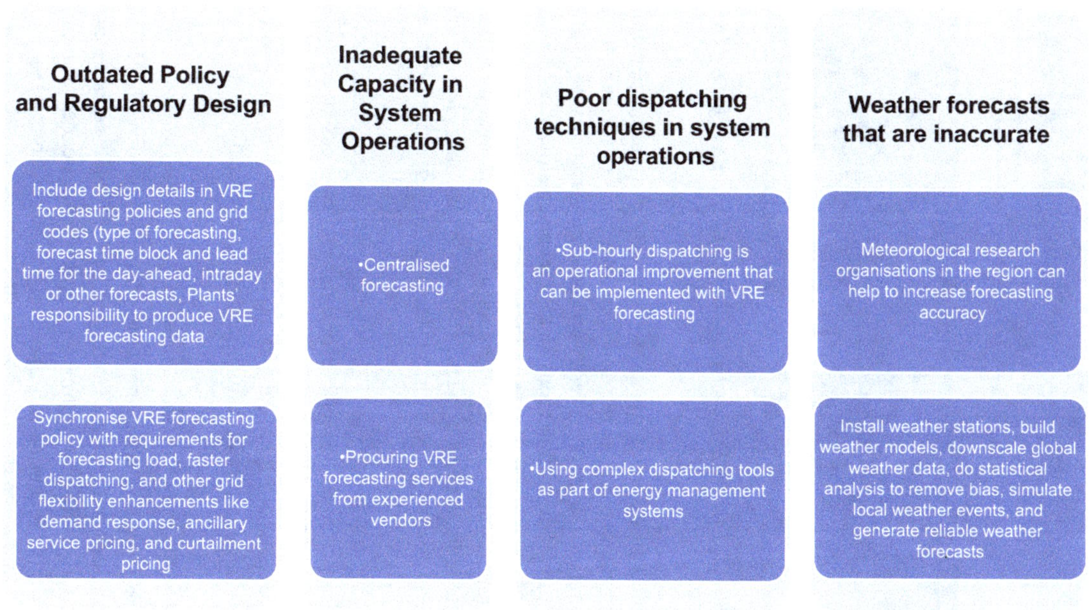

Figure 7.14 Developmental challenges and the way forward, adapted from [19]

7.2 Regulatory aspects on the increased integration of renewable energy

7.2.1 Introduction

The increasing share of variable renewable energy has made it necessary for TSOs to develop specific guidelines, more precisely described as Grid Codes, for IBR's incorporation to ensure the dependability of the BPS in the event of contingency conditions. Figure 7.15 illustrates this process. A thorough analysis of the global Grid Codes, both international and regional, was completed in [20], and this section's introduction provides a condensed version of that analysis.

The European electric system is structured similarly to the North American Reliability Corporation (NERC). ENTSO-E is an umbrella organisation representing 43 energy TSOs from 36 European countries. It is in charge of creating network Grid Codes and in aiding with their implementation. A harmonised grid connection code establishes a minimal requirement that each member must follow in their own Grid Codes. Some of its member's Grid Codes may need to be lightly modified accordingly. This standard is a minimal requirement rather than a suggested set of standards, and generally speaking, national Grid Codes are more extensive and stringent in their permissible ranges for technical parameter ranges [20].

In the United States, The Federal Energy Regulatory Commission (FERC) and NERC are the principal organisations in charge of standards and requirements. The interstate transmission of electricity, natural gas, and oil inside the United States is governed by FERC, a federal organisation. In July 2006, FERC approved NERC as the Electric Reliability Organization (ERO) for the United States; in this capacity, NERC develops reliability standards through an open stakeholder process. These standards are then submitted to FERC for approval after being completed by NERC. NERC will be in charge of verifying adherence to these requirements upon approval. Historically primary frequency response (PFR) generator requirements have not been the focus of these standards and criteria [20].

In addition to interconnection-wide standards and regulations, a Balancing Authority (BA) can request FERC approval for a standard that applies purely to a particular region. NERC BAL-001-TRE-1, adopted by FERC in 2014 for the Texas

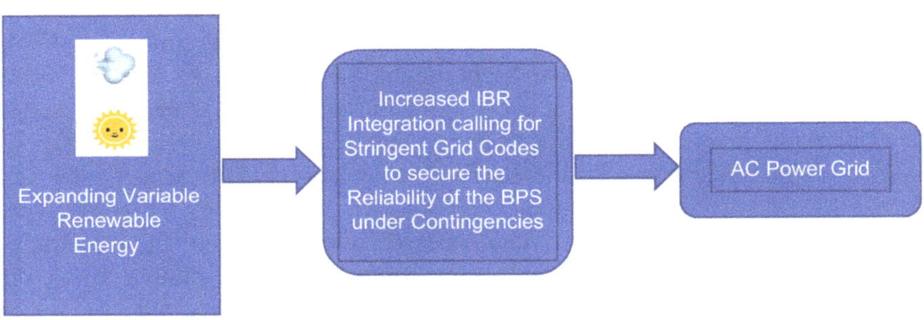

Figure 7.15 Coping with increased shares of Variable Renewable Energy

Interconnection, run by the Electric Reliability Council of Texas, Inc. (ERCOT), is an example of a regional standard.

The FERC has authorised Western Electricity Coordinating Council (WECC) to be Western Interconnection's regional entity. In this capacity, WECC Regional Standards command that every BA and Reserve Sharing Group must provide backup spinning generation for at least 50% of its contingency reserve commitments. These are defined as the greater of a single contingency loss and the total of the 3 per cent of hourly load loss and the 3 per cent of hourly generation. Further, the scheme should be swift enough to respond to frequency deviations via a control scheme or a speed governor [20].

The most important PFR standard in the US, the NERC BAL-003-1, was authorised by FERC in 2014 and revised as NERC BAL-003-1.1 in 2015. Each BA must meet a yearly minimum Frequency Response Measure (FRM) requirement under this standard to demonstrate that it can prevent UFLS for a given generation loss. Instead of being prescribed by NERC, the specifics of how BAs achieve this goal are left to the BAs themselves. To promote compliance with BAL-0033, NERC has provided suggested speed governor settings for each of the three US interconnections [20].

To achieve short-term compliance with NERC BAL-003-1.1, CAISO is now rethinking its method of acquiring PFR utilising a two-phase strategy. The first is enabling CAISO to transfer PFR from other BAs in the Western Interconnection, ensuring compliance with BAL-003-1.1 obligations. The second is increasing the spinning reserve requirements for generators offering PFR to enhance those provisions. The development of a maximum permitted dead band is the key difference between CAISO's amended criteria and WECC's system-wide Standard BAL-002-WECC-29 [20].

In Ontario, the bulk electrical system and the energy market are managed by the Independent Electricity System Operator (IESO). Following the August 2003 blackout, the Ontario Energy Board and NERC signed a Memorandum of Understanding in 2006 that designated NERC as the ERO for IESO. This happened after NERC regulations had already been applied to IESO [20]. Northeast Power Coordinating Council, Inc. (NPCC), which oversees the IESO's adherence to NERC dependability criteria, is also a member of the IESO.

Regarding PFR, the IESO requires that generators, including wind and solar power, meet particular capabilities as part of its connection requirements, explicitly concerning the maximum dead band, changeable droop range, and delivery rate. Units delivering the IESO's frequency regulation ancillary service must operate with a working governor in-service meeting these technical requirements [20].

In addition to these specifications, IESO now requires wind turbines to be able to give the system a synthetic inertia response after a frequency variation. If the frequency has not adequately recovered, this reaction must be initiated within 1 s and sustained for at least 10 s. This criterion prevents wind generators from reducing their output to contribute to the system's PFR. IESO alerts wind generators when this service should be enabled [20].

Other regional Grid Codes that can be cited are those in Brazil, Great Britain (National Grid), Finland, Italy (Terna), Ireland (EirGrid), New Zealand, Singapore, Spain, and Switzerland. It is not within the scope of this narrative to make a detailed analysis of all grid codes worldwide; therefore, to portray specific relevant aspects in this book, reference will be made only to some national or regional grid codes [20].

7.2.2 Grid Codes Relevant Regulations

The International Renewable Energy Agency (IRENA) issued a report [21] on the technical specifications for connecting to the electric power grid. It explains the basic criteria VRE power plants and new VRE facilitators must satisfy to access the power grid. It also describes the emerging trends seen in the most advanced technical specifications to cope with higher VRE shares and the related changes of decentralisation, digitalisation, and electrification of the final energy demand.

One of the oldest grid code requirements for conventional generating units is the frequency and voltage ranges that must be maintained during regular operation and contingencies.

Power System operational requirements have evolved throughout the years to characterise the behaviour of VRE plants during faults and contingencies. Some of the most recent changes to grid codes address the loss of inertia from the reduced use of synchronous generators following higher shares of renewable generation. The addition of VRE lowers inertia and raises the initial Rate of Change of Frequency (ROCOF), which is the rate at which frequency changes after an event and a metric that can activate system protective mechanisms, as will be described in the next section. As a result, newer limitations and operational restrictions, operational measures, and novel ways to counteract inertia loss are to be expected.

Figure 7.16 depicts the grid codes and innovation trends envisaged by IRENA [21]. It is imperative to agree with this viewpoint, especially when transmission systems are planned toward 100 per cent of VRE, where meeting Grid Code regulations for black start needs, for example, is especially important. Note the significant roles of Grid codes in achieving a high share of the VRE goal.

Grid Codes are also used by TSOs to secure a reliable and economic operation of transmission grids under steady state and dynamic system performance, markedly after the increasing incorporation of variable renewable resources in the power grid. These directives address voltage, reactive power, frequency, and power quality under steady-state conditions and frequency, active power ramps, frequency and dynamic control of voltage and reactive power, voltage ride through, and inertia, to mention some, under dynamic conditions.

The first step in developing grid codes is to settle an agreement among all relevant stakeholders aimed at facilitating private developers' access to a friendly integration of VRE generation plants, with straightforward requirements that are simple to assess through simulation studies and possible to achieve while keeping the operation of the electric system safe. As with any other regulatory documents, it is expected that grid codes will undergo frequent revisions to integrate new

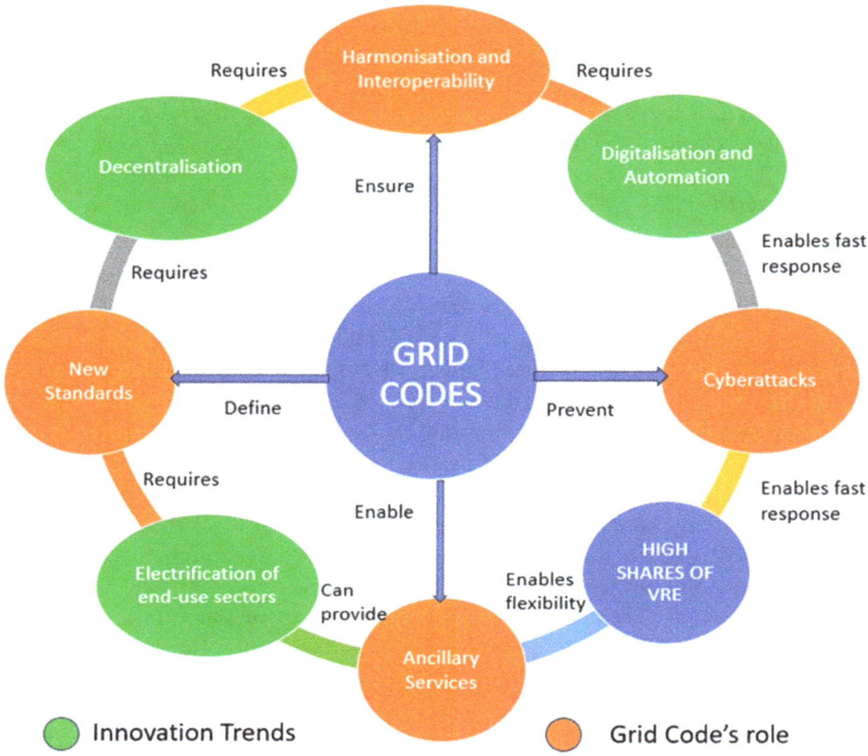

Figure 7.16 Grid codes and innovation trends, adapted from [21]

developments and power utility requirements to optimise the economical operation of electric grids.

As more generation capacity and flexibility are available in distribution systems due to decentralisation, operational control limits to keep them within limits become harder to achieve due to the intermittent nature of variable renewable energy. New strategies to optimally use the grid assets and energy resources are expected.

7.2.3 Voltage and frequency requirements

The first requirements related to VRE generators focused on requesting the same performance as conventional synchronous generators. Perhaps some of the first instances occurred when the German Affiliated Company Deutsche Verbundgesellschaft solicited that they should endure the first low-voltage ride-through (LVRT), followed by some more specific duties regarding voltage and frequency control under repeating fault sequences as shares of VRE became substantial. By then, it was clear to grid code originators that IBRs could safely integrate into conventional power systems and that their distinctive functionality could, in some

Frequency Hz:	Requirement upon feed–in from generating units for 110 kV Systems		Requirement upon feed–in from generating units for 220 kV Systems		Requirement upon feed–in from generating units for 480 kV Systems	
47.5–48	87–115% for t ≥ 10 min	At Network voltage ≤ 84.5% the generating units may disconnect from the network	87–115% for t ≥ 10 min	At Network voltage ≤ 84% the generating units may disconnect from the network	85.5–115.8% for t ≥ 10	At Network voltage ≤ 85.5% the generating units may disconnect from the network
48–48.5	87–115% for t ≥ 20 min		87–115% for t ≥ 20 min		85.5–115.8% for t ≥ 20	
48.5–49	87–115% for t ≥ 30 min		87–115% for t ≥ 30 min		85.5–115.8% for t ≥ 30	
49–49.5	90.1–111.8% for t ≥ 2 h		90.1–111.8% for t ≥ 2 h		92.1–110.5% for t ≥ 2 h	
49.5–50	90.1–111.8% for t ≥ 2 h		90.1–111.8% for t ≥ 2 h		92.1–110.5% for t ≥ 2 h	
50–50.5	90.1–111.8% for t ≥ 2 h		90.1–111.8% for t ≥ 2 h		92.1–110.5% for t ≥ 2 h	
50.5–51.5	87–115% for t ≥ 30 min		87–115% for t ≥ 30 min		85.5–115.8% for t ≥ 30	

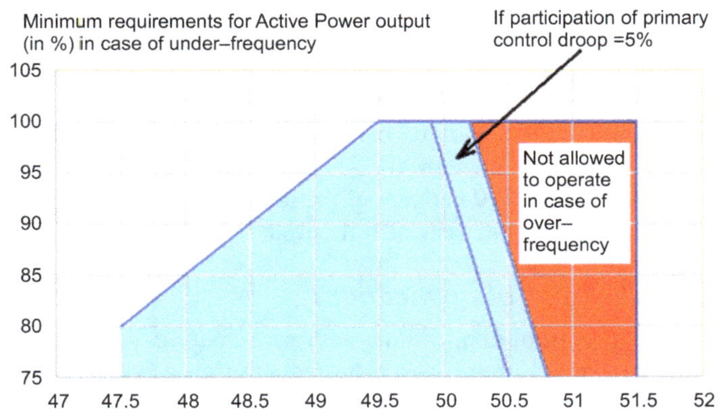

Figure 7.17 Voltage and frequency requirements from the German Transmission Code, from [23]

respects, transcend that of their synchronous counterparts. This was shown early in this century when LVRT ranges and after-fault active power restoration times were set by German and Danish grid codes for wind turbine IBRs, which would have been impossible for conventional synchronous generators to achieve [22–24].

The voltage and frequency requirements imposed by the German Grid Code for transmission system operators are shown in Figure 7.17 [23].

In the event of network disturbances caused by frequency or voltage deviations (over or under fluctuations), or following the resolution of an isolated event, automatic synchronisation of the various generators with the network is allowed only if the lowest of the phase-to-phase voltages at the grid connection point is higher than 105 kV in the 110 kV network, higher than 210 kV in the 220 kV network, and higher than 370 kV in the 380 kV network [23]. After the disturbance, the active power supplied to the network should not increase at more than 10 per cent of the network connection capacity per minute.

The added capabilities of the VRE technology introduced some conflicting issues to pursuing unbiased regulations regardless of technology. The grid codes handled this by requesting the best that every technology could offer to concur in a friendly working environment. This led to some more relaxed requirements for synchronous generators.

Since 2021, voltage and frequency have stood among the foremost issues in grid codes. This arises from the fact that all generators connected to the grid have to work harmoniously. Regarding voltage, it is important to establish the operating range. Generators are needed to operate continuously at nominal voltage under full load conditions. Typically, operation ranges consider voltages under-to-over nominal values for particular periods like those shown in Table 7.1 for the Danish Grid Code [24] and those illustrated in Figure 7.18 that the Nordic Grid Code [25] imposes for categories C (connection below 110 kV and capacity at or above the threshold identified by the individual TSOs) and D (connection above 110 kV and capacity at or above the threshold identified by the individual TSOs) wind power plants. These include frequency variations in the 49–51 Hz range and voltage tolerances between 90 and 105 per cent of rated voltage. A valid operation for 1 h within the frequency range of 49.7–50.3 Hz at voltages between 85 and 90 per cent or between 105 and 110 per cent of the rated voltage is allowed. An output power reduction of up to 10 per cent of full output is acceptable.

7.2.4 Frequency control applications

Regarding frequency control applications with participating VRE resources, most grid codes contemplate requirements on the reduction or a boost of active power following over-frequency or under-frequency events, respectively. The active power increase is most relevant for synchronous or VRE generators working underloaded, i.e. below their rated conditions or when energy storage resources are available. However, for fast frequency or inertial response support, some limited operation times are often considered for situations involving under-generation or over-rated generation.

According to the German Grid Code for high and extra high voltage [26], the offshore Zertifizierung Erzeugungsanlagen (EZA) Certification of Production Plants must have the technical capability to change the active power output within the reach of frequency restoration and replacement reserve using the parameters shown in Table 7.2.

Table 7.1 Requirements of the Danish Grid Code for Category D power plants, from [24]. Protective functions with associated operating settings and trip time must match the values in the table below.

Protective function	**Symbol**	**Setting**		**Trip time**		**Standard value**
Overvoltage (step 3)	$U_{>>>}$	$1.20 \cdot U_n$	V	0–100	ms	100 ms
Overvoltage (step 2)	$U_{>>}$	$1.15 \cdot U_n$	V	100–200	ms	200 ms
Overvoltage (step 1)	$U_{>}$	$1.10 \cdot U_n$	V	60	s	60 s
Undervoltagc (step 1)	$U_{<}$	$0.90 \cdot U_n$	V	10–60	s	10 s
Overfrequency	$f_{>}$	52	Hz	200	ms	200 ms
Underfrequency	$f_{<}$	47	Hz	200	ms	200 ms
Change of frequency	df/dt	±2.5	Hz/s	50–100	ms	80 ms

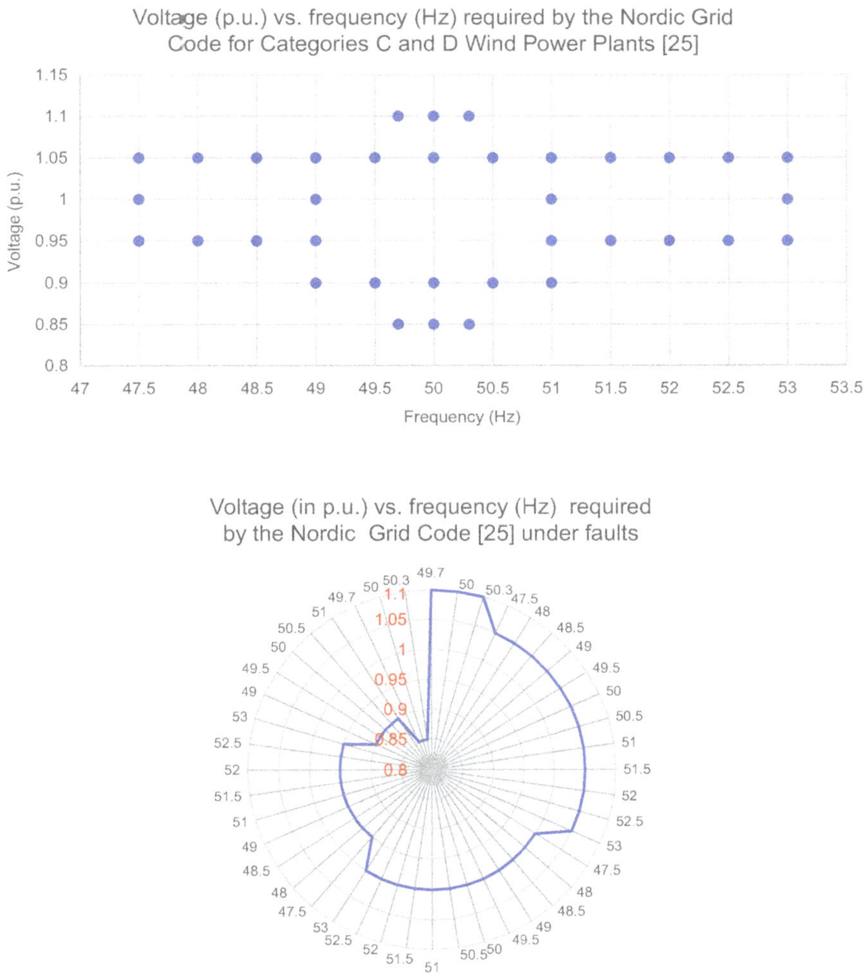

Figure 7.18 Voltage and frequency performance requirements for the Nordic countries, adapted in different arrangements from [25]. For categories C (connection below 110 kV and capacity at or above the threshold identified by the individual TSOs) and D (connection above 110 kV and capacity at or above the threshold identified by the individual TSOs) wind power plants.

The TSO provides these system functions, and the prerequisite for the offshore EZA is the capacity for signal-controlled active power adjustment.

Almost all grid codes impose reactive power provision and the ability to enforce diverse control strategies related to reactive power engagement on transmission, sub-transmission, and distribution systems for synchronous and IBR resources.

Table. 7.2 Adjustment required to the offshore EZA for active power output, adapted from [26]

Parameter:	Minimum control band (not simultaneous)	Max. activation delay	Time until full activation
Frequency restoration reserve	±10% PAV	30 s	2.0–7.5 min
Replacement reserve	±20% PAV	7.5 min	7.5–15 min

The activation time and the ability to supply reserves separately for upward and downward control are other essential characteristics for wind power operators.

The activation time enters these criteria because of the wind's intermittent nature. Wind generators may be unable to offer the needed reserves if the service is required for a lengthy period. Regulating upwards may result in less wind power output than is available. Furthermore, the ability to separate offers for downward and upward regulation allows wind operators to make more cost-effective provisions, as downward regulation costs are always lower than spot pricing [27].

Today's IBRs are almost all grid-following (GFL) devices, which sense the grid's voltage and frequency and inject current to supply the right amount of active and reactive power. The key premise underlying the design of GFL IBRs is that there are still enough synchronous generators on the grid to produce a somewhat robust and steady voltage and frequency signal that GFL IBRs can "follow". However, because GFL levels are rising, there will eventually be a limit to how far GFL controls can be implemented. Further sophisticated inverter controls, known as grid-forming (GFM), will be required to keep the system stable.

As clouds travel over a plant, they may impact the output of one section, leaving other sections free of obstruction. Therefore, clouds produce a variety of fluctuations in PV production within plants and between different plants. Therefore, output aggregation of many independent PV plants might result in a comparative decrease in ramps for the combined effect of several plants compared to a single plant.

According to [28], a sizable quantity of generation connected to a distribution system might significantly influence the BPS's dependability. The absence of disturbance tolerance, which calls for LVRT and fault ride-through (FRT) capacity, is especially concerning. The Distributed Energy Resources (DER) must trip offline when the grid frequency or voltage is already out of balance due to strict statutory trip restrictions and a lack of ride-through capacity. Multiple DER units can scale up to form a very large resource when seen collectively. For instance, the California Independent System Operator (CAISO) reported in 2016 that the BA had 4,900 MW of DER. The Public Utility Commission of California adopted CA rule 21 to address these issues in response to their concerns about DER. This rule identified and implemented many of the functions that are now part of IEEE 1547-2018 [29], and it required that all solar PV DER installed beginning on September 9, 2017, use smart inverters.

The increased participation of VRE in modern electric systems poses technical and operational challenges, including under and overvoltage issues, more frequent tap changes in distribution transformers, and higher power losses. However, one of the most critical challenges is the increased negative impact on stability as systems inertia decreases, brought about by the replacement of conventional synchronous generation by VRE resources.

Most grid codes have imposed the need for VRE resources connected to transmission, sub-transmission, and distribution systems to offer the required flexibility, given their harder predictability. This includes their capability to perform active and reactive power control, often fulfilled by enabling VRE remote setpoint adjustments, as well as improving power system flexibility through participation in automatic generation control (AGC) to provide reliability services. The anticipated active power control action requirements in grid codes are the minimum ramp rates for conventional synchronous generation and battery units and the maximum ramp rate requirements for VRE, which are mostly pertinent to start-up, shutdown, or setpoint adjustments with the appropriate kilowatt resolution – see Energinet Technical Regulation for Wind Power Plants 2016 [24]. These capabilities make up the PV units' task to bolster grid stability during events leading to overvoltages and over frequencies.

Droop control actions to reduce VRE curtailment are under consideration in the Hawaii solar power grid [21], although this approach can also be applied to wind plants. Figure 7.19 depicts the Danish regulation regarding frequency control, showing the upward and downward active power regulation as a function of frequency variations and the corresponding droops [24].

7.2.5 Power quality

Another relevant topic on VRE operation is power quality. Anything that impacts the integrity of a clean sinusoidal output or becomes a source for overvoltages, resonances, additional heating, maloperation of protecting devices, etc., is carefully reflected and regulated by operational standards, grid codes, guidelines, or recommended operational practices.

Power electronics are a source of harmonic distortion due to waveform frequency components, integer multiples of nominal frequency. This occurs in current and voltage signals during conversion processes like rectifying and inversion. Therefore, IBRs are subjected to due regulatory directives in grid codes to keep within limits the individual and total harmonic distortion at the point of common coupling (PCC) of individual plants. For example, Tables 7.3 and 7.4 show the

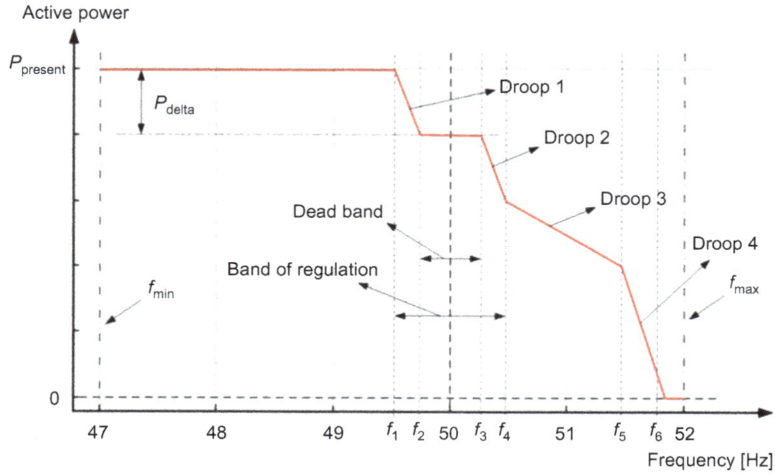

Figure 7.19 Frequency control curve for a PV power plant, from [24]. https://energinet.dk/regler/el/nettilslutning/

Table 7.3 Current harmonic distortion limits on the PV systems, adapted from [21]

Standards	Type	Harmonic Order (h)	Distortion Limit	THD (%)
IEEE 1547 AS 477.2 (Australia), GB/T (China), and ECM (Malaysia)	Odd	$33 < h$	<0.3%	5%
		$23 \leq h \leq 33$	<0.6%	
		$17 \leq h \leq 21$	<1.5%	
		$11 \leq h \leq 15$	<2%	
		$3 \leq h \leq 9$	<4%	
	Even	$10 \leq h \leq 32$	<0.3%	
		$2 \leq h \leq 8$	<1%	
UK (EREC G83 Stds.)	Odd	$h = 3$, 5 and 7	< (2.3, 1.14 and 0.77)%	<3%
		$h = 9$, 11, 13	< (0.4, 0.33 and 0.21)%	
		$11 \leq h \leq 15$	0.15%	
	Even	$h = 2$, 4 and 6	< (1.08, 0.43 and 0.3)%	
		$8 \leq h \leq 40$	<0.23%	
IEC 61000-3-2	Odd	$h = 3$, 5 and 7	< (3.45, 1.71 and 1.15)%	<5%
		$h = 9$, 11 and 13	< (0.6, 0.5 and 0.3)%	
		$15 \leq h \leq 39$	<0.225%	
	Even	$h = 2$, 4 and 6	< (1.6, 0.65 and 0.45)%	
		$8 \leq h \leq 40$	<0.345	

current and voltage distortion limits, respectively, allowed by different grid codes in the world [21].

Likewise, inter-harmonics, frequency components with non-integer multiples of nominal frequency, will likely become evident in applications where AC

Table 7.4 Voltage harmonic distortion levels on the PV systems, adapted from [21]

Standards	Voltage Bus	Max. Individual Harmonics	THD (%)
IEEE 519	$(V \leq 1)$ kV	5.0%	8.0%
	$(1 \leq V \leq 69)$ kV	3.0%	5.0%
	$(69 \leq V \leq 161)$ kV	1.5%	2.5%
	(>161) kV	1.0%	1.5%
IEC 61000-3-2	$(2.3 \leq V \leq 69)$ kV	3.0%	<5%
	$(69 \leq V \leq 161)$ kV	1.5%	2.5%
	(>161) kV	1.0%	1.5%

systems are interconnected through a DC link, as with VREs. These harmonics will likely excite flicker and initiate harmonic instabilities [29,30]. IEEE-519 [29] and IEEE-517 [29] Standards guide recommended inter-harmonics limits from inverters connected to the AC grid based on flicker excel measurements.

7.2.6 Low-voltage ride-through

Conventional synchronous generators are associated with inherent large fault currents. At the same time, VRE resources, due to their power being carried through power electronics, are characterised by smaller but more controlled short-circuit currents. These characteristics define the stiffness of conventional systems. They are also associated with the capability of conventional and VRE systems to manage active and reactive power at the point of common coupling (PCC). VRE output power to the AC grid impacts converter, cables, transformer, and switched reactive power losses, which are relevant elements in grid codes. Providing the proper overcurrent protection for VRE resources is an elaborate topic, as discussed in Chapter 11.

It is also important to mention that as the share of IBRs in the power systems increases, they become weaker since their Short Circuit Current decreases given the reduced fault current that IBRs can provide; this, in turn, lowers the Short Circuit Ratio, which is a direct function of short circuit current and the stiffness of the system can potentially be impacted, leaving it more disposed to perturbations. Although for all systems where renewable generation is a significant supplier, it is vital for weak systems that appropriate models for the representation of IBR controllers and angular instability of synchronous condensers, among others, are used.

Grid Codes are expected to enforce stiffer requirements as IBRs enter the grid to provide power-balancing capabilities. This is remarkable when referring to enough periods for TSOs to devise and apply remedial actions required under contingencies to restore operation to normal conditions. Also, all power electronic inverter operations must meet voltage sag immunity standards or guidelines. Voltage dips during fault conditions can create several nuisance-tripping problems. Therefore, voltage tolerance curves for wind farms are a relevant subject addressed by grid codes. Figure 7.20 portrays the grid voltage amplitude

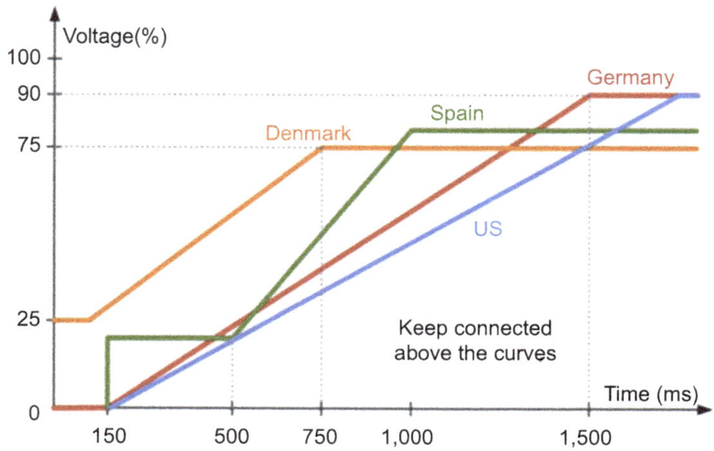

Figure 7.20 Voltage tolerance curves for wind farms under fault conditions, from [31]

under fault conditions versus the permitted times required in Denmark, Spain, Germany, and the USA [31]. Under fault conditions, wind generators can remain connected to the BPS for grid voltages falling above the curves. Still, they can be pulled out from the BPS for voltages falling below the curves for the indicated specific times.

In the case of symmetrical and asymmetrical events, that is, faults in three phases, two phases, or a single phase in the Danish system, the following rules must be followed: Area A: The wind power facility must sustain regular output and remain linked to the grid, Area B: The wind farm must maintain its grid connection but if several successive faults sequences occur, disconnection is allowed. This is illustrated in Figure 7.21. To assist in stabilising voltage within the design parameters made possible by the present wind power plant technology, the wind power plant must provide maximum voltage support by delivering an additional amount of regulated reactive current, as displayed in Figure 7.22, which depicts the Wind Turbine generator requirements under voltage sags created during fault conditions in the grid [32]. TSOs have the challenging task of specifying the delivery of reactive power from renewable resources and coordinating it with existing reactive power resources in the grid to optimise the grid response to voltage sag events. Disconnecting the wind farm is permitted in Area C of Figure 7.20 and Figure 7.21.

The microelectronics industry has created voltage sag ride-through capability in Standards SEMI 10 [33]. This Standard defines the SAG ride-through capability. Figure 7.23 depicts the SEMI voltage tolerance curve overlaid with the CBEMA curves [34]. Essentially, the equipment has to run through voltage sags down to 50 per cent of nominal for 200 ms (12 cycles).

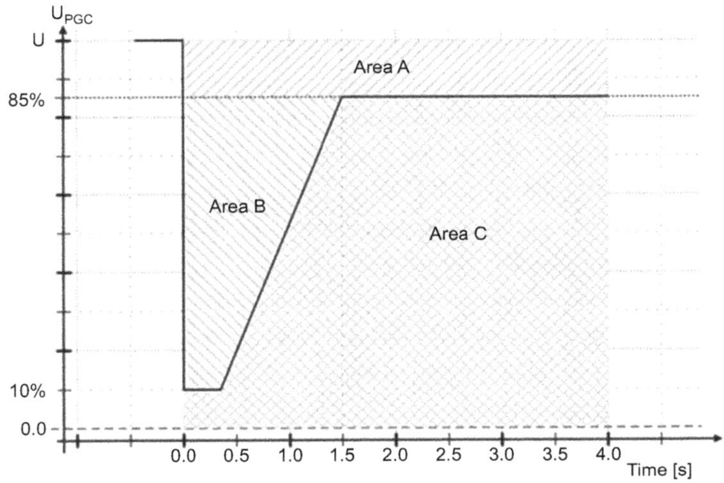

Figure 7.21 Voltage dip tolerance requirements for category C and D PV power plants, from [32]

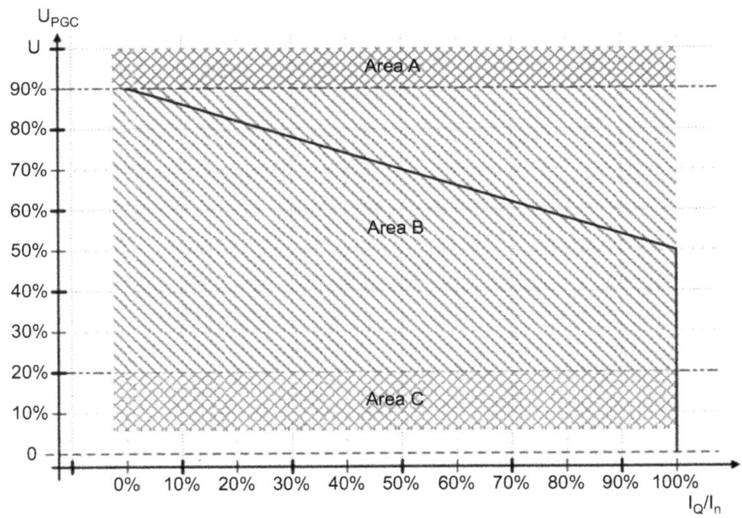

Figure 7.22 Requirements for the delivery of additional reactive current IQ during voltage dips for category C and D wind power plants, from [32]

Standards [35] and [36] are also part of the SEMI Standards. Standard F47-0200 specifies the required voltage sag tolerance for semiconductor fabrication equipment, and SEMI F42 explains the method for testing compliance with SEMI F47 [36].

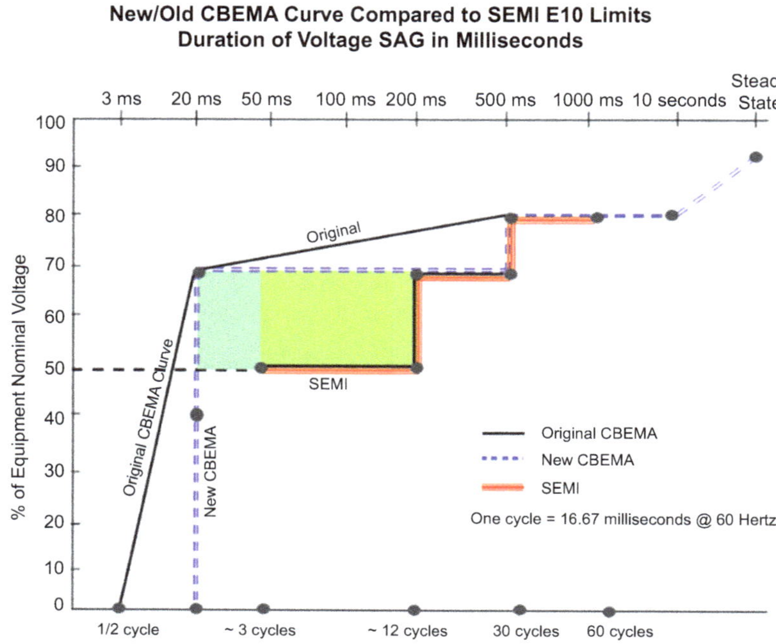

Figure 7.23 Overlay of SEMI curve superimposed on the CBEMA curves for semiconductor equipment, from [34]

7.3 Power electronics and control functions for IBRs

7.3.1 Introduction

With the rapid growth in electricity demand and the ageing of fossil fuel-based energy, VRE offers several favourable elements: (i) better cost-effectiveness, (ii) improvement of voltage profile, (iii) higher reliability, (iv) energy efficiency, active and reactive power control, and sustainable endurance. However, the fast-tracked integration of IBR poses an extraordinary challenge to the electric grid and power system protection planners and transmission and distribution system operators, to mention a few. Therefore, the IBRs and plant controls must offer a high degree of reliability during power system disturbances, or a limited number of them must be established to keep the power system stable and resilient.

7.3.2 High tech requirements

Variable renewable generation is created as variable DC voltage in PV generation and AC voltage in Wind generation. DC/DC converters are used in PV generation to maximise the DC power output per module. Wind generators use an AC/DC, or rectifying, and a DC/AC, or inversion, processes and some control features for coupling to the AC grid. This can include the Grid-Following Inverters (GFI) and GFM attributes when they are supplemented. Figure 7.24 sketches a high-level view of the structural design of a modern RES power generation system. Intelligent control functions are needed to ensure that power transfer is performed consistently with the established and expected operation planning targets, considering the variable nature of renewable energy resources [37].

The central control actions required for inverter-based resources to transfer renewable power to the electric AC power grid are depicted in Figure 7.25. The key features are reactive power management, dynamic support, and system monitoring. However, protection, communications, and virtual inertia are other significant components needed to achieve reliability in the whole arrangement. More insight into virtual inertia is afforded in the Inertia Loss and Synthetic Inertia section of this chapter.

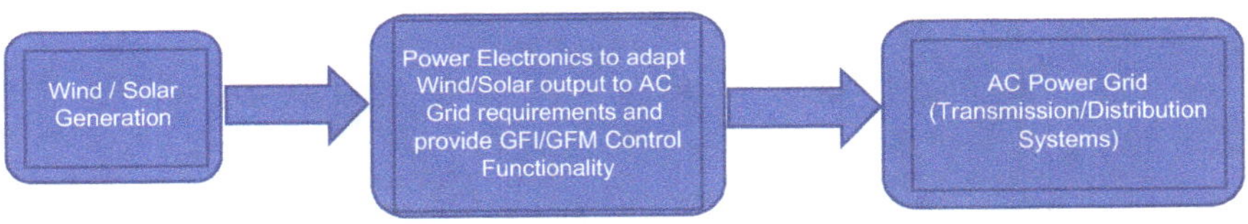

Figure 7.24 High-level view of RES integration

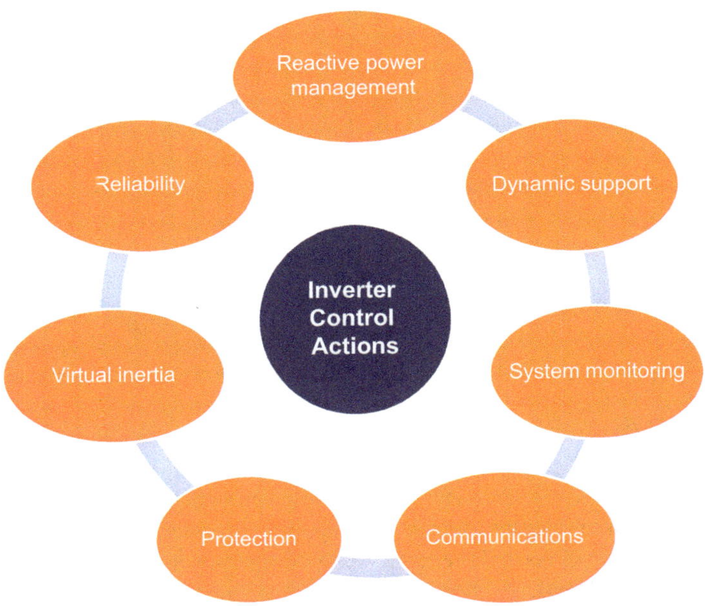

Figure 7.25 Variable Inverter-Based Resource (IBR) control actions

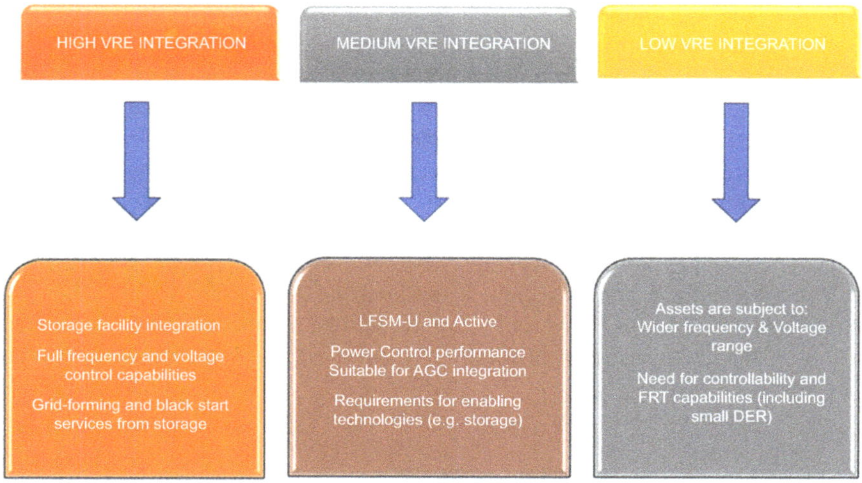

Figure 7.26 VRE technological requirements for Small Systems, adapted from [21]

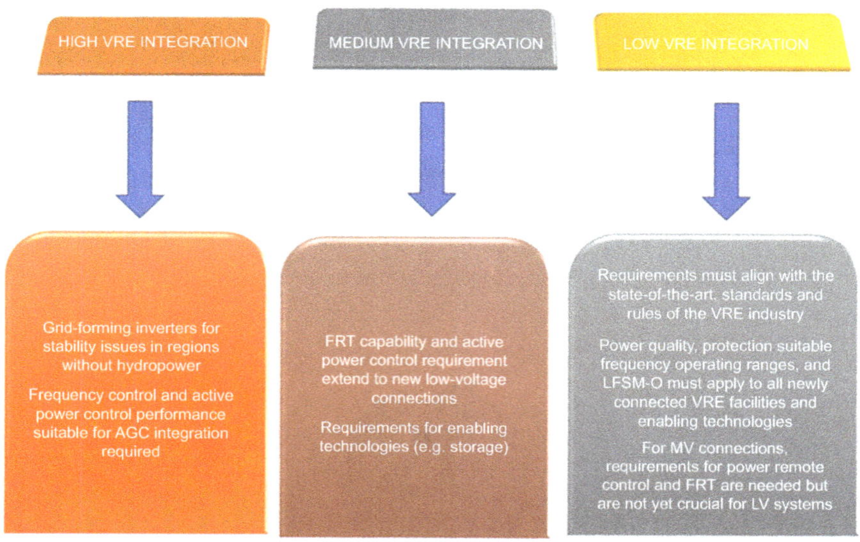

Figure 7.27 VRE technological requirements for Medium Systems, adapted from [21]

Based on the anticipated penetration of VRE generation and the size of the system, Figures 7.26–7.28 illustrate the various technological requirements necessary for power systems of different extents and levels of VRE penetration [21]. For all cases, LFSM-U stands for Limited Frequency Sensitive Mode-Underfrequency, a power generating module or HVDC system operating mode that will increase active power output in

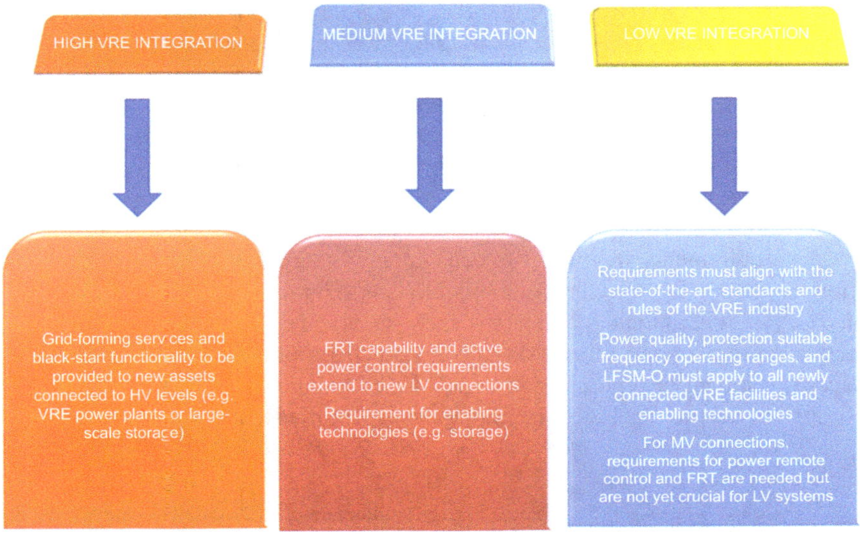

Figure 7.28 VRE technological requirements for Large Systems, adapted from [21]

response to a change in system frequency below a certain value. LFSM-O means Limited Frequency Sensitive Mode-Over frequency, used in the UK grid code and seeks automatic generator module action for frequencies above a certain threshold.

7.3.3 Power converter requirements

The generation, transmission, distribution, and delivery to end loads of electrical power involve a great deal of power conversion. The various kinds of generation assets generate electric power, which in most cases is transported long or very long distances to reach the final power consumption sites. Power is usually conducted at high voltages to diminish technical transmission and distribution losses since currents are decreased when transmitted at a higher voltage. This can bring losses, which vary by the square of the current, to moderated levels.

Power voltage is stepped up at generator sites to connect to transmission lines and stepped down at remote industrial and medium – or low-voltage customer installations, which are usually commercial developments and residential areas. However, an increasing share of renewable power, like PV solar and wind power, which are intermittent, is set to inject larger amounts of power into the grid at constant voltage, as required by grid code legislation.

Carrying out this power conditioning uses diverse power converters depending on the type of renewable energy involved.

Solar PV power requires a DC/DC converter to harness the voltage from the photovoltaic cells and then guide it to the inverter's DC bus, from where a second power conversion through a central DC/AC inverter is carried out to extract that power and inject it into the AC grid through suitable power transformers. Similarly, wind turbine generators produce a variable power that depends on the speed of the wind, which in turn depends on the atmospheric pressure.

Voltage source inverter (VSI) and current source inverter (CSI) inverters provide the required inverter capability to harness solar PV generation. However, VSI inverters are more commonly used in integrating this kind of renewable power generation because they provide good voltage regulation, keeping power quality indices within allowable limits.

The requirements of power dispatch operators essentially determine the control actions in their effort to maintain a reliable operation of the electric grid to ensure that relevant operational parameters such as voltage, frequency, and real and reactive power, among others, do not deviate from predefined thresholds established by national or regional grid codes. Considering the grid code requirements described in the previous section, among the capabilities that PV and Wind Turbine

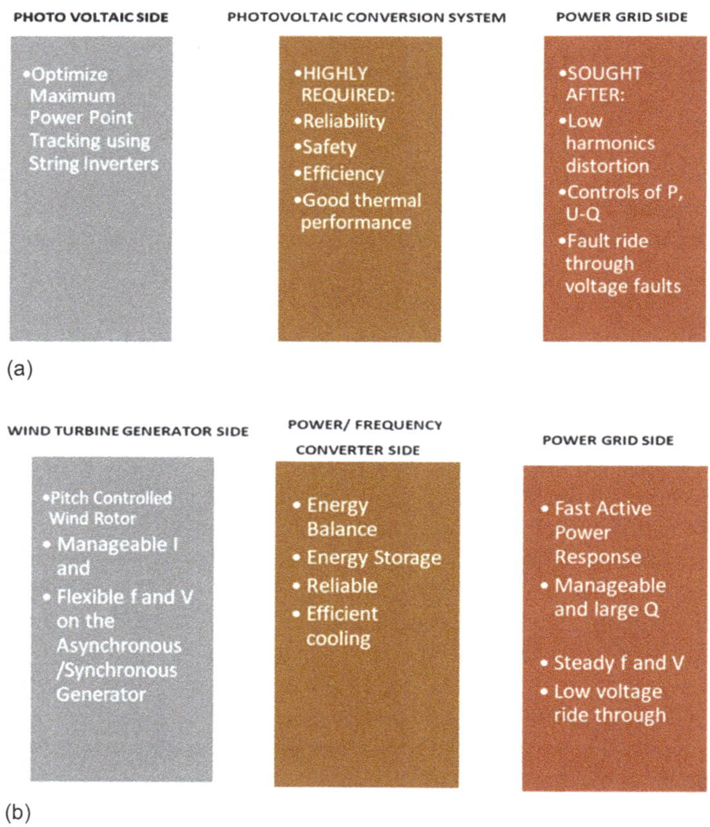

Figure 7.29 Requirements for renewable power, adapted from [31]. (a) For PV and (b) for wind turbine generators.

Systems must guarantee to ensure good performance under steady state and contingency conditions are those depicted in Figure 7.29 [31].

As described earlier, for system voltages above 1 kV, the output current's total harmonic distortion (THD) ought to be kept at values lower than 5 per cent at the grid side of PCC. The AC grid side also requires that under large PV penetration, inverters must ride through voltage faults and maintain the grid voltage in response to grid faults [38,39]. As illustrated in Figure 7.29, PV systems have the same requirements and must abide by the same grid standards as wind turbine systems [31].

Due to the PV technology's relatively low power capacity per producing unit and relatively high energy costs, there is a considerable need for high-efficiency power conversion to reach a reasonable cost per produced kWh. Aiming at this, PV inverters directly coupled to the AC grid without transformers are under development in order to increase conversion efficiency. Therefore, it is important to consider the consequent absence of isolation between the DC and AC circuits and consider all safety concerns [59].

Finally, dependability is crucial for power electronics-based PV systems, just like wind power conversion systems, because it increases total energy output (service time) and lowers energy costs. Also, PV converters must be more temperature insensitive due to exposure or a smaller housing chamber, which is advantageous for reliability and performance [31].

7.3.4 Control tasks to meet the requirements for renewable power integration

Integrating variable renewable energy into the electric grid also requires smooth synchronisation. This is achieved through the monitoring of phase, amplitude, and frequency in the AC grid, which can be attained with a Phase-Locked Loop (PLL) employing an algorithm that can deliver fast and accurate synchronisation and that can afford disturbance rejection. This will best contribute to a dynamic response suitable for wind and solar IBR integration, supporting overall power system stability, especially under LVRT conditions.

It is very important to understand the role of power electronics in the operational performance of electric power systems. This is especially significant due to the natural high-frequency switching that takes place in the operation of power electronic valves. Apart from creating harmonics, which are high-frequency current oscillations that impact the voltage waveform and the frequency at the source and load sides of the inverters, some events in the power system can create high-frequency oscillations. This may occur, for example, under switching, transformer energisation, or lightning events, which can lead to power frequency faults with ensuing voltage sags that can trigger IBR's Undervoltage protection, causing them to shut down.

Figure 7.30 displays some control tasks expected from the converters' power electronics to be executed at different levels in integrating variable renewable energy into the AC power grid [37].

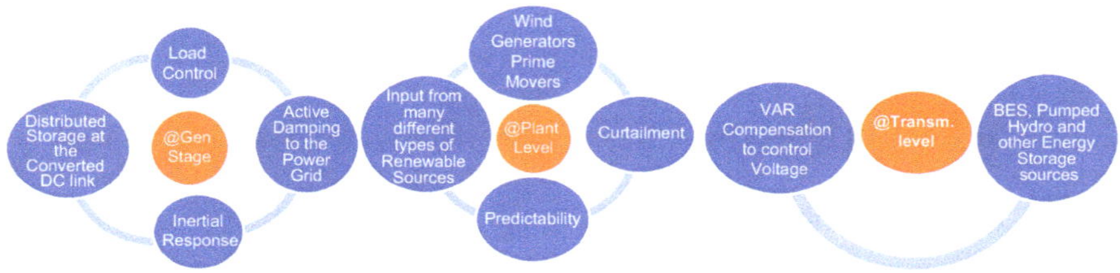

Figure 7.30 Expected control tasks to meet the requirements for renewable power integration, adapted from [37]

These routine tasks require using highly efficient and exceptionally reliable power electronics, which allow transmission and distribution operators to fine-control weather-dependent complex bursts of energy from wind and solar variable renewable resources to meet the specific customer's demand. Furthermore, energy transfer from these renewable technologies may require advanced ancillary services like grid support through reactive power provision and low-voltage ride-through [37].

Additional control actions include regulating electromagnetic torque in wind turbines to fully exploit the generated power and balancing energy under inertia mismatch events between mechanical and electrical power. This leads the converter to emulate the behaviour of conventional synchronous generation notwithstanding the speed of the wind and even contributes to grid fault recovery. In some circumstances, namely during a power mismatch between the wind farms and the grid, balancing and energy storage may become an expected requirement at generation wind plants.

Wind turbine systems must also comply with specific requirements imposed by TSOs to support power system events such as voltage sags during fault conditions. This emphasises the importance of power electronics technologies in fulfilling the requirements of power utilities.

7.3.5 Grid-following inverters (GFI)

Once converted into AC voltage, there are still several challenges for the integration of these renewable resources to be overcome: The voltage supplied at PCC to the AC power grid to deliver reliable power under dynamic demand conditions during the day, during the week, during weekends, and even during seasons of the year. Also, voltage sags during remote fault clearance can cause changes in the power frequency. The voltage variations occur within the limits allowed by the TSO.

This led to the development of the so-called Grid-Following or Grid-Feeding Inverters, which must adjust to the existing voltage and frequency in a well-regulated AC grid. They are essentially current source inverters with PLL controllers to extract the grid voltages' phase angle to synchronise. Inverters use voltage and frequency from the AC grid and can receive active and reactive power setpoints from the power plant controller. With this information, inverters are set to perform as controllable current sources synchronised with the grid to inject current with a unit power factor and keep track of changes in the grid voltage and frequency, acting therefore on a grid-following control action, providing the AC grid with the demanded amounts of active and reactive power. Grid-following inverters quickly respond to intermittences, enabling them to achieve smaller deviations from nominal power frequency.

7.3.6 Grid-forming (GFM) inverters

GFMs can lead to establishing the system voltage rather than relying on synchronous generators (SGs) to give a consistent voltage. GFM resembles a voltage source inverter where voltage and frequency are regulated via active and reactive power control. The information of the fundamental grid voltage frequency phasor at the PCC is not required for GFMs in an islanded system or a microgrid, in which case a GFM functions as a slack bus. The voltage value is independent of the active power control, and it is possible to adapt a GFM to contribute to voltage and frequency support.

Some control options designed to modify GFIs to behave like synchronous machines have recently been adapted [38,40–43].

Reactive power and system load influence the amplitude and frequency of the voltage. As a result, the output power is changed almost instantly for load balancing, local voltage management, and frequency control. For energy storage applications, a power conversion system is utilised to preserve the state of the battery charge and even boost power quality.

GFMs connecting wind and solar energy to the power grid can act to restore power to the grid following a blackout. This is a strategy known as a "black start".

After a blackout, electricity may be restored to the grid using Grid-forming inverters that connect wind and solar energy.

When installed, active and reactive powers of charging and discharging BESSs interact via the GFMs with the AC grid as power is absorbed or inserted into it. The converter performs this through a bidirectional DC/DC control scheme.

A comparison between electric systems based on synchronous generation and modern and future systems can be summarised as follows:

- Synchronous generators provide rotational inertia, which is well used to counterbalance system disturbances and load fluctuations.
- Innovative generation comprising wind and solar VRE is run by solid-state electronics Inverter-based Resources (IBR) in the conversion process since they lack rotating inertia and cannot support the connecting power grid under contingency conditions.
- Ways to afford the diminished short circuit capability from synchronous generators lost as VREs take over conventional generation must be found. Losing short circuit capability can translate into an increased instability risk during disturbances in the network. In other words, the weaker stiffness of the power grid with high shares of VRE has to be handled by adding counterbalancing measures. This can include synchronous condensers and stored energy, among others.
- Control capabilities of IBR technologies are the key attributes to ensure that they will be able to be incorporated into the electric grid without

compromising the stability, security, and efficiency of the system, considering their intrinsic qualities like:

- Lack of or insufficient power system inertia can potentially give rise to substantial frequency deviations, leading to power shutdowns or increasing the chance for load shedding.
- Variable in character due to weather dependency, IBRs are bound to uncertainty to the delivery of power to the AC grid, potentially producing power and voltage fluctuations,
- Enhanced supervision of power grid parameters, which leads to better and more accurate forecasting, control, and participation of VRE.
- Added control features, including virtual inertia (VI) enhanced frequency and demand response, provision of ancillary services, and support features for the AC grid.

7.3.7 Some noteworthy aspects of VRE integration

Energy extraction from renewable sources is prioritised in the RES management approach known as maximum power point tracking (MPPT). As RES penetration increases, specific grid restrictions have been implemented for these systems. The RESs should provide ancillary services like LVRT, reactive power control, and frequency management through active power control to provide a dependable and efficient power conversion in harnessing renewable energy [31].

The Distribution/Transmission System Operators have prioritised creating solutions to guarantee the RESs' dependability and integrate additional renewable energy. The progression of wind turbine size and electronics has also evolved enough so that wind turbine propellers can reach as much as 150 m in diameter, yielding 10 MW power output. This requires the development and utilisation of power electronics that can handle the increased power levels. Additional functionality implicates integrating proportional-integral (PI) and proportional-resonant (PR) controllers to properly handle grid synchronisation, DC bus stability, and current regulation [31].

As seen in Figure 7.31, sophisticated control capabilities of RESs are required for wind turbines (WTs) and are the same as those required for PV systems. These capabilities include operating through grid faults and delivering grid-support services. The turbine's rotational speed may be modified in the variable speed wind turbine idea to produce the most power possible using the available wind power. Typically, the current in the generator is varied by adjusting the generator side converter. To operate under a grid fault, coordinated management of many wind turbine subsystems, including the generator/grid side converters, brake chopper/crowbar, and pitch angle controller, is required [31].

Wind Turbines require power injection to cope with Low-Voltage-Ride-Through events. This can be managed in several ways, including constant active or reactive power control, unity power factor control, and positive and negative sequence control. However, the controlled power option can produce oscillations following the interaction between sequence currents and voltages created during faults in the electric grid.

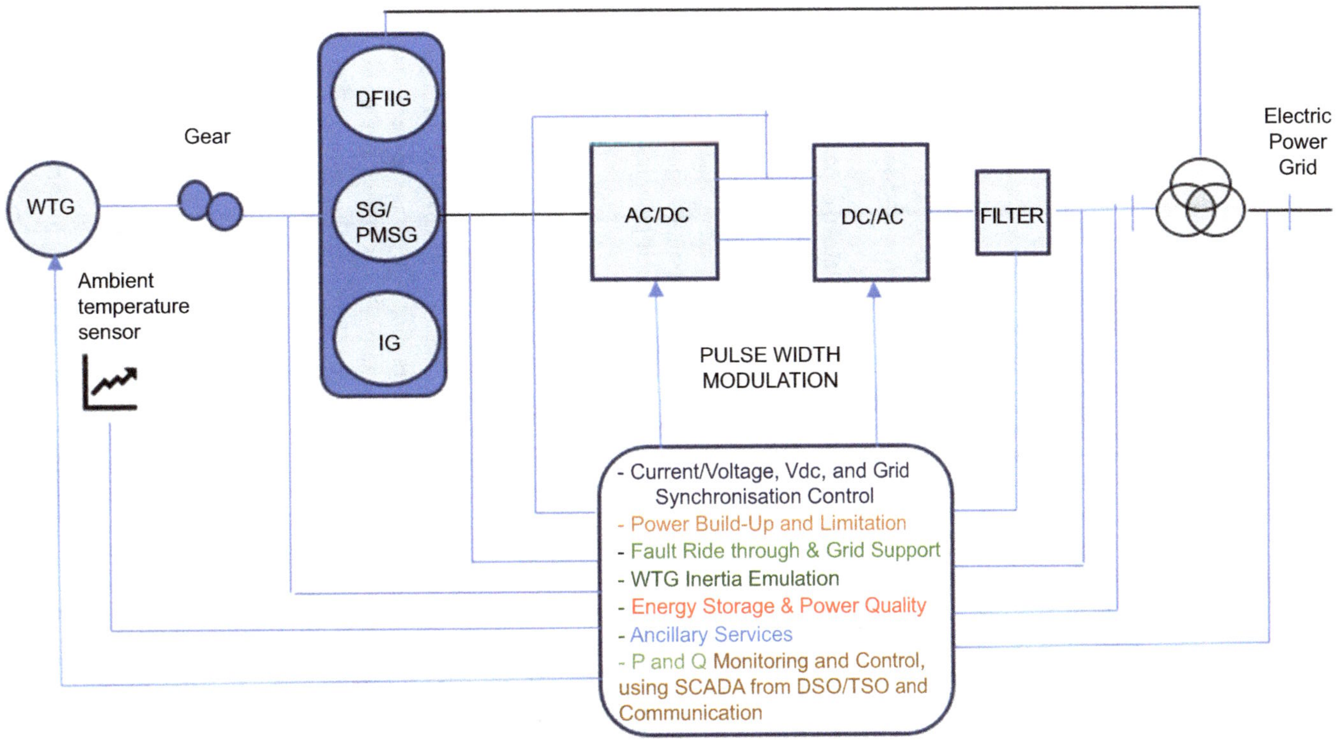

Figure 7.31 Advanced RES control functions, adapted from [31]

For single-phase PV systems, which exist in residential concentrations, although there is less control freedom for grid voltage and current during grid faults, anti-islanding protection is often required to protect them by interrupting supply to local loads in the event of grid faults. Nevertheless, with higher shares of PV generation, the need to boost reactive power injection to achieve LVRT builds on, and the right approach must be implemented, considering the over-current protection and specific inverter requirements for reactive power support.

Considering the fast-moving progress in the integration of megawatt-size wind and solar resources into the electric power grid, it is vital to acknowledge the need for increasingly reliable power converters and control functionality to meet the rigorous requirements of the power industry, which keeps devising new and more cost-effective control strategies to improve the coexistence of new IBRs with the waning synchronous generating technologies. Also important is to realise that the rising influx of distributed Renewable Generation systems is leaving behind the concept of centralised large generation plants and that systems protection, communication, and energy management issues need to be reassessed and approached accordingly.

In this context, the traditional studies to calculate the performance of the electric power system must be taken to a new stage where they consider the presence of elements that were not part of them, like the power IBRs that are the essence of all solar and wind integration along with all of their sophisticated control functions. This becomes increasingly important with rising shares of renewable power generation in the grid. NERC has already pinpointed the need for electromagnetic transient (EMT) studies in interconnection and planning studies of BPS-connected IBRs [44].

7.4 Inertia loss and synthetic inertia

7.4.1 Introduction

This section addresses key issues relating to the loss of inertia following a sudden increase in load or a significant loss of generation, as well as the needed action to rapidly inject active power into the power system. This is known as Fast Frequency Response (FFR), which prevents frequency from moving into the area of Underfrequency Load Shedding (UFLS) levels. The ensuing remarks are based on the insightful white paper put together by NERC, on the subject [44].

After all, neither wind nor solar generators can prevent the power system's inertia from decreasing, with larger shares of renewable energy in a typical power system powered by synchronous generation. Additionally, as the amount of renewable energy in the system increases, TSOs are figuring out ways to compensate for the loss of inertia, which indicates a definite desire to push the percentage of renewable output to its maximum level. This illustrates the increasing importance of power electronics in the renewable energy integration process.

PV systems connected to the AC grid are quickly becoming more advanced and may eventually supply a substantial amount of the energy needed in some areas. Thus, as PV systems grow in power capacity, laws governing PV systems are increasingly stringent. However, fast frequency response is essential for any power system to support stability when a large generation unit drops or after a sudden increase in loads occurs. Consequently, it is significant to recognise that renewable power will play an essential role in this process.

7.4.2 Fast frequency response

The BPS's inertia decreases as more inverter-based resources are incorporated, which will affect small interconnections. Therefore, we need to take suitable measures to maintain frequency stability. Fast frequency response is essential for reliable operation. Responsible entities should consider system characteristics when dealing with inverter-based resources.

Balancing generation and load is vital for reliable BPS. Three subjects are key to avoiding frequency control issues. These are usually referred to as Primary, Secondary, and Tertiary controls. Power System characteristics are important to consider when dealing with these resources.

In the event of an unexpected generation loss, it is important to be aware of the possible frequency decrease. The energy stored in the rotating masses of machines connected and synchronised with the grid will naturally be extracted, causing a decline in machines' speeds and resulting in a change in system frequency. To maintain reliability in the power system, it is essential to utilise Primary, Secondary, and Tertiary frequency controls and consider the characteristics of inverter-based resources.

As shown in the previous section, during a power imbalance, the system's frequency reaches the lowest point, which is best described as the NADIR. During the frequency arresting period and the subsequent recovery period, the synchronous generator's turbine-governors and the inverter-based resource active power-frequency control systems start responding to the deviation of system frequency by increasing their active power outputs. This is achieved through the use of Primary frequency response and secondary frequency controls, specifically automatic generation control (AGC).

Figure 7.32 depicts different frequency response provisions under contingency conditions. Typically, a high ROCOF (red curve) delays frequency response actions, risking the frequency falling into UFLS thresholds. It is important to assess the time taken to reach the UFLS threshold without responsive actions. This determines the timeframe for deploying adequate frequency response measures, as indicated in Figure 7.32.

The Primary Frequency Control (PFC) must act during the arresting period, as the frequency starts to drop owing to a contingency, such as a sudden generation loss or load increase, and heading toward the UFLS region (red curve). To compensate for the generation/load imbalance, the remaining generators use their mass inertia to generate active power to arrest the fall of frequency in what is commonly referred to as inertial response. Since this causes a speed reduction, the frequency starts falling. Power frequency response increases generation output as the net

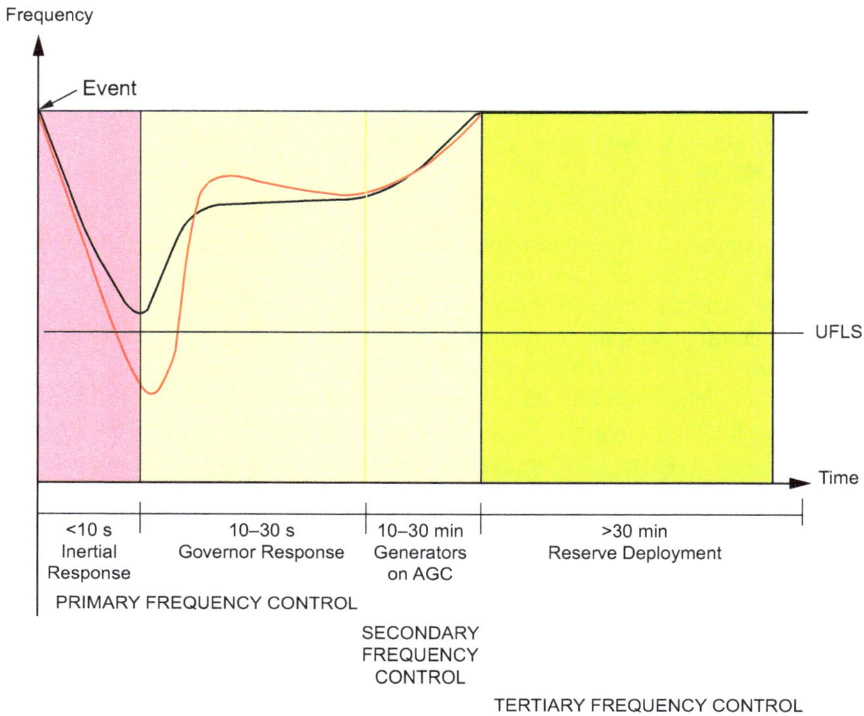

Figure 7.32 Frequency NADIR and subsequent fast frequency control actions

imbalance approaches zero; the frequency is arrested at its lowest frequency drop, or NADIR, to stay above the UFLS level (black curve) to prevent further generator loss effects.

Note that the Power Frequency Response actions increase the Generator's output and halt the continuous fall of frequency but do not return the frequency to pre-contingency values. This is ultimately carried out by AGC systems and by routine Scheduled Time Out's (STO's) pre-established actions to achieve an ensuing generation-demand balance.

Remaining wind generators and other IBR resources like PV and certain energy storage technologies can be used to provide power support to restore the imbalance for a rapid and sustained response. This is identified as a Fast-Frequency Response (FFR).

Certainly, RES do not participate in Primary Frequency Control due to their natural unpredictability. As with any other non-schedulable resource, they need to hold the active power supply as a function of frequency. For over-frequencies, a power decrease is required, and this could be manageable. However, given the uncertainty of resource availability, this could be unachievable the other way around, i.e., for under-frequency conditions, battery energy storage systems could be considered a resource to cope with this limitation.

PV and wind power are naturally non-dispatchable since they lack a continuous driving resource to deliver energy. However, when operating at maximum capacity, they function as deployable resources and can quickly modify their output level in response to the weather. For wind, this calls for decreasing the wind turbine's output, which can be accomplished by altering the blade pitch angle and lowering the quantity of energy captured from the wind. Regarding solar generation, power electronics can be used to command the inverter to limit output for solar generation before swiftly increasing it again. This action results in a trade-off between the volume of energy generated and the ability to respond to imbalances. As a result, the amount of generated power can be decreased.

When a massive contingency like the loss of a large generation unit occurs, the remaining generators' rotating inertia or kinetic energy is transformed into actual power to compensate for the supply/demand unbalance by the action of inertial response. This process slows down the generators, leading to a drop in frequency that eventually triggers the power frequency response (PFR) action, which prompts the increase of active power in the grid until power and demand come into balance. At this point, the frequency decay is arrested, and PFR helps restore the frequency. Fast Frequency Response is specifically significant in some interconnections where increases in ROCOF may potentially increase the risk of network instability.

7.4.3 Virtual inertia (VI) based inverters

The insertion of fast-responding RES generation inverters in the AC system can give rise to some instabilities in the AC system they are hooking up to, related primarily to frequency, voltage, and phase angles. This may happen more frequently in grid segments with low short circuit ratios. To address this, the VRE site can simulate a synchronous generator-like response by instrumenting suitable IBR control strategies to mimic the behaviour of synchronised spinning masses. This is called virtual inertia (VI) in technical terminology. This resource may not offer the same response as a traditional synchronous generator, so it is important to evaluate their response time [45]. Consequently, if this event occurs, the ROCOF will display steeper ramps [39].

The NADIR, or, as previously stated, the lowest point on frequency drop under a contingency, is calculated following a frequency change over 0.5 seconds following a sudden loss in generation.

A greater NADIR occurs during events such as faults or swift load fluctuations. Therefore, the ROCOF will show steeper ramps when this type of event occurs, and under-frequency relay protection may consequently be triggered to provide dynamic frequency support to grids with high wind and solar power penetration.

The reduction of frequency NADIR achieved during a contingency enabling virtual inertia is shown in Figure 7.33, which was adjusted from [46]. This is equivalent to providing an inertial response in the primary control frequency containment process, which is very important to achieve when integrating large shares of wind and solar energy into the AC electric grid.

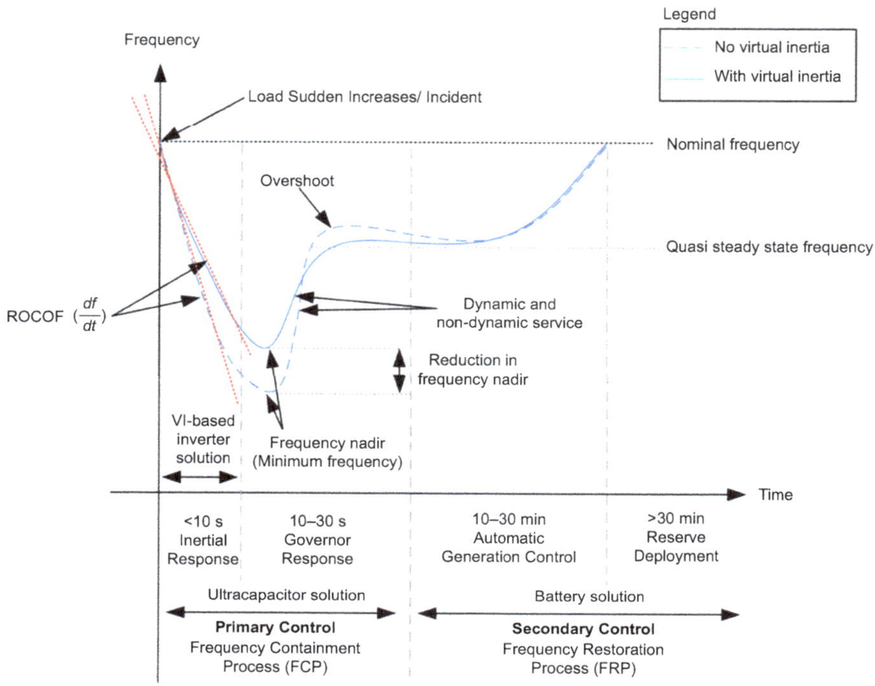

Figure 7.33 The effect of virtual inertia, adapted from [46]

7.5 Energy storage

7.5.1 Introduction

The interconnection of renewable resources to the AC electric grid requires that the operation of the electric system follow demand patterns. Under the operation of the systems with only synchronous generation, the systems must be ready to involve additional generation from gas, diesel, or carbon-fuelled plants to cover peak demands and, conversely, to shut or slow down the participation of those generation resources at times of low demand. This usually follows daily, weekly, and seasonal consumption patterns.

When renewable variable resources come into play, the process becomes more complicated due to their intermittent nature, which does not necessarily align with the dynamic electric demand fluctuations. It is, therefore, necessary to realise that these resources will also have to stand on and off when merging with the electric grid. An alternative to this is to store the energy that is produced and not brought into the grid so that it can be optimally leveraged.

It is noteworthy that one of the most widely used and well-established forms of energy storage technologies, pumped hydropower, currently accounts for 93% of all utility-scale energy storage capacity in the United States. It is projected that pumped storage hydropower will remain the most cost-effective solution (together with compressed air energy storage) for energy arbitrage and other applications involving extended charge/discharge cycles through 2025 [47].

Conceptually, a BESSs can be conceived as a power generator: A positive generator when it provides energy to the grid and a negative generator during the charging cycles. They can be used as energy resources with charging and discharging capabilities, like Pumped Storage Hydropower. Under low electricity demand, water is pumped from a lower reservoir to an upper reservoir. When demand increases towards peak values, water from the upper reservoir is released into the lower reservoir, flowing through a turbine and increasing power generation as needed as it descends. BESS can be conceived as going through similar energy storage charging and discharging cycles governed by daily solar irradiation levels and electricity demand [47]. Utility-grade battery storage can help during periods of high demand, store renewable energy during periods of low demand, or manage energy curtailments as determined by dispatch systems or STOs [48].

7.5.2 Current and potential battery-based energy storage trends to support the electric grid

The most developed grid codes, such as those in Europe and the United States, which have specific requirements for new users, mention storage and other consumer-producer connections. For some requirements, such as frequency, robustness, LVRT, voltage stability, and reactive power capacity, Belgium, for instance, treats storage as a generation asset. German regulations for low-voltage facilities for consumption, generation, storage, and EV charging are differentiated by DER connections [48].

The rising shares of inverter-based resources in electric power systems degrade their inertia and impose the need to find support for power frequency containment to maintain the power balance. This situation calls for converter-interfaced BESSs as a potential grid frequency regulation solution. The good news is that battery systems are expected to become cost-competitive for energy storage applications that require rapid cycles and quick responses. It has also been suggested that incorporating batteries could be a promising option for improving system frequency control capabilities in pumped storage hydropower plants [49].

On the other hand, other potential utility options like flywheel energy storage systems (FESS) have drawn considerable attention as pollution-free short or medium-term energy storage devices worldwide [50]. This technology is typically known for its long lifespan, high cycle efficiency, little pollution, and low maintenance costs, which makes it suitable for short-term applications. FESS is common in electric vehicles, space projects, military equipment, and power system applications. FESS is appropriate for load frequency regulation applications, in which the power frequency and power output are adjusted to match power system demand, ensuring system stability and reliability.

Superconducting Magnetic Energy Storage (SMES) Systems deserve to be mentioned as a potential option for utilities to store energy in the magnetic field of a coil maintained in a cryogenic environment below the critical temperature at which it exhibits superconductivity. At this condition, the magnetic field can flow indefinitely even after the voltage across the coil is eliminated [51]. The power conditioning system must generate a positive voltage across the coil to store energy during the charging phase and a reverse voltage that causes the coil to discharge in the discharging phase. SMES systems are capable of swift action. They are capable of switching between charge and discharge modes in seconds. Additional benefits include the absence of moving components and high cycling efficiency. In locations where other energy storage technologies are impractical, SMES systems can be used. SMES applications for renewable energy integration have gained attention in recent times. SMES is a viable and inexpensive choice for applications such as minimising power fluctuation, frequency control, transient stability enhancement, and power quality enhancements of grid-connected renewable energy systems, such as wind energy conversion and solar photovoltaic systems [51].

Figure 7.34 shows the yearly cost of various power storage technologies based on data from the US Department of Energy [52]. The results suggest that the Lithium Cadmium power source has the lowest annualised price among all power sources.

As illustrated in Figure 7.35, typical battery energy storage systems require thermal handling, thermal management, and thermal supervision of the power electronics supervision functionality, as well as rigorous monitoring and control of all relevant parameters.

Batteries can provide operational support during power system events to maintain safety and reliability in situations involving significant changes in generation or load, like generation tripping or load shedding. Energy Storage

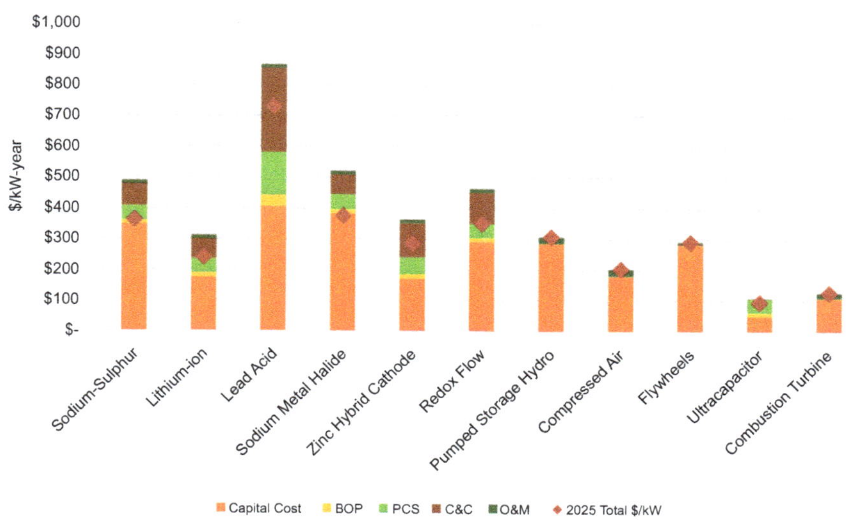

Figure 7.34 Annualised cost per kWh of all technologies, from [50]

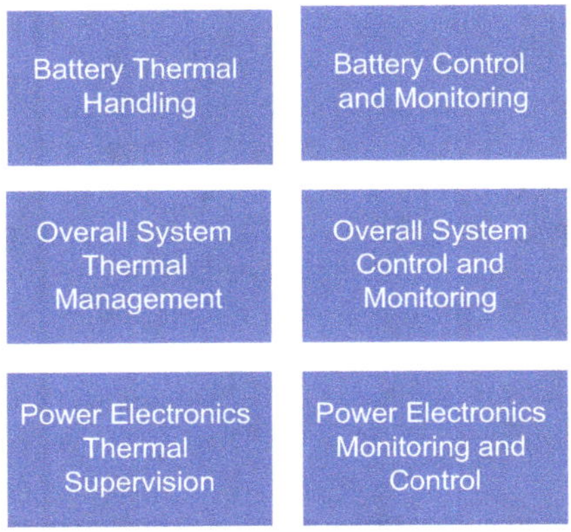

Figure 7.35 Typical BESS control functions

Resources (ESR) are intensively employed in ERCOT's Energy and Ancillary Service markets. Almost 68 per cent of ERCOT's Regulation Up and Responsive Reserve Service from Primary Frequency Response is provided by ESRs. ERCOT anticipates a further increase in the share of Ancillary Services from ESRs as the number of ESRs within its footprint rises [53].

According to the REF study by the National Renewable Energy Laboratory, approximately 140 GW of installed storage capacity would be needed to achieve a 90 per cent renewable energy scenario, including 48 per cent wind and solar energy [54].

At the highest VRE penetration levels, energy storage becomes an important player in enabling significant energy shifting and maintaining grid availability if the VRE is unavailable.

As VRE penetration rises, the amount of energy that must be stored and restricted due to overproduction at specific times of the day typically needs to increase. Another crucial issue is that to guarantee system stability, ESR will need to be afforded to provide the full spectrum of necessary grid reliability services [54].

In Australia, the Dalrymple Battery Energy Storage System (BESS), or ESCRI-SA, is a grid-forming BESS with microgrid automation consisting of a 30 MVA/8 MWh capability. It was put in place in 2018 on the lower Yorke Peninsula, close to the conclusion of a protracted 132 kV single-circuit radial feeder. The Dalrymple BESS, based on Virtual Synchronous Generator technology, is the first (and only) large-scale Grid-forming BESS connected to the Australian NEM (National Electricity Market). It provides synthetic inertia and high fault current to enable the operation of more renewable energy sources [55].

Additionally, the system offers services related to flexibility and dependability, including quick power injection, seamless islanding, and black starting of the local distribution network. When faults on the upstream feeder occur, the system automatically islands in coordination with the nearby Wattle Point Wind Farm, which has a capacity of 91 MW and distributed solar PV to maintain the operation of a local islanded power system and ensure the local customers' supply is never interrupted. The Dalrymple BESS is thus more than just an energy storage system; it is also the biggest autonomous microgrid in the world.

Results from the project and its operation have shown for the first time on the NEM how crucial grid-forming converters – as opposed to grid-following converters – can be in bolstering the grid and making it possible to meet ambitious renewable energy targets. In addition, unlike comparable power system support technologies like synchronous condensers, the Dalrymple BESS offers competitive market services to the NEM, generating a financial return for the operator.

Battery storage may help existing power generation units enable the black-start capability of decentralised generators via BESS support. Due to its high nominal power and low self-discharge, Lithium-Ion battery storage is a good fit for this application [55].

A comprehensive study in [56] assessed grid-forming and grid-following converter-interfaced battery energy storage systems. It revealed that the grid-forming control strategy outperforms its grid-following converter counterpart, achieving better frequency containment and a lower relative ROCOF, as demonstrated quantitatively by the results using appropriately defined frequency metrics. The study carried out quantitative simulations that showed that large-scale BESSs could significantly enhance system frequency containment, and the degree of improvement was proportionally related to the degree of frequency-power control gain.

The Finish Grid Code follows a Compliance Process for type D Grid Energy Storage Systems. Type D grids have a point's voltage level of at least 110 kV, and the grid energy storage system's rated capacity in production mode is at least 30 MW [57]. The planning stage includes compliance with: general data, technical data, operating voltage and frequency range, fault-ride-through capability, active power control and frequency control, house load and changes in production power, reactive power capacity, voltage, reactive power control (including voltage control step response calculation), protection settings of the power generating facility and impact on power quality, data required for dynamic modelling, real-time measurement data and instrumentation, specific study requirements, and grid energy storage system project's schedule and commissioning and statement of compliance [57].

Where IBR penetration is on the rise, GFM controls can be considered an option to offer grid-stabilising backup towards BPS's reliability. This array exploits the existing DC support, leading the task to be completed through user interface design that can lead to reduced costs to achieve stability improvements without the need for minimising much more expensive enhancements since GFM controls can be implemented on any type of IBR, including new solar photovoltaic and wind plants with some limitations [58].

Based on findings from small or island system studies, it has also been anticipated that as the total share of IBRs in the power systems nears 100 per cent, the required amount of GFMs to ensure system stability will be more related to the diminishing inertia constant rather than to the percentage of IBR connected generation. Preliminary estimates suggest that around 30 per cent of IBRs on the horizon need to be implemented with GFM abilities [58].

Using smart technologies and controls is crucial to making the most out of power electronics, battery storage [57], and FACTS technologies. These advancements are necessary to position our electric grids as reliable and resilient resources for optimal operation. It will take a team effort from different partners to achieve this goal. The electric utility industry, public organisations, private companies, renewable energy experts, integration specialists, manufacturing corporations, universities, and technology developers are all important in this work. Efforts are underway in various parts of the world and are thoroughly outlined in the relevant NERC publication [58].

Appendix A: List of acronyms

AC	Alternating Current
BA	Balancing Authority
BESS	Battery Energy Storage System
BPS	Bulk Power System
CAISO	California Independent System Operator
CBEMA	Computer Business Equipment Manufacturers Association
CSI	Current Source Inverter
DSO	Distribution System Operator
ENTSO-E	European Network of Transmission System Operators for Electricity
ERCOT	Electricity Reliability Council of Texas
ERO	Electric Reliability Organization
EU	European Union
EZA	Zertifizierung Erzeugungsanlagen - (EZA) Certification of Production Plants
FESS	Flywheel Energy Storage Systems
FERC	Federal Energy Regulatory Commission
FRT	Frequency Ride-Through
GFL	Grid-following
GFM	Grid-forming
IAM	Integrated Assessment Model
IESO	Independent Energy System Operator
IBR	Inverted-Based Resources
LVRT	Low-Voltage Ride-Through
MPPT	Maximum Power Point Tracking
NEM	National Electricity Market in Australia
NERC	National Electric Reliability Corporation
NPCC	Northeast Power Coordinating Council

PCC	Point of Common Coupling
PFR	Primary Frequency Response
PI	Proportional-Integral
PLL	Phase Locked Loop
PR	Proportional-Resonant
PV	Photo Voltaic
REF	Renewable Electricity Futures
RES	Renewable Energy Source
ROCOF	Rate of Change of Frequency
SEMI	Semiconductor Equipment and Materials International
TSO	Transmission System Operator
UFLS	Under Frequency Load Shedding
VR	Variable Resources
VRE	Variable Renewable Energy
VSI	Voltage Source Inverter
WECC	Western Electricity Coordinating Council

References

[1] Blaabjerg, F., Teodorescu, R., Liserre, M., and Timbus, V., 'Overview of control and grid synchronization for distributed power generation systems', *IEEE Trans. Ind. Electron.*, vol. 53, pp. 1398–1409, 2006.

[2] IEA *Global Energy Review* April 2021.

[3] IEA *Renewables 2022, Analysis and Forecast to 2027*. This article is available under the terms of the Creative Commons Attribution License (CC BY) 2022.

[4] IEA *Renewable Energy Market Update: Outlook for 2021 and 2022* (2021).

[5] Energy Efficiency & Renewable Energy, Office of Energy Efficiency & Renewable Energy, US Department of Energy, *Renewable Energy Resource Assessment Information for the United States*, Washington, DC, 2022, Available from March https://www.energy.gov/eere/renewable-energy [accessed 7 Nov 2023].

[6] International Energy Agency, *Renewable Energy Market Update, Outlook for 2022 and 2023*, IEA, Paris, 2022.

[7] U.S. Energy Information Administration, Annual Energy Outlook 2023. (https://www.eia.gov/outlooks/aeo/narrative/index.php#TheElectricityMix-inth) EIA, Washington, DC, March 2023 [accessed 3 Jul 2024].

[8] Publications Office of the European Union, *Communication from the Commission to the European Parliament, the Council, the European Economic and Social Committee, and the Committee of the Regions, EU Solar Energy Strategy*, Brussels, 18 May 2022, COM (2022) 221, final.

[9] International Energy Agency, *Africa Energy Outlook 2019, World Energy Special Report*, IEA, Paris, 2019.

[10] Cronin, J., Anandarajah, G., and Dessens, O., 'Climate change impacts on the energy system: a review of trends and gaps', *Clim. Change*, vol. 151, no. 2, pp. 79–93, 2018, doi:10.1007/s10584-018-2265-4. Available online from: https://link.springer.com/content/pdf/10.1007/s10584-018-2265-4.pdf[accessed 7 Nov 2023].

[11] Parkinson, S., and Djilali, N., 'Robust response to hydro-climatic change in electricity generation planning', *Clim. Change*, vol. 130, no. 4 pp. 475–489, 2015.

[12] de Lucena, A., Schaeffer, R., and Szklo, A., 'Least-cost adaptation options for global climate change impacts on the Brazilian electric power system', *Global Environ. Change*, vol. 20, no. 2, pp. 342–350, 2010.

[13] Seljom P., Rosenberg, E., Fidje, A., *et al.*, 'Modelling the effects of climate change on the energy system—a case study of Norway', *Energy Policy*, vol. 39, no. 11, pp. 7310–7321, 2011.

[14] The World Bank, funded by ESMAP, *Global Solar Atlas 2.0 Technical Report*. Washington, DC, 2019.

[15] The World Bank, funded by ESMAP, *Global Wind Atlas, Version v3.3*. Wahington, DC, 2023.

[16] The World Bank, funded by ESMAP, *Global Wind Power Density Potential*, Washington, DC, 2023.

[17] Mazzeo, D., Baglivo, C., Matera, N., De Luca, P., Congedo, P., and Oliveti, G., *Energy and Economic Dataset of the Worldwide Optimal Photovoltaic-Wind Hybrid Renewable Energy Systems*, Elsevier, Amsterdam, Data in Brief 33, 2020. This article is available under the terms of the Creative Commons Attribution License (CC BY). Available online from: Energy and economic dataset of the worldwide optimal photovoltaic-wind hybrid renewable energy systems (sciencedirectassets.com) [accessed 3 Jul 2024].

[18] US Agency for International Development, 'Grid integration series: variable renewable energy forecasting', *Scaling Up Renewable Energy Project*, Prepared for Energy and Infrastructure Office U.S. Agency for International Development, submitted by: Tetra Tech ES, Inc., Washington, DC, 2020.

[19] Pearre, N., and Swan, L., 'Spatial and geographic heterogeneity of wind turbine farms for temporally decoupled power output', *Energy*, vol. 145, pp. 417–429, 2018.

[20] Roberts, C., 'Ernest Orlando Lawrence Berkely National Laboratory', *Review of International Grid Codes*, Berkeley, CA, 2018, LBNL-2001104.

[21] IRENA, *Grid Codes for Renewable Powered Systems*, International Renewable Energy Agency, Abu Dhabi, 2022.

[22] Troester, E., 'New German grid codes for connecting PV systems to the MV power grid', 2nd International Workshop on Concentrating PV Power Plants, Optical, Design, Production, Grid Connection, Germany, 2008.

[23] European Network of Transmission System Operators for Electricity, *Network Code for Requirements for Grid Connection Applicable to All*

Generators Requirements in the Context of Present Practices, Working Document, 26 June 2012.
[24] Energinet/DK, *Technical Regulation 3.2.2 for PV Power Plants above 11 kW*, Published UK Edition, 14 July 2016.
[25] Nordel, Nordic Grid Code *2007 (Nordic Collection of Rules)*, Jan. 2007.
[26] TenneT TSO GmbH, *Grid Connection Requirements – High and Extra-High Voltage*, Germany, June 2023.
[27] Ivanova, A., Domínguez-García, J., and Corcher, C., 'Frequency support markets and wind power integration', *Energies*, vol. 14, p. 7450, 2021, https://doi.org/10.3390/en14217450 [accessed 6 Nov 2023].
[28] NERC, *A Report by the Integration of Variable Generation Task Force (Task 1-7), Performance of Distributed Energy Resources During and after System Disturbance Voltage and Frequency Ride-Through Requirements*, USA, Dec 2013.
[29] IEEE, *IEEE Standard for Interconnection and Interoperability of Distributed Energy Resources with Associated Electric Power Systems Interfaces*. IEEE Std. 1547-2018, April 2018. [Online]. Available: IEEE Standards Coordinating Committee 21 (SCC21) – IEEE Std 1547-2018 (Revision of IEEE Std 1547-2003) [accessed 6 Nov 2023].
[30] IEEE 519-1992, *Recommended Practices and Requirements for Harmonic Control in Electric Power Systems*, USA, 1993.
[31] M. M. Mahmoud, B. S. Atia, Y. M. Esmail, *et al.*, 'Evaluation and comparison of different methods for improving fault ride-through capability in grid-tied permanent magnet synchronous wind generators', *Int. Trans. Electrical Energy Syst.*, vol. 2023, Article ID 7717070, 2023.
[32] Energinet, *Technical Regulation 3.2.5 for Wind Power Plants with a Power Output Greater than 11 kW*, Denmark, July 2016.
[33] SEMI E10, *Specifications for Definition and Measurement of Equipment Reliability, Availability, and Maintainability (RAM) and Utilization*, 2022 edn, California, USA, 1999.
[34] Pacific Gas & Electric, Voltage Tolerance Boundary, California, USA, 1999.
[35] *Semiconductor Equipment and Materials International, Specification for Semiconductor Processing Equipment Voltage Sag Immunity*, SEMI F47-0200 Standard.
[36] 1999 Semiconductor Equipment and Materials International, *Test Method for Semiconductor Processing Equipment Voltage Sag Immunity*, SEMI F42-0600 Standard, 1999.
[37] Muljadi, E., 'Power electronics roles in renewable energy generation: challenges and opportunities panel session: challenges/issues for smart grid', *National Renewable Energy Laboratory, IEEE PES General Meeting*, July 17–21, 2016, Boston, MA, USA.
[38] Li, B., and Zhou, L., 'Power decoupling method based on the diagonal compensating matrix for VSG-controlled parallel inverters in the microgrid', *Energies*, vol. 10, no. 12, p. 2159, 2017.

[39] Johnson, S., Rhodes, J., and D., Webber, M., 'Understanding the impact of non-synchronous wind and solar generation on grid stability and identifying mitigation pathways', *Appl. Energy* vol. 262, p. 114492, 2020.

[40] Driesen, J., and Visscher, K., 'Virtual synchronous generators', 2008 IEEE Power and Energy Society General Meeting – Conversion and Delivery of Electrical Energy in the 21st Century, 2008, pp. 1–3.

[41] Liu, J., Miura, Y, and Ise, T., 'Comparison of dynamic characteristics between virtual synchronous generator and droop control in inverter-based distributed generators. *IEEE Trans. Power Electron.*, vol. 31, no. 5, pp. 3600–3611, 2016.

[42] Beck, H., and Hesse, R., 'Virtual synchronous machine', 9th International Conference on Electrical Power Quality and Utilisation, 2007, pp. 1–6.

[43] Yu, M, Roscoe, A. Booth, C, *et al.*, 'Use of an inertia-less Virtual Synchronous Machine within future power networks with high penetrations of converters', *Power Systems Computation Conference, PSCC*, 2016, pp. 1–7.

[44] NERC, *Reliability Guideline: Improvements to Interconnection Requirements for BPS-Connected Inverter-Based Resources*, USA, 2019.

[45] Miller N, Lew D, and Piwko R., 'Technology capabilities for fast frequency response', *Tech. Rep.*, GE Energy Consulting, Schenectady, NY, 2017.

[46] Yap, K.Y., Sarimuthu, Ch., and Mun-Yee, L.J., 'Virtual inertia-based inverters for mitigating frequency instability in grid-connected renewable energy system: a review', *Appl. Sci.*, vol. 9, Article No. 5300, 2019. This article is available under the terms of the Creative Commons Attribution License (CC BY).

[47] DOE, *U.S. Hydropower Market Report*, Office of Energy Efficiency & Renewable Energy, Water Power Technologies Office, Washington, DC, 2021.

[48] International Renewable Energy Agency, *Grid Codes for Renewable Powered Systems*, IRENA, Abu Dhabi, 2022.

[49] Pérez-Díaz, J., Lafoz, M., and Burke, F., 'Integration of fast acting energy storage systems in existing pumped-storage power plants to enhance the system's frequency control', *WIREs Energy Environ.*, vol. 9, no. 2, p. e367, 2019.

[50] Li, X. and Palazzolo, A., 'A review of flywheel energy storage systems: state of the art and opportunities', *J. Energy Storage*, vol. 46, p. 103576, 2022.

[51] Adetokun, B, Oghorada, O, and Abubakar, S., 'Superconducting magnetic energy storage systems: prospects and challenges for renewable energy applications', *J. Energy Storage*, vol. 55, p. 105663, 2022.

[52] Mongird, K, Fotedar, V, Viswanathan, V, *et al.*, 'Energy storage technology and cost characterization report', *Hydro Wires*, US Department of Energy, Washington, DC, 2019.

[53] Mago, N., 'ERCOT'S operational experience with battery energy storage', *G ST/ESIG Webinar Series*, 5 April 2023.

[54] Kroposki, B., 'Integrating high levels of variable renewable energy into electric power systems', *J. Modern Power Systems Clean Energy*, vol. 5, no. 6, pp. 831–837, 2017.

[55] Cherevatskiy, S., Sproul, S., Zabihi, S., *et al.*, 'Grid forming energy storage system addresses challenges of grids with high penetration of renewables: a case study', *CIGRE Session 48*, C2-C6-322, Paris, 2020.
[56] Zuo, Y, Yuan, Z., Sossan, F., Zecchino, A., Cherkauoi, R., and Paolone, M., 'Performance assessment of grid-forming and grid-following converter-interfaced battery energy storage systems on frequency regulation in low-inertia power grids', *J. Sustainable Energy, Grids Netw.*, vol. 27, p. 100496, 2021.
[57] FINGRID, *Grid Code Specifications for Grid Energy Storage Systems, SJV2019*, Helsinki, Finland, Feb. 2020.
[58] NERC, *White Paper: Grid Forming Functional Specifications for BPS-Connected Battery Energy Storage Systems*, Atlanta, Georgia, Sep. 2023.
[59] Blaabjerg, F., Ma, K., and Yang, Y., 'Power electronics for renewable energy systems – status and trends', *CIPS 2014*, Nuremberg/Germany, Feb. 2014.

Chapter 8

Enhanced dynamic and transient operation in power transmission using VSC-HVDC links

Luis M. Castro[1]

8.1 Introduction

The frequency control assignment in large-scale electrical power systems is one of the most arduous tasks to be carried out by a Transmission System Operator (TSO) at an energy control centre. The exercise carries a great deal of uncertainty because the energy demanded by most users changes randomly in the short, medium, and long-term time spans. It involves domestic, commercial, and industrial consumers. This is compounded by the nature of the electrical power grid, which comprises hundreds of thousands of components and pieces of equipment that are expected to work as a cohesive whole. However, it is susceptible to failures in one or several of its parts due to internal and external stresses, such as adverse weather conditions, human errors in operation, and even sabotage. Internal stresses may include continuous operation with distorted voltage and current waveforms produced by large power converters and saturated power transformers, and spikes produced by the switching of capacitor reactor banks and transformers. More recently, incorporating rather large blocks of renewable power generation connected to the grid through power electronic inverters introduces a random behaviour to the generation dispatch required to meet the electrical demand.

Due to the time-varying and random nature of demand and generation, both planning and operation of power networks become an increasingly challenging undertaking, particularly with the integration of large blocks of onshore and offshore wind power plants and the many megawatt-level solar PV plants. Besides, the power grid is not in a steady state but in a perpetual transient state. TSOs would hope for small changes to take place, but this is not warranted because network disturbances caused by fortuitous events can cause sudden changes in the operating state of the system, affecting nodal voltages, line currents, active and reactive power flows, and electrical frequency. At best, power system faults cause transient power outages, temporarily affecting the continuity of electricity service to users. In worst-case scenarios, critical contingencies can damage electrical equipment or

[1]Department of Electrical Energy, National Autonomous University of México (UNAM), México

even compromise the stability of the system, leading to wide-area blackouts, time-consuming restorations and important social and economic losses.

It goes without saying that power system dynamic assessments covering a wide range of abnormal operating conditions are key to the well-being of the electrical power grid. The electricity supply industry has a wide range of simulation tools with which comprehensive studies of power systems can be carried out, but the electrical power grid is becoming larger and more complex, incorporating new technologies which may not yet be fully covered in the existing commercial software used by power utility planners and operators. This presents an opportunity to carry out timely research and development in academic institutions and industrial centres.

This chapter addresses the impact of power imbalances on the dynamic and transient operation of the electrical power system and the positive impact that state-of-the-art power electronics equipment, methods and software models may have in deterring the worst effects of power imbalances. The material presented in this Chapter takes a research-oriented approach, which is based on recent original work published by the author [1]. The state-of-the-art equipment is HVDC systems using voltage source converters (VSC). The point-to-point VS-HVDC link is addressed first and then moves onto the multiterminal VDC-HVDC system. The former case has immediate practical application since, as discussed in Chapter 2, there are many VSC-HVDC links which have been installed around the world over the last two decades. The latter case looks toward the future as multiterminal VSC-HVDC systems forming DC grids currently exist only at the prototype stage (at the present time – see Chapter 2). The kind of power imbalances addressed are the demand-generation mismatches, leading to slowly varying transient responses. This is with reference to those taking place in large power system areas interconnected by tie-lines. The so-called Automatic Load Frequency Control (ALFC), which is a classical control method in inter-area power systems, is very relevant to this study.

First, the reader is presented with a concise review of ALFC. This comprises the primary and secondary frequency controls available in conventional generating power plants using synchronous generators. Key characteristics of the ALFC response of single and multi-area AC power systems are highlighted. These points are addressed with reference to a simplified dynamic modelling framework characterised by its linear models and responses, which is useful for performing ALFC simulations involving slow electromechanical transients. Next, it is shown that by replacing the traditional AC tie line interconnecting a two-area network with two-terminal VSC-HVDC links, the power systems dynamic operation shows a much-improved response; power oscillations are mitigated due to the fast action of the VSC station controls. It is also emphasised that the system flexibility increases because one of the fundamental tasks of the secondary control is carried out locally by the VSC-HVDC link, i.e., the tie-line power flow control is exerted by the VSC converters. Multiterminal VSC-HVDC systems are also studied in this chapter within the context of ALFC analyses. The fundamental control schemes of VSC stations are discussed, i.e., grid-forming, grid-following, and fixed-power control schemes. Contingencies related to large power imbalances in the AC systems, outage of VSC stations, and DC grid loads are discussed using ALFC dynamic simulations. It is established how VSC control schemes govern the dynamic operation of the hybrid AC/DC power grid, particularly those of the interconnected AC systems. In summary, this chapter focuses on enhancing the dynamic and transient operation of power systems using VSC-HVDC links as tie-lines.

8.2 One-area and two-area AC systems

The frequency control of electric power systems is perhaps the most important assignment carried out by TSOs in the control centre. This is a real-time operation problem where the energy balance between consumption and generation must be continuously guaranteed. The ever-increasing electrical energy requirements motivated by population growth and the sophistication of human activities make matters worse for the already complex task of achieving the energy balance in the power system. This task is further complicated when renewable intermittent generating plants, such as wind and solar energy resources, are involved in expanding and improving the power generation matrix. These aspects lay new challenges to TSO's doors in matters related to frequency control. From the system operation point of view, the more complex patterns of generation and demand, together with the unforeseen contingencies that occur in the transmission grid, can lead to frequency variations difficult to mitigate when the power network is more vulnerable, e.g., during high-demand periods, low-inertia operating conditions, etc. Poor system frequency control may result during outages of critical tie lines in interconnected power systems or areas of a large power system. This may result in the disconnection of zones with high power demand. In worst-case scenarios, these are often the starting point for the power system to experience partial or total blackouts, leading to great economic losses and socio-political disquiet.

When addressing the equilibrium issue between demand and generation in electrical power networks, the key measure of success is the frequency behaviour. It is well known that even relatively small frequency deviations, in the range of 1%–2% of the nominal operating frequency (f_s = 50 Hz or f_s = 60 Hz), can jeopardise power system stability. This is critical in power systems dominated by thermal power plants, as these are more sensitive to changes in the mechanical speed of their turbine rotors, affecting the ongoing internal thermal balance processes inside the turbines. If these power plants were to be disconnected from their primary protections, ensuing cascading failures may occur, putting further strain on the power system and its stability. To overcome this situation, there are mechanisms designed to help grid-connected power plants to remain connected and help to maintain system frequency. These are divided into primary and secondary frequency control mechanisms [2]. Both mechanisms, which complement each other, are generally available in large power plants driven by thermal and hydro turbines [2,3]. Integrating large blocks of renewable power resources into electrical networks, particularly wind and solar power, is a reality. Hence, these power plants must now participate in frequency control provision, as mandated by most grid codes worldwide. This issue has been addressed in Chapter 7. Accordingly, closed-loop frequency controls like those employed in conventional power plants have been developed for photovoltaic and wind power plants, mainly aimed at primary frequency control. These control schemes often incorporate supplementary control loops that emulate a virtual inertia response to mimic, as much as practicable, the behaviour of conventional power plants, as discussed in Chapter 7.

8.2.1 Fundamentals of primary and secondary frequency control

Conventional thermal and hydropower plants employ three-phase synchronous generators (SG). The voltage/reactive power and active power/frequency regulations are carried out by two fundamental control loops, namely, the AVR (Automatic Voltage Regulator) and the ALFC. These are schematically shown in Figure 8.1 for the case of a power plant with a hydro turbine and a power plant with a three-stage reheating steam turbine. The control loop responsible for regulating

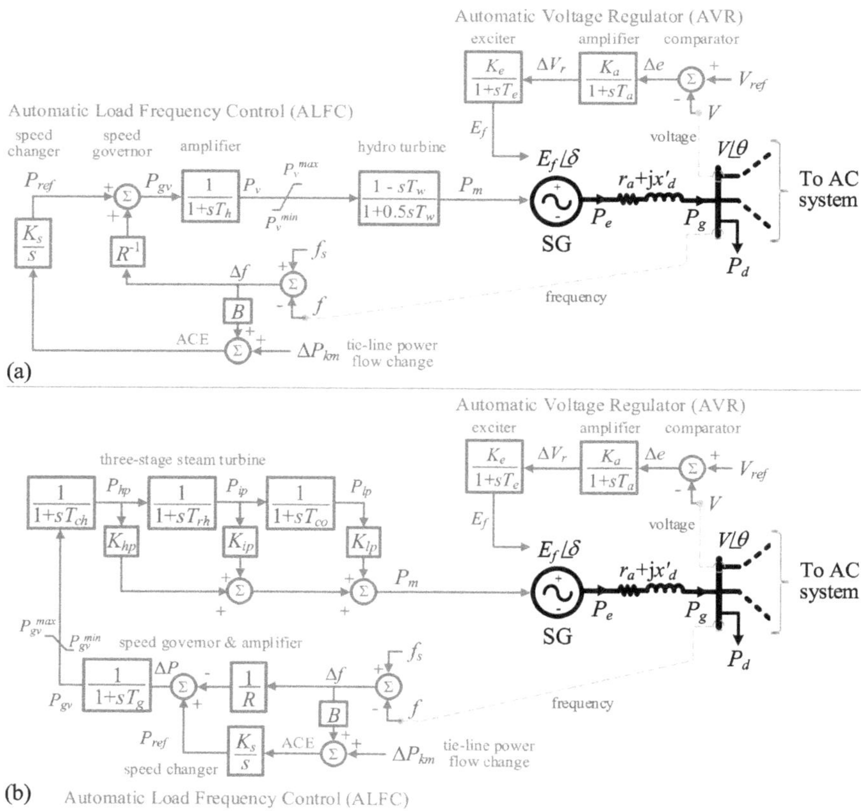

Figure 8.1 Conventional SG power plants with AVR and ALFC. (a) Hydroelectric power plant. (b) Thermoelectric power plant.

the voltage at the power plant terminals is the AVR, associated with the generator exciter control. The AVR continuously measures the voltage level of the synchronous generator, and upon rectification, it uses these signals to control the field winding current of the generator's rotor. In connection with Figure 8.1, the generator voltage magnitude V is compared with a reference value $V_{ref,}$ and the net difference is the input signal to the exciter. The exciter adjusts the generator field windings voltage, E_f, thereby regulating the generator's terminal voltage. The AVR indirectly affects the frequency control of the electrical power grid, although these are only second-order effects [2,3].

The second fundamental control loop is the ALFC. It directly regulates the turbine's mechanical power P_m to control the mechanical speed ω_m at its nominal value $\omega_s = 2\pi f_s$. In turn, this action regulates the synchronous frequency of the power system f_s. The frequency regulation action exerts direct control over the mechanical speed of the turbine-generator group of the power plant. As illustrated in Figure 8.1, the ALFC comprises two closed-loop controls: the primary control is related to the turbine speed governor, and the secondary control is related to the turbine speed changer. The primary control is a relatively fast control loop that operates on the direct measurement of the electrical frequency of the generator. This exercises control over the turbine speed governor, which regulates the valves through which steam flows in thermal turbines or the blades and gates through which water flows to the turbine in hydro turbines.

The response of this droop control is characterised by the generation plant's statism, or regulation constant, R. The main objective of the primary control loop is to counteract rapid load changes and power deviations in the power system, trying to maintain a continuous balance between the generated power and the customers' load demand. The secondary control loop, on the other hand, acts more slowly than the primary control. Its main task is to make fine adjustments on the frequency (i.e., $\Delta f \rightarrow 0$) by regulating the reference power of the speed changer, P_{ref}, additionally establishing actions to maintain the pre-established power exchange, P_{km}, with other areas of the same electrical power grid, or between interconnected power grids. Indeed, secondary control plays an important role in reducing the so-called Area Control Error (ACE) when two or more AC systems are interconnected. One of the control strategies that have prevailed over the years in the operation of interconnected power systems is the widely employed tie-line bias control, which establishes that all power plants must contribute to reducing frequency deviations, Δf, while at the same time enabling a net power exchange between control areas: $\text{ACE} = \Delta P_{km} + B \times \Delta f$, where B is the bias parameter that represents the frequency response characteristic of the control area [2,3].

Hence, conventional plants are equipped with suitable mechanisms to adjust their mechanical powers to achieve a balance between load and generation as quickly as possible. The time response of the turbine-governor group is of particular interest to TSOs because the system demand varies over the course of a day and because random contingencies do occur, such as transmission line and generator outages, thus disrupting the energy balance of the power system. To be ready to counteract such complex and unforeseen operating scenarios, exhaustive offline dynamic analyses are extensively carried out at control centres, considering a complete set of credible power imbalances that are likely to occur to assess the impact of such scenarios on the system frequency stability. Based on these studies, power system analysts design suitable corrective actions in subsequent steps. For instance, ones that enable the implementation of remedial action schemes relating to load curtailment and generator tripping or the design of under-frequency load shedding schemes.

The second-order oscillation represented by Equations (8.1)–(8.2) that characterise the electromechanical response of synchronous generators can be the starting point for studying the slow system dynamics originating from load/generation imbalances [4]. This encompasses both the action of the mechanisms associated with the ALFC and the electrical response of the transmission system. By referring to Figure 8.1, it is inferred that the ALFC operation is exerted on the mechanical power, P_m. In contrast, the response of the electric power of the generator is characterised by a non-linear function, $P_e = \mathrm{Re}\{\mathbf{E}_f \times \mathbf{I}^*_g\}$ with $\mathbf{I}_g = (\mathbf{E}_f - \mathbf{V}) \times (r_a + \mathrm{j}x'_d)^{-1}$. This relationship is dependent on the internal voltage of the generator $\mathbf{E}_f = E_f e^{\mathrm{j}\delta}$ and the system voltage $\mathbf{V} = Ve^{\mathrm{j}\theta}$. The generator's electric power P_g is always consistent with the system demand P_d, where $P_g = -\mathrm{Re}\{\mathbf{V} \times \mathbf{I}^*_g\}$ is the power injected into the electrical network. In the case of lossless generators (i.e., $r_a = 0$), the following relationship applies: $P_e = P_g$. It follows that in a lossless generator, the load changes would be immediately reflected upon the electrical power generation, instantaneously satisfying the system load, i.e., $\Delta P_d = \Delta P_e$. However, even when the generator's winding resistances are neglected owing to their rather small value, there are time constants associated with the synchronous generators in the power plant that prevent the instantaneous setting of ΔP_e induced by changes in the mechanical power ΔP_m. The associated delays cause the generator rotor to accelerate (i.e., $d\omega/dt > 0$) when there are negative changes in demand (i.e., $\Delta P_d = \Delta P_e < 0$). Conversely, a deceleration in the generator rotor occurs (i.e., $d\omega/dt < 0$) when changes in system demand are positive (i.e., $\Delta P_d = \Delta P_e > 0$).

In summary, system load changes cause mechanical speed variations with respect to the synchronous speed: $\Delta\omega = \omega - \omega_s \neq 0$, with ensuing electrical frequency deviations: $\Delta f = f - f_s \neq 0$. The primary control action yields a new equilibrium point: $P_m = P_e = P_d$, having a different frequency value: f. A new equilibrium point occurs several seconds after the change in system load occurred and once the ALFC loops of the power plants have had time to counteract the system power imbalance: ΔP_d.

The following classic equations encapsulate well the dynamics of such speed, frequency and power variations:

$$\frac{d\omega_k}{dt} = \frac{\omega_s}{2H_k}(P_{mk} - P_{ek} - D_k\Delta\omega_k) \quad \forall k = 1\ldots\gamma \tag{8.1}$$

$$\frac{d\delta_k}{dt} = \omega_k - \omega_s \tag{8.2}$$

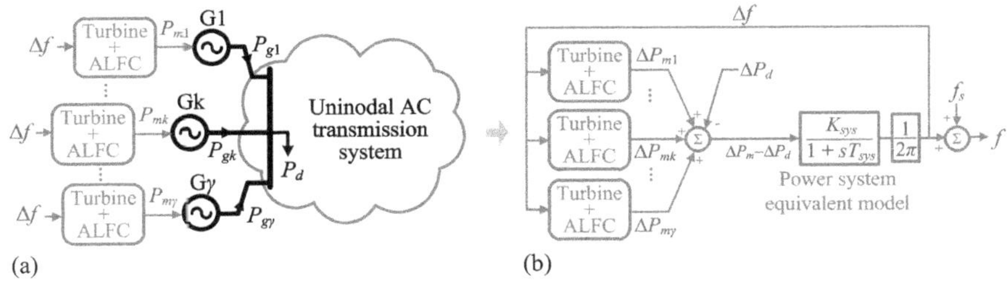

Figure 8.2 (a) Power system with SG power plants participating in frequency regulation. (b) System frequency response model for ALFC studies.

where H is the generator inertia constant, D is the generator damping constant, $\Delta\omega = \omega - \omega_s$, $\omega = 2\pi f$ is the mechanical angular speed, $\omega_s = 2\pi f_s$ denotes the synchronous angular speed, f is the shaft frequency, f_s is the synchronous frequency and k is the generator counter.

The set of equations (8.1)–(8.2) applies to each SG in the power plant and to each power plant in the power network, as outlined in Figure 8.2(a). While conducting ALFC studies, the full set of equations must be solved simultaneously to assess the system frequency response and to determine the contribution of each power plant to system frequency. For utility-level power grids containing several tens of generating units, ALFC dynamic analyses are quite taxing, owing to the very long simulation times involved, ranging from tens of seconds to minutes, to observe both primary and secondary control loop frequency responses. To overcome this, ALFC studies are often separated from the analysis of the AVR response since its dynamics are much faster compared to the slow ALFC's electromechanical dynamics. This justifies that the voltage magnitudes are assumed to remain constant throughout the ALFC study. This simplifies ALFC studies a great deal, enabling the analyst to concentrate on obtaining only the useful information for the design of operating strategies that best serve the system for coping with potential power imbalances, such as those caused by sudden load changes or generator outages. Furthermore, since the electromechanical oscillations produced by power imbalances are relatively slow, it is often assumed that generators operate in unison and that the system dynamics can be grouped. In other words, it is assumed that all generation units in one area experience the same frequency deviations $\Delta\omega_1 = \Delta\omega_k = \Delta\omega_\gamma$. This means that for small changes and after some algebra, (8.1) can be expressed as (8.4):

$$\frac{2H_{sys}}{\omega_s}\frac{d\Delta\omega}{dt} - D_{sys}\Delta\omega = \Delta P_m - \Delta P_d \tag{8.3}$$

This can be solved for $\Delta\omega$ using its Laplacian form:

$$\Delta\omega(s) = \left(\frac{K_{sys}}{1 + sT_{sys}}\right)[\Delta P_m(s) - \Delta P_d(s)] \tag{8.4}$$

where $K_{sys} = 1/D_{sys}$, $T_{sys} = 2H_{sys}/(\omega_s\, D_{sys})$, $H_{sys} = \Sigma H_k\ \forall k = 1, \ldots, \gamma$, whereas the power system equivalent mechanical power change is $\Delta P_m = \Sigma\Delta P_{mk}\ \forall k = 1, \ldots, \gamma$.

Equation (8.4) is schematically shown in Figure 8.2(b), which represents the so-called system frequency response (SFR) model.

This simplified SFR model in Figure 8.2(b) is advantageous because the system frequency response to power imbalances can be obtained efficiently. However, it should be noted that the AC transmission system is not considered in this one-node representation, and the total system demand is being grouped into a single node. This implies that the effect of the transmission lines that interconnect different AC systems cannot be directly estimated. Also, the secondary control loop of the ALFC cannot be included, see Figure 8.1.

This is one important reason why simplified linear models such as the SFR are not widely used. Most electrical utilities continue to use the transient stability models of commercial software packages to carry out Full Dynamic Simulations (FDS) for ALFC studies. These are non-linear models solved by injecting nodal currents into the system nodal admittance matrix [4,5]. These simulation studies yield detailed results of the performance of generators and transmission systems behaviours for both steady-state and transient operating regimes.

However, the computational burden of the models increases with the size of the network, resulting in time-consuming ALFC simulations in utility-size power grids. A suitable compromise must be grasped between efficiency and accuracy in the calculations. Aiming to resolve this outstanding issue in ALFC application studies, a simplified network formulation is developed below. It combines seamlessly the ALFC control loops of power plants with a simplified representation of the electrical power grid. It uses linear power flows for the internal generation powers P_e, given in (8.5a) and for the transmission lines P_{km}, given in (8.5b). Simplifying assumptions are that the generators and transmission lines contain no resistance and that the associated nodal voltage magnitudes remain at their nominal values throughout the ALFC study, e.g., $E_{fi} = V_k = V_m = 1$ pu.

$$(a)\ P_{ei} = \frac{\delta_i - \theta_k}{x'_{di}} \quad \text{and} \quad (b)\ P_{km} = \frac{\theta_k - \theta_m}{x_{km}} \tag{8.5}$$

Figure 8.3 shows the one-line diagram of an equivalent two power plants, feeding into nodes k and m, respectively, and interconnected by an AC transmission line. Using (8.5) to formulate the nodal power balances in the equivalent power system of Figure 8.3, including the internal nodes of the generators, the linear

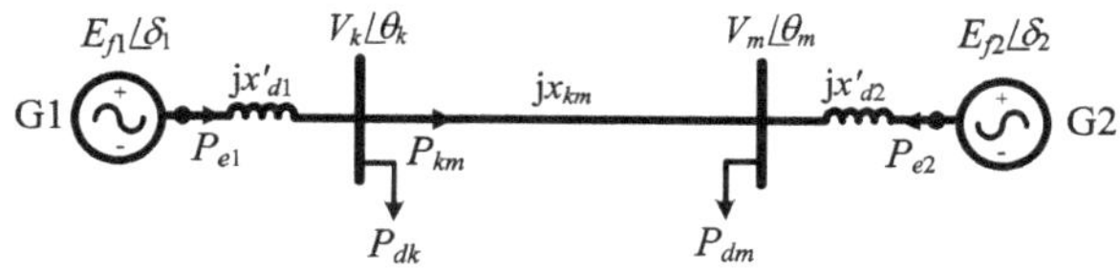

Figure 8.3 Two SG power plants interconnected by a tie-line reactance

power flow model is written down in (8.6).

$$\begin{pmatrix} -P_{dk} \\ -P_{dm} \\ P_{e1} \\ P_{e2} \end{pmatrix} = \begin{pmatrix} x'^{-1}_{d1} + x^{-1}_{km} & -x^{-1}_{km} & -x'^{-1}_{d1} & 0 \\ -x^{-1}_{km} & x'^{-1}_{d2} + x^{-1}_{km} & 0 & x'^{-1}_{d2} \\ -x'^{-1}_{d1} & 0 & x'^{-1}_{d1} & 0 \\ 0 & -x'^{-1}_{d2} & 0 & x'^{-1}_{d2} \end{pmatrix} \begin{pmatrix} \theta_k \\ \theta_m \\ \delta_1 \\ \delta_2 \end{pmatrix}$$

$$\Rightarrow \quad \begin{aligned} -\mathbf{P}_d &= \mathbf{B}\,\boldsymbol{\theta} + \mathbf{B}_g \delta \\ \mathbf{P}_e &= \mathbf{B}_g^T \boldsymbol{\theta} + \mathbf{B}_d\, \delta \end{aligned} \tag{8.6}$$

The two-power plant model is easily extendable to the case of n_{ac} buses and γ SG power plants, as shown by the equations in bold typeface in (8.6). Here, $\mathbf{P}_d$ is a n_{ac}-by-1 vector that contains the nodal power demands, and $\mathbf{P}_e$ is a γ-by-1 vector that accommodates the SG internal powers; $\mathbf{B}$ is a n_{ac}-by-n_{ac} nodal susceptance matrix made up of the series transmission lines reactances x_{km} and SG transient reactances x'_d; $\mathbf{B}_g$ is a n_{ac}-by-γ matrix containing the internal SG transient reactances in locations corresponding to connection nodes, and $\mathbf{B}^T_g$ being its transposed matrix; $\mathbf{B}_d$ is a γ-by-γ diagonal matrix that accommodates the generator reactances x'_d; θ is a n_{ac}-by-1vector with the nodal phase angles; δ is a γ-by-1 vector with the internal SG load angles.

To determine the full operating state of the power plants, the ALFC loops must be incorporated into the formulation, just like in the SFR model, see Figure 8.2(b). This requires the numerical integration of the associated set of differential-algebraic equations (DAE) [6,7]. The numerical solution may be carried out using a suitable numerical method, such as the implicit trapezoidal rule, widely used in dynamic simulations of utility-scale power systems [7]. In particular, the set of DAE resulting from the SG oscillation, equations (8.1)–(8.2) and the ALFC loops of the power plants are shown in Figure 8.1, are as follows:

$$\begin{aligned} F_H &= y_{(t-\Delta t)} + 0.5\Delta t\dot{y}_{(t-\Delta t)} - (y_{(t)} - 0.5\Delta t\dot{y}_{(t)}), \quad \forall y \in \{\omega, \delta, P_m, P_v, P_{ref}\} \\ F_T &= y_{(t-\Delta t)} + 0.5\Delta t\dot{y}_{(t-\Delta t)} - (y_{(t)} - 0.5\Delta t\dot{y}_{(t)}), \quad \forall y \in \{\omega, \delta, P_{lp}, P_{ip}, P_{hp}, \Delta P, P_{ref}\} \end{aligned} \tag{8.7}$$

where F_H denotes the DAE for hydroelectric power plants, F_T represents those of thermoelectric power plants, and Δt is the integration time step. Both sets of equations can be solved for time (t) as follows,

$$\begin{aligned} \Delta\mathbf{y}_{H(t,i)} &= -\mathbf{J}^{-1}_{H(t,i)}\ \mathbf{F}_{H(t,i)}; \quad \mathbf{y}_{H(t,i+1)} = \mathbf{y}_{H(t,i)} + \Delta\mathbf{y}_{H(t,i)} \\ \Delta\mathbf{y}_{T(t,i)} &= -\mathbf{J}^{-1}_{T(t,i)}\ \mathbf{F}_{T(t,i)}; \quad \mathbf{y}_{T(t,i+1)} = \mathbf{y}_{T(t,i)} + \Delta\mathbf{y}_{T(t,i)} \end{aligned} \tag{8.8}$$

where i is the iteration counter, $\mathbf{J}_H$, $\mathbf{J}_T$ are Jacobian matrices, the mismatch vectors are $\mathbf{F}_H = [F_\omega, F_\delta, F_{Pm}, F_{Pv}, F_{Pref}]^T$ and $\mathbf{F}_T = [F_\omega, F_\delta, F_{lp}, F_{ip}, F_{hp}, F_{\Delta P}, F_{Pref}]^T$, whereas $\mathbf{y}_H = [\omega, \delta, P_m, P_v, P_{ref}]^T$ and $\mathbf{y}_T = [\omega, \delta, P_{lp}, P_{ip}, P_{hp}, \Delta P, P_{ref}]^T$ are the vectors of state variables of both SG power plants. Upon successful solution of $\mathbf{y}_{H(t)}$

and $\mathbf{y}_{T(t)}$ for time (t), the internal SG load angles are known $\delta_{(t)}$. Moreover, the system nodal phase angles can be calculated from (8.6), as follows,

$$\boldsymbol{\theta}_{(t)} = \mathbf{B}^{-1}\left(-\mathbf{P}_{d(t)} - \mathbf{B}_g \delta_{(t)}\right) \tag{8.9}$$

The dynamic ALFC modelling framework is summarised in Figure 8.4(a) and (b), where γ-SG and n-power nodes are included in the generic, schematic representation of the multi-SG, multi-node power grid. The solution algorithm is shown in flowchart form in part (b) of Figure 8.4. The transmission network is now included in the new ALFC formulation. This enables the assessment of the secondary frequency control response, which deals with the ACE that is affected by AC tie lines linking the power systems. Equally significant is the fact that this opens the door to conducting ALFC studies for HVDC-connected AC power grids using this dynamic modelling framework, as it will be shown below.

8.2.2 Primary and secondary control in two-area AC systems

Before addressing the more complex issue of AC systems interconnected by VSC-HVDC links, let us look at the simplified two-area power system in Figure 8.5 to illustrate some key issues on frequency regulation arising from the ALFC loops of SG power plants.

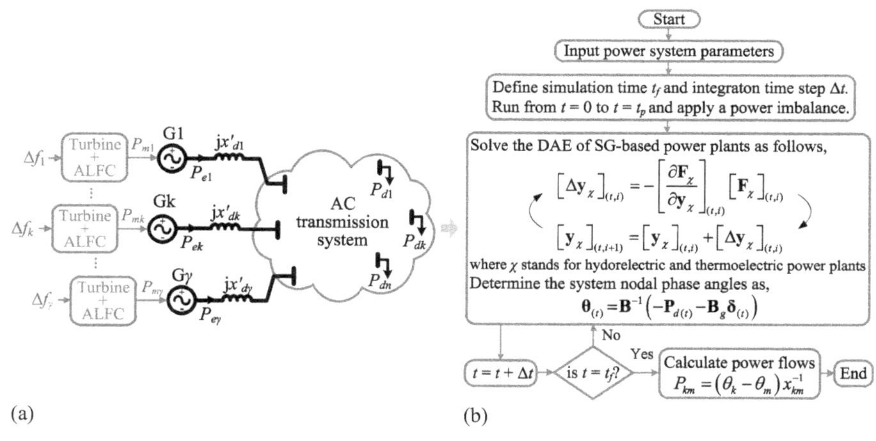

Figure 8.4 (a) Power system with SG power plants participating in frequency regulation. (b) Dynamic modelling framework for ALFC studies.

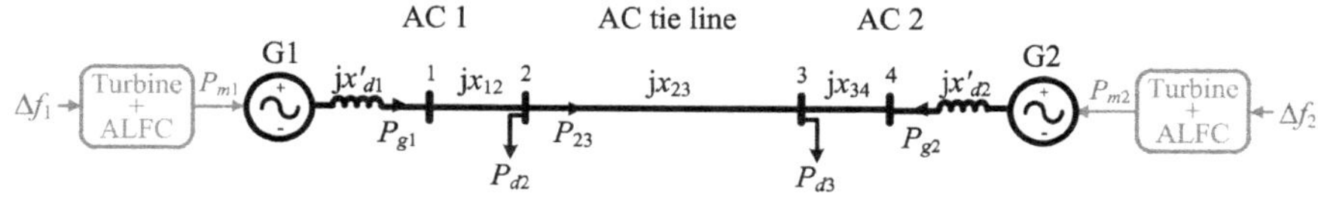

Figure 8.5 Two-area power system interconnected by an AC tie-line

The tie-line reactance x_{23} links AC 1 and AC 2. Both equivalent system areas operate at their nominal frequency f_s = 50 Hz. The power network AC 1 represents a 1,200-MW equivalent system, whereas AC 2 is rated at 1,000 MW. The power generation dynamics of both areas correspond to thermoelectric power plants (see Figure 8.1). Both AC 1 and AC 2 have similar characteristics, with AC 1 feeding into a load of P_{d2} = 1,000 MW and AC 2 supplying a load of P_{d3} = 750 MW. At steady state, G1 in AC 1 generates P_{g1} = 1,100 MW, exporting 100 MW to AC 2. Hence, G2 in AC 2 generates P_{g2} = 650 MW to cover the demand in its own area. The power flow through the AC tie line is P_{23} = 100 MW. Table 8.1 gives the parameters of the power plants and transmission lines, noting that both SG power plants have primary and secondary spinning power reserves, SR_p and SR_s, respectively, with respect to their power dispatched in a steady state. The ALFC study uses the dynamic modelling framework summarised in Figure 8.4. The integration time step used in the simulation study below is Δt = 1 ms. The simulation study assumes a power imbalance contingency involving a sudden demand increment in AC 2 of 10% (i.e., 75 MW) at a time t_p = 3 s. The dynamic simulation runs for t_f = 150 s.

To assess the responses of the SG power plants' primary and secondary control loops separately, two operating scenarios are considered below: (i) the interconnected systems depend only on primary frequency control, and (ii) both primary and secondary ALFC controls are active. The ALFC responses of the two-area power system are provided in Figures 8.6 and 8.7, respectively.

It is observed that the frequencies at AC 1 and AC 2 behave similarly, both reaching a frequency nadir of approximately $f_1 \approx f_2 \approx$ 49.4 Hz. The reason for this is that both AC systems share the same inertial response, with the power imbalance causing slow electromechanical transients at both SG terminals, which are propagated through the nodal phase angle variations.

Nevertheless, the system frequencies are quite different when the primary and secondary ALFC loops are active, the system frequencies are better controlled throughout the simulation run due to the action of the two control loops; see f_1 and f_2 at t = 150 s for (i)–(ii) in Figures 8.6(a) and 8.7(a), respectively. This point is clearer in the AC tie-line power flow P_{23}. In the first scenario (i), when the primary frequency control acts unaided, the power flow reaches a peak nearing 160 MW

Table 8.1 Parameters of the two-area power system equipment

Transmission systems AC 1 and AC 2 100-MW base power		**SGs driven by thermoelectric turbines**	
Loads	P_{d2} = 10 pu P_{d3} = 7.5 pu	G1: 1,200 MW SR_p = 5%, SR_s = 4.1%	G2: 1,000 MW SR_p = 5%, SR_s = 10%
Generators	P_{g1} = 11 pu P_{g2} = 6.5 pu	$x'_{d1} = x'_{d2}$ = 0.314 pu; $D_1 = D_2$ = 0 s/rad $H_1 = H_2$ = 5 s; $R_1 = R_2$ = 0.08 pu; $T_{g1} = T_{g2}$ = 3 s	
Transmission Lines	x_{12} = 0.08 pu x_{23} = 0.10 pu x_{34} = 0.07 pu	$K_{s1} = K_{s2}$ = 0.1; $B_1 = B_2$ = 12.5 pu; $T_{ch1} = T_{ch2}$ = 0.3 s $T_{rh1} = T_{rh2}$ = 7 s; $T_{co1} = T_{co2}$ = 0.2 s; $K_{hp1} = K_{hp2}$ = 0.3 $K_{ip1} = K_{ip2}$ = 0.4; $K_{lp1} = K_{lp2}$ = 0.3.	

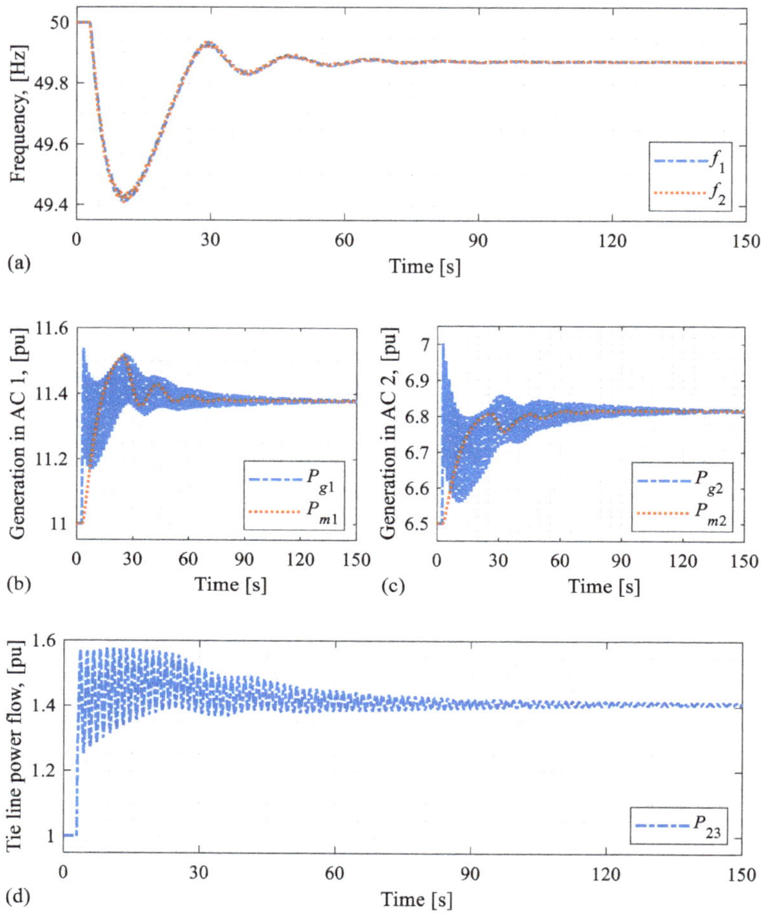

Figure 8.6 Two-area power system with primary frequency control only: (a) generators' frequencies, (b) electrical and mechanical powers from G1, (c) electrical and mechanical powers from G2, and (d) tie line power flow

during the first 20 s following the disturbance. In contrast, in scenario (ii) it decreases more rapidly.

In scenario (i), the tie line power settles at P_{23} = 140 MW at t = 150 s, with 40 MW more than in its pre-contingency, steady-state value, see Figure 8.6(d). In contrast, when both ALFC loops are active – scenario (ii), the AC tie-line power flow settles to $P_{23} \approx$ 100 MW at t = 150 s. Nonetheless, transient power oscillations are present, but they nearly disappear at t = 90 s, as observed in Figure 8.7(d).

As expected, with the 10% demand increase at node 3, the transient and steady-state power flows through the tie-line are strongly linked with the power generations at both systems AC 1 and AC 2, as shown in Figures 8.6(b) and (c) and 8.7(b) and (c) for scenarios (i) and (ii), respectively. Note that since the rated

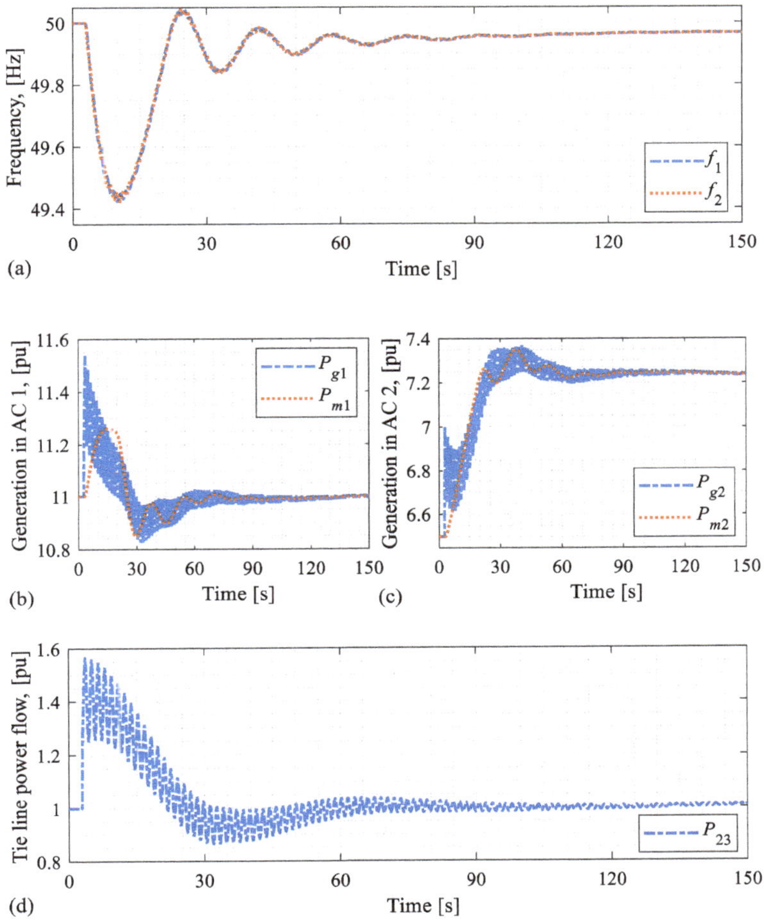

Figure 8.7 Two-area power system with both ALFC loops active: (a) generators' frequencies, (b) electrical and mechanical powers from G1, (c) electrical and mechanical powers from G2, and (d) tie line power flow

capacity of AC 1 and AC 2 stand at 1,200 MW and 1,000 MW, respectively, there is sufficient generating capacity to cover the transient power peaks at scenarios (i) and (ii), respectively.

It should be noticed that when the primary and secondary controls are active, G1 considerably reduces its output power P_{g1}, while assisting the overall system in bringing back the tie line power flow to its scheduled value in steady state (i.e., 100 MW); see Figure 8.7(b) and (c). The opposite happens for G2, whose power generation P_{g2} must increase further to be responsible for supplying the additional demand increase of 10%, and node 3, totalling 725 MW, see Figure 8.6(b) and (c).

In conclusion, AC 1 and AC 2 regulate the frequency of the two-area system while simultaneously exercising the tie-line power flow control. That is, both systems bear a shared responsibility of acting together to reduce the related area control error, i.e., $ACE_1 = \Delta P_{23} + B_1 \times \Delta f_1$ for AC 1 and $ACE_2 = \Delta P_{32} + B_2 \times \Delta f_2$ for AC 2, see Figure 8.1.

To conclude this section, relating to ALFC of SG power plants interconnected by AC tie lines, it should be remarked that tie-line bias control is much more complicated to achieve than what has been proclaimed by this rather contrived test case, particularly in highly interconnected AC systems. In part, this is due to reverse power flows that take place between interconnected areas, some of which are difficult to manage in practice. Examples of these are quite common in European countries whose electrical power networks are highly interconnected between regions and between countries. It is due to these complex operating conditions that VSC-HVDC-based interconnections are rapidly gaining adepts among power grid planners. Particularly within bilateral energy transactions between countries involving inland and submarine power transmission.

This trend is well illustrated by the many VSC-HVDC links that have been built recently or are in the planning stage, in Europe: the 2,000 MW INELFE link between France and Spain, the 1,200 MW Fréjus Road Tunnel between France and Italy, the 1,000 MW Nemo link between Belgium and England, the 1,000 MW ALEGrO link between Belgium and Germany, the 1,000 MW IFA-2 link between France and England, the 1,400 MW North Sea Link between Norway and England, the 1,400 MW FAB link between France and England, the 1,400 MW Viking link between Denmark and England, the 700 MW Celtic Interconnector between Ireland and France, the 500 MW Green link between Ireland and Wales, the 1,400 MW NeuConnect link between England and Germany, and the 700 MW Harmony link between Lithuania and Poland. Refer to Table 2.9 for further details.

In this tenor, key aspects of power/frequency regulation in VSC-HVDC-connected AC systems are discussed in the following section.

8.3 Two-area VSC-HVDC connected AC systems

The interconnection of power systems through conventional AC transmission lines is not amenable to power flow controllability in the link. The dynamics of the interconnected power systems are governed by physics laws, such as Ohm's and Kirchhoff's laws. One way to circumvent this problem is through secondary controls in SG power plants, as discussed above, or by using the unified power flow controller (UPFC), as reported in Chapter 2. Within this context, the former option, which is used in practice, requires additional infrastructure to be deployed throughout the transmission network, which is coordinated at energy control centres. Examples of these auxiliary equipment are telemetry, measurement devices, communication channels, and signal processing tools, all of which are required by TSOs with a high degree of reliability.

Improved voltage and power regulation are continuously pursued by TSOs in the quest for enhanced system stability, particularly when one of the interconnected AC systems is weak. For instance, lacking sufficient voltage support or frequency regulation means. This situation certainly aggravates further under stressing dynamic operating conditions following non-planned outages of critical equipment. In this sense, much of the effort to deal with frequency control problems is associated with shunt-connected generating sources, which are vastly reinforced with frequency control loops. Cases in point are conventional and renewable power plants and, more recently, battery energy storage systems (BESS). Over the last two decades, transmission systems have been reinforced with shunt-connected or series-connected FACTS equipment. Particularly, VSC-HVDC links have been very popular because this technology offers wide-range voltage and power controllability in a decoupled manner [8], just as is the case with SG power plants at their interconnection points with the power grid. Regarding frequency regulation issues, another key factor is the transmission infrastructure – even though this is not so evident in many cases – whose dynamic response to power imbalances impacts the system frequency stability.

AC tie lines have traditionally been used for interconnecting AC electrical power systems; however, power fluctuations can flow freely back and forth in response to changes in the transmission line terminal voltages. On the other hand, the use of power electronics equipment, particularly VSC-HVDC links, play a very effective role in mitigating power and frequency fluctuations. VSC-HVDC links are not exposed to the imbalances taking place in the AC power grids that they interconnect because the power oscillations are not propagated through the link but decoupled due to the intrinsic AC-to-DC energy conversion process taking place in the VSC stations. The enhancement of the system dynamic response in view of load/frequency control issues using power electronics equipment is addressed below. Key technical aspects regarding the slow transients related to ALFC in VSC-HVDC-connected AC power systems are discussed in detail.

8.3.1 Basic VSC-HVDC link elements, modelling, and controls

Figure 8.8 shows the schematic representation of the point-to-point VSC-HVDC link with fundamental voltage and power control loops. The HVDC link is

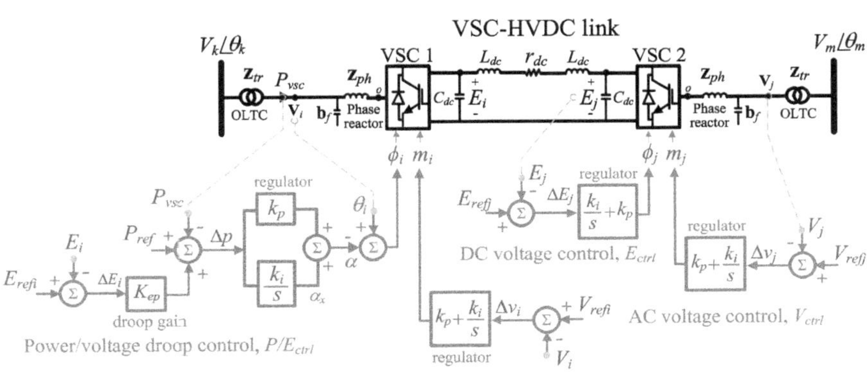

Figure 8.8 Schematic representation of two-terminal VSC-HVDC link with AC and DC voltage/power control loops

integrated into the electrical network by step-up voltage transformers [8]. The converters' AC output voltage $\mathbf{V}_o$ is a direct function of the DC voltage E and the modulating variables $\mathbf{m} = me^{j\phi}$ driven by a sinusoidal PWM technique, i.e., the converter AC-to-DC voltage relationship is given by $\mathbf{V}_o = \sqrt{(3/8)} \times E \times me^{j\phi}$. Each VSC station uses AC and DC filtering equipment ($\mathbf{z}_{ph}$, $\mathbf{b}_f$, L_{dc}, C_{dc}) to improve the quality of the input/output signals while stabilising the voltages at both sides of the converter.

AC voltage control is carried out independently at nodes i and j by converter stations VSC 1 and VSC 2 using the modulation ratio control variables m_1 and m_2. Using both power converters, the aim is to reduce the errors between the reference and actual voltages, V_{ref} and V, at their corresponding AC voltage terminals, V_i and V_j. A parallel can be drawn between the action of this voltage control and the AVR control of a conventional SG power plant, exerting voltage regulation at their corresponding interconnection points. It should be noted that for stable VSC-HVDC link operation, there must be at least one VSC station exerting voltage control in the DC link or DC grid in the case of multiterminal VSC-HVDC system. The associated VSC control loop continuously compares the reference and actual DC voltages, E_{ref} and E, driving the resulting error to zero throughout time. This is achieved by adjusting the converter phase-shifting angle ϕ, see VSC 2 in Figure 8.8. A VSC station performing DC voltage control is often called a grid-following converter. Grid-following converters guarantee stable power flow transfers through the DC link, enabling overall AC/DC system stability.

From the perspective of power systems operation, VSC-HVDC links are primarily charged with carrying a fixed power flow independently of the external operating conditions of the interconnected AC systems. Hence, regulating power flow to a reference value, $P_{vsc} = P_{ref}$, is the key task of the power control loop in the DC transmission links (see VSC 1 in Figure 8.8).

The TSOs set this scheduled power following economic and technical criteria. For instance, to fulfil a bilateral contract between parties (interconnected AC systems) or to act to benefit the power system operation under certain loading conditions directly. Concerning the latter, HVDC links may be equipped with VSC stations with grid-forming control schemes characterised by power/voltage droop controls. This gives the DC link improved system dynamic responses when undergoing perturbations. In this operating scenario, the VSC reference power P_{ref} is allowed to dynamically change from $P_{vsc} = P_{ref}$ to $P_{vsc} = P_{ref} + K_{ep}(E_{refi} - E_i)$, according to voltage variations at its corresponding DC voltage E_i, where K_{ep} is the constant droop control gain. The usefulness of this control strategy is more evident when applied to multiterminal VSC-HVDC systems, where power sharing between VSC stations occurs following a contingency in the AC/DC system.

Table 8.2 gives the VSC controls and variables that characterise the dynamic response of VSC-HVDC links. A close examination of the power/voltage droop control of VSC 1 in Figure 8.8, reveals that fixed-power and grid-forming control strategies are equivalent in steady-state operation given that the VSC-HVDC follows the voltage restriction: $E_i = E_{refi}$, which yields: $P_{vsc} = P_{ref}$.

Table 8.2 VSC control strategies and associated variables in VSC-HVDC links

Control strategy	Acronym	Controlled variables	Reference values
Grid-following E_{ctrl}	E_{ctrl}	E, V	E_{ref}, V_{ref}
Fixed power P_{ctrl}	P_{ctrl}	P_{vsc}, V	P_{ref}, V_{ref}
Grid-forming E/P_{ctrl}	P/E_{ctrl}	E, P_{vsc}, V	E_{ref}, P_{ref}, V_{ref}

This would establish the fundamental contribution of VSC dynamics to ALFC actions when a VSC-HVDC links two AC systems. However, it should be noted that the VSC control loops for AC and DC voltage regulation are several times faster than those performing power control in an SG power plant [9]. In fact, the combined action of the VSC station and its voltage controls mimic well the behaviour of an ideal voltage source. Also, it is worth recalling that the overall ALFC system response is governed by slow electromechanical transients related to conventional power plants' power/frequency controls. From a practical standpoint, this signifies that for ALFC studies, the AC and DC voltages, V and E, may be taken to be constant during power imbalance events.

Bearing in mind the above points and with reference to Figure 8.8, the dynamics that characterise grid-forming VSC stations P/E_{ctrl} may be encapsulated in (8.10) and (8.11).

$$\frac{d\alpha_x}{dt} = k_i\left(P_{ref} + K_{ep}\Delta E_i - P_{vsc}\right) \tag{8.10}$$

$$\alpha = \alpha_x + k_p\left(P_{ref} + K_{ep}\Delta E_i - P_{vsc}\right) \tag{8.11}$$

Notice that converters with only a fixed power control strategy, P_{ctrl}, can be represented by these dynamic equations by just setting K_{ep} to zero.

In (8.10) and (8.11), $\alpha = \theta_i - \phi_i$ is the angular aperture between the phase angles of voltage $\mathbf{V}_i = V_i e^{j\theta i}$ and the converter's output voltage $\mathbf{V}_o = \sqrt{(3/8)} \times E \times me^{j\phi}$; $\Delta E_i = E_{refi} - E_i$; K_{ep} is the constant droop control gain; constants k_p and k_i are proportional and integral gains of the PI regulator; P_{ref} is the reference power; P_{vsc} is the actual power of the converter exercising power control, that is, a fixed-power or grid-forming VSC station.

Following the ideas put forward in the previous section, a linear power flow model may be developed for the VSC-HVDC links aimed at assessing ALFC responses [1]. As before, a lossless power approach will be pursued: the resistive elements of the transformers and phase reactors are taken to be zero. Also, as before, the AC voltage magnitudes will be taken to be at their nominal values $V_k = V_i = V_j = V_m = 1$ pu. It follows that $|\mathbf{V}_o| = 1$ pu, i.e., $\sqrt{(2/3)} \times E \times m = 1$ pu. The VSC power flow is: $P_{vsc} = \mathrm{Re}\{\mathbf{V}_i \times \mathbf{I}_{io}^*\}$, with $\mathbf{I}_{io} = (\mathbf{V}_i - \mathbf{V}_o) \times (\mathrm{j}x_{phi})^{-1}$.

Equations (8.12a) and (8.12b) give the linear power flows in the transformer and VSC 1, P_{tr} and P_{vsc}, respectively. Similar equations are set up for VSC 2 and its associated transformer. Concerning the internal HVDC power flow, the link resistance, r_{dc}, can be calculated from $P_{dc} = E_i \times I_{ij}$ and $I_{ij} = (E_i - E_j) \times (r_{dc})^{-1}$. The DC power flow expression is linear under the assumption of the sending-end voltage,

E_i, operating at nominal voltage. The DC power equals the line current, $P_{ij} \approx I_{ij}$, as given in (8.12c):

$$\text{(a) } P_{tr1} = \frac{\theta_k - \theta_i}{x_{tr1}}; \quad \text{(b) } P_{vsc1} = \frac{\theta_i - \phi_i}{x_{ph1}}; \quad \text{(c) } P_{ij} = \frac{E_i - E_j}{r_{dc}} \tag{8.12}$$

Examining (8.12b) reveals that the VSC power flow directly depends on the angular difference between the phase angle i and converter phase-shifting angle i. This explains why the power control exerted by VSC 1 in the VSC-HVDC arrangement of Figure 8.8 acts on the angular difference $\alpha = \theta_i - \phi_i$. This control action follows a linear model to great effect.

Equation (8.12) may be combined with those of the interconnected AC systems to obtain the nodal power balances at nodes k and m, including those of the VSC-HVDC link at nodes i, 1_{dc}, 2_{dc}, j. We follow a similar approach as for the case of the AC tie line in (8.6). The ensuing linear power flow model is given in (8.13).

$$\begin{pmatrix} -\mathbf{P}_d \\ \mathbf{P}_e \\ -\mathbf{P}_{dc} \end{pmatrix} = \begin{pmatrix} \mathbf{B} & \mathbf{B}_g & \mathbf{B}_v & \mathbf{0} \\ \mathbf{B}_g^T & \mathbf{B}_d & \mathbf{0} & \mathbf{0} \\ \mathbf{B}_v^T & \mathbf{0} & \mathbf{B}_p & \mathbf{G}_{dc} \end{pmatrix} \begin{pmatrix} \boldsymbol{\theta} \\ \delta \\ \boldsymbol{\varphi} \\ \mathbf{E} \end{pmatrix} \tag{8.13}$$

The vector entries in (8.13) are detailed below:

$$\mathbf{P}_d = \begin{pmatrix} -P_{dk} \\ -P_{di} \\ -P_{dj} \\ -P_{dm} \end{pmatrix}, \mathbf{P}_e = \begin{pmatrix} P_{e1} \\ P_{e2} \end{pmatrix}, \mathbf{P}_{dc} = \begin{pmatrix} -P_{1dc} \\ -P_{2dc} \end{pmatrix}, \boldsymbol{\theta} = \begin{pmatrix} \theta_k \\ \theta_i \\ \theta_j \\ \theta_m \end{pmatrix},$$
$$\delta = \begin{pmatrix} \delta_1 \\ \delta_2 \end{pmatrix}, \boldsymbol{\varphi} = \begin{pmatrix} \phi_i \\ \phi_j \end{pmatrix}, \mathbf{E} = \begin{pmatrix} E_i \\ E_j \end{pmatrix} \tag{8.14}$$

The matrix entries in (8.13) are given below,

$$\mathbf{B} = \begin{pmatrix} x'_{d1}-1 + x_{tr1}^{-1} & -x_{tr1}^{-1} & 0 & 0 \\ -x_{tr1}^{-1} & x_{tr1}^{-1} + x_{ph1}^{-1} & 0 & 0 \\ 0 & 0 & x_{tr2}^{-1} + x_{ph2}^{-1} & -x_{tr2}^{-1} \\ 0 & 0 & -x_{tr2}^{-1} & x'_{d2}-1 + x_{tr2}^{-1} \end{pmatrix}, \mathbf{B}_g$$
$$= \begin{pmatrix} -x'_{d1}-1 & 0 \\ 0 & 0 \\ 0 & 0 \\ 0 & -x'_{d2}-1 \end{pmatrix}, \mathbf{B}_v = \begin{pmatrix} 0 & 0 \\ -x_{ph1}^{-1} & 0 \\ 0 & -x_{ph2}^{-1} \\ 0 & 0 \end{pmatrix},$$
$$\mathbf{B}_d = \begin{pmatrix} x'_{d1}-1 & 0 \\ 0 & x'_{d2}-1 \end{pmatrix}, \mathbf{B}_p = \begin{pmatrix} x_{ph1}^{-1} & 0 \\ 0 & x_{ph2}^{-1} \end{pmatrix}, \mathbf{G}_{dc} = \begin{pmatrix} r_{dc}^{-1} & -r_{dc}^{-1} \\ -r_{dc}^{-1} & r_{dc}^{-1} \end{pmatrix}, \tag{8.15}$$

where the vectors and matrix terms $\mathbf{P}_d$, $\mathbf{P}_e$, θ, δ, $\mathbf{B}$, $\mathbf{B}_g$, and $\mathbf{B}_d$ are similar to those obtained in the case of power systems interconnected by conventional AC tie lines, see (8.6).

Furthermore, the n_{dc}-by-1 vector $\mathbf{P}_{dc}$ contains the nodal demands of the DC link; ϕ is a ρ-by-1 vector with the VSC phase-shifting angles, e.g., $\rho = 2$ for a two-terminal VSC-HVDC link; vector $\mathbf{E}$ contains the n_{dc} DC nodal voltages of the VSC-HVDC link. $\mathbf{B}_v$ is a n_{ac}-by-ρ matrix containing the negative of the inverted series reactances of the VSC coupling reactors x_{ph}. $\mathbf{B}_p$ is a n_{dc}-by-ρ matrix that contains the inverted reactances of the coupling reactors. $\mathbf{G}_{dc}$ is the n_{dc}-by-n_{dc} nodal conductance matrix of the DC link.

Figure 8.9 shows the one-line diagram of two equivalent AC systems interconnected by a point-to-point VSC-HVDC link, which is useful to illustrate the effectiveness of the linear power system model just derived in carrying out efficient ALFC studies.

The algebraic model (8.13) defines the power flows within the VSC-HVDC. They are dependent on the dynamic variables of the VSC stations, which are governed by the corresponding control loops acting on the converter phase-shifting angle ϕ_i that allows the power flow regulation. The DAE must be resolved at each time (t), just as in the case of SG power plants' power/frequency control loops. The DAE (8.10) is expressed in the following generic form:

$$F_V = y_{(t-\Delta t)} + 0.5\Delta t\dot{y}_{(t-\Delta t)} - \left(y_{(t)} - 0.5\Delta t\dot{y}_{(t)}\right), \ \forall y \in \{\alpha_x\} \tag{8.16}$$

where F_V denotes the DAE for VSC stations that exert power flow control, and Δt is the integration time step, which is solved for time (t), as follows:

$$\Delta\mathbf{y}_{V(t,i)} = -\mathbf{J}^{-1}_{V(t,i)}\ \mathbf{F}_{V(t,i)}; \quad \mathbf{y}_{V(t,i+1)} = \mathbf{y}_{V(t,i)} + \Delta\mathbf{y}_{V(t,i)} \tag{8.17}$$

where $\mathbf{J}_V$ is the Jacobian matrix, $\mathbf{F}_V = [F_{\alpha x}]$ is the mismatch vector, $\mathbf{y}_V = [\alpha_x]$ is the dynamic state variable of the VSC carrying out power flow control, and i is the iteration counter.

Solving $\mathbf{y}_{V(t)}$ at time (t), yields the calculation of the VSC phase-shifting angles $\varphi_{(t)}$, using $\phi_{i(t)} = \theta_{i(t)} - \alpha_{(t)}$ with $\alpha_{(t)}$ obtained by (8.11) at time (t). Since the SG load angles contained in the vector $\delta_{(t)}$ are also known at the time (t) through the solution of (8.8), the nodal voltage phase angles of the VSC-HVDC can be determined from (8.13) using the expression (8.18). Similarly, once the system nodal phase angles θ

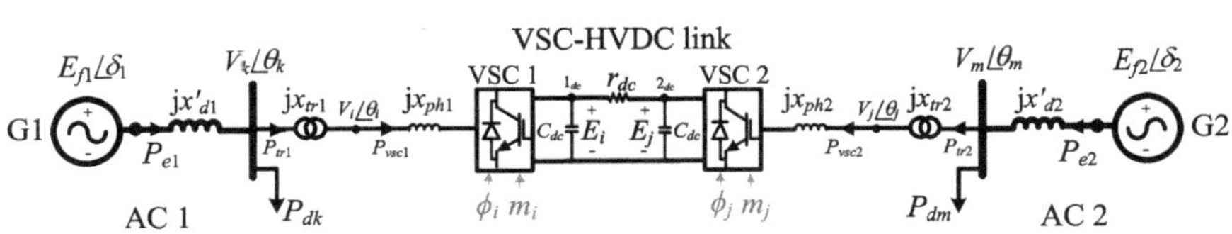

Figure 8.9 Schematic representation of a power system for ALFC studies comprising two AC systems interconnected by a VSC-HVDC link

and converter phase-shifting angles φ are known, the DC nodal voltages $\mathbf{E}$ are determined using (8.19).

$$\boldsymbol{\theta}_{(t)} = \mathbf{B}^{-1}\left(-\mathbf{P}_{d(t)} - \mathbf{B}_g\delta_{(t)} - \mathbf{B}_v\boldsymbol{\varphi}_{(t)}\right) \tag{8.18}$$

$$\mathbf{E}_{(t)} = \mathbf{G}_{dc}^{-1}\left(-\mathbf{P}_{dc(t)} - \mathbf{B}_v^T\boldsymbol{\theta}_{(t)} - \mathbf{B}_p\boldsymbol{\varphi}_{(t)}\right) \tag{8.19}$$

It should be borne in mind that the equation associated with the VSC station exerting DC voltage control (grid-following converter) must be deleted and removed from (8.19) for a correct solution. This is so because the associated voltage is known *a priori* and remains constant during the ALFC study: $E_j = E_{refj}$, as previously stated.

The differential-algebraic model (8.13)–(8.18) may be extended quite straightforwardly for a n_{ac}-bus power system with γ SG power plants and ρ VSC stations, forming a multiterminal HVDC system. A particular case is when one VSC-HVDC link interconnects two AC transmission systems, as illustrated in Figure 8.10(a). The solution algorithm is presented in Figure 8.10(b). This formulation is useful to assess the ALFC response of VSC-HVDC connected AC systems. This is a natural progression of the framework shown in Figure 8.4 for power systems linked by conventional AC tie lines. All the advantageous characteristics of the original framework for assessing ALFC response in conventional

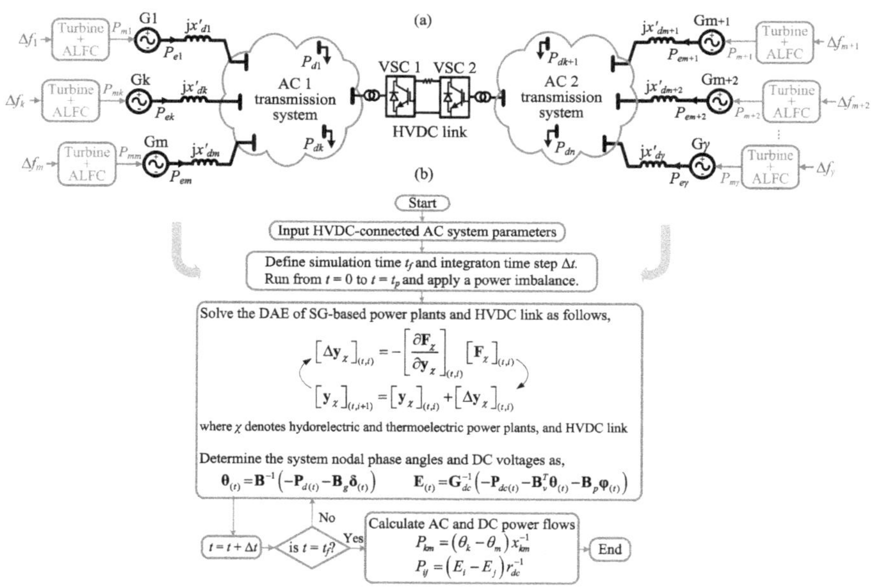

Figure 8.10 (a) Schematic representation of a point-to-point, VSC-HVDC link connecting two equivalent AC transmission systems. (b) Dynamic modelling framework for ALFC studies.

AC power systems are passed on to this extended framework, including VSC-HVDC systems.

8.3.2 Frequency control in two-area AC systems interconnected by a VSC-HVDC link

The contrived power system shown in Figure 8.11 may be used to discuss the impact of VSC-HVDC links on power/frequency regulation in interconnected AC systems. The two-area electrical power network in Figure 8.5 has been modified by replacing the AC tie line with a point-to-point, 200-MW VSC-HVDC link. It should be recalled that AC 1 and AC 2 represent 1,200 MW and 1,000 MW equivalent power networks, which operate at f_s = 50 Hz. The AC transmission systems and synchronous generators have the same parameters as before. They are given in Table 8.1. The parameters of the VSC-HVDC link are new and given in Table 8.3. In this two-terminal HVDC link, the converter VSC 1 is taken to exert fixed power control, P_{ctrl}, with a setting P_{ref} = 1 pu. Hence, it establishes a constant power transmission of 100 MW through the DC link. VSC 2 is taken to be a grid-following converter, E_{ctrl}, responsible for regulating voltage E_{2dc} at the set-point value E_{ref} = 2 pu. Under these conditions, the steady-state power generated by G1 and G2 and power flows in the HVDC link are the same as those in the two-area power system of Figure 8.5.

A sudden demand increase of 10% in AC 2 is assumed to occur at node 3. That is, P_{d3} changes from 750 MW to 825 MW at t_p = 3 s. The ALFC study is carried out for t_f = 150 s to fully observe the primary and secondary responses of the two-area AC power system linked by the VSC-HVDC system.

Figure 8.12 provides the ALFC responses of the VSC-HVDC-connected AC systems. It should be emphasised that the operating scenarios in the two-area system interconnected by a conventional AC tie line – with results given in Figure 8.7

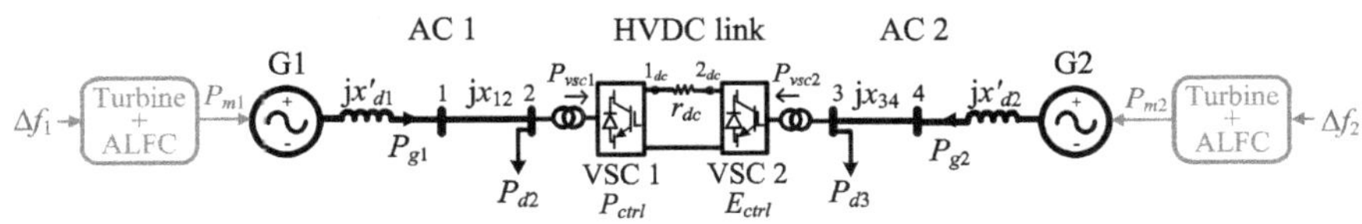

Figure 8.11 Two-area power system formed by a two-terminal VSC-HVDC link

Table 8.3 Parameters of the two-terminal VSC-HVDC link

VSC-HVDC link 100-MW base power	
VSC power rating	P_{nom} = 200 MW
DC line	r_{dc} = 0.0278 pu
Phase reactors	$x_{ph1} = x_{ph2}$ = 0.075 pu
Transformers	$x_{tr1} = x_{tr2}$ = 0.075 pu
VSC 1 – P_{ctrl}	k_p = 1e-4; k_i = 50e-4; K_{ep} = 0
VSC 2 – E_{ctrl}	E_{ref} = 2.0 pu

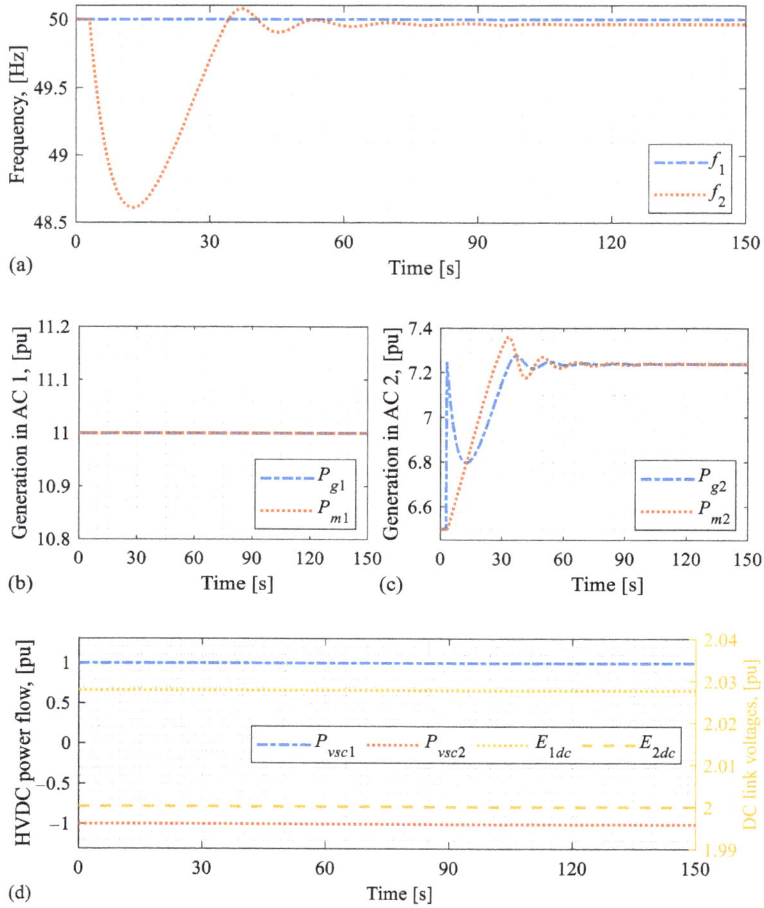

Figure 8.12 Two-area HVDC system with both ALFC loops active: (a) generators' frequencies, (b) electrical and mechanical powers from G1, (c) electrical and mechanical powers from G2, and (d) powers and voltages in the HVDC link

– and this one are the same. However, the dynamic responses of the two systems are strikingly different.

It is observed from Figure 8.12(a) that the system frequency of AC 1 remains unchanged, i.e., $f_1 = 50$ Hz. The reason is that the power imbalance occurs at AC 2, and the power perturbation does not propagate from AC 2 towards AC 1 owing to the frequency conversion process of the VSC converters of the DC link, namely, f_1–0–f_2, conducted by both VSC stations of the VSC-HVDC. It is concluded that both power systems, AC 1 and AC 2, are electrically decoupled.

However, the frequency drop experienced by AC 2 reaches a minimum frequency of $f_2 = 48.6$ Hz during the transient period. This frequency value is lower than $f_2 \approx 49.4$ Hz in area 2 for the case when the two-area system is linked by a conventional AC tie line, see Figure 8.7(a). This is explained by the fact that now AC 2 is taking care of f_2 independently. That is, without the inertia support and primary frequency regulation afforded previously by AC 1 to AC 2. This is a direct consequence of the electrical decoupling brought about by the HVDC link sitting between the two AC systems. Also, the power generated by G1, P_{g1}, remains unperturbed, leaving the power plant G2 to carry the full burden of frequency regulation at its own control area. This is satisfactorily returned to its nominal value $f_2 \approx 50$ Hz at $t = 80$ s, owing to the primary and secondary spinning power reserves available in AC 2, which are delivered to the electrical network through the operation of the ALFC loops of the power plants.

Comparing the dynamics of the two systems, shown in Figures 8.12(d) and 8.7(d), these could not be more different. In the case of the VSC-HVDC link, power oscillations are non-existent. Qualitatively speaking, this aspect is directly transferred into the dynamic responses of frequency f_2 and power generation P_{g2} at AC 2 since both variables present much smaller oscillations than those observed in Figure 8.7(a) and (c). Such dynamic behaviour results from the power and voltage controls of the VSC stations. For one, VSC 1 appropriately controls its input power $P_{vsc1} = 1$ pu to the reference value $P_{ref} = 1$ pu. Note that in the case of a point-to-point VSC-HVDC link, the power flow of VSC 2 is opposed to that of VSC 1, $P_{vsc2} = -P_{vsc1}$. The other key point is that converter VSC 2 controls its DC bus voltage E_{2dc}, keeping it constant at the voltage setpoint $E_{ref} = 2.0$ pu. Since the DC link power is kept constant by VSC 1, its voltage also remains constant at $E_{1dc} = 2.0278$ pu throughout the contingency period. The DC link voltages related to both converters are shown in Figure 8.12(d).

An important aspect that should not be overlooked when point-to-point VSC-HVDC links replace AC tie lines is that one of the main functions of the secondary frequency control carried out by SG power plants becomes unnecessary.

It should be reminded that the area control errors in the conventional two-area power system, as shown in Figure 8.5, are: $\text{ACE}_1 = \Delta P_{23} + B_1 \times \Delta f_1$ for AC 1 and $\text{ACE}_2 = \Delta P_{32} + B_2 \times \Delta f_2$ for AC 2. However, in the two-area system joined by the HVDC link, the power plants are no longer required to minimise the power flow errors in the tie line, ΔP_{23} and ΔP_{32}. The reason is that it is the VSC-HVDC link that regulates the power flow locally. This leads to new area control errors in AC areas linked by HVDC systems, namely, $\text{ACE}_1 = B_1 \times \Delta f_1$ and $\text{ACE}_2 = B_2 \times \Delta f_2$ for AC 2. Under these operating conditions, some favourable technical and economic consequences can be readily identified:

(i) Power flows are better controlled in quality and quantity, and time. In terms of both quantity and quality, power flow overshoots above the transmission line ratings are prevented, and power oscillations are filtered out by the AC-to-DC conversion process carried out by the VSC stations, along with their ancillary equipment and controls. Concerning time, the power flow regulation is exerted by the power electronic converters, which have shorter time responses than the electromechanical controls of SG power plants.
(ii) VSC-HVDC links carry out power flow control locally, so costly measurement infrastructure is not needed along the tie-line linking the power plants.
(iii) SG power plants selected to carry out secondary control would experience a much-reduced mechanical stress on their generator and turbine shafts.

In this context, when the power flowing through the VSC-HVDC link is modified following a technical-economic action, the converter charged with power control strategy, P_{ctrl} or P/E_{ctrl}, receives a change signal to modify the VSC power set point P_{ref}, either from the HVDC operator, or the TSOs, or automatically, according to the operating criterion established.

To illustrate these points, a simulation is carried out assuming that VSC 1 exerts power control, P_{ctrl}, in the VSC-HVDC link of Figure 8.7. A power change takes place at t = 3 s of the simulation. The power setpoint P_{ref} changes from 100 MW to 120 MW with a rate of change of 20 MW/min. Figure 8.13 shows the dynamic results for the change in VSC-HVDC power flow. It is observed that the impact on the electrical power system is not taxing at all.

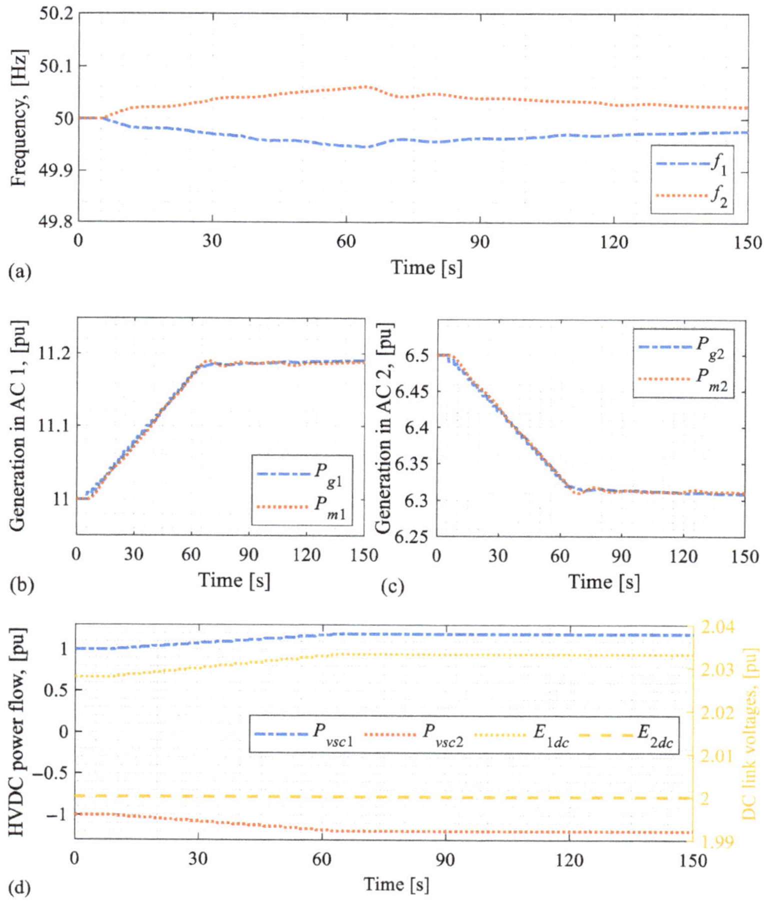

Figure 8.13 ALFC results with the VSC-HVDC link changing its reference power: (a) generators' frequencies, (b) electrical and mechanical powers from G1, (c) electrical and mechanical powers from G2, and (d) powers and voltages in the HVDC link

It should be noted that since more power is taken from AC 1 and injected into AC 2 through the VSC-HVDC link, f_1 reduces and f_2 increases, albeit in a transient manner. However, they do not deviate much from their nominal values – both staying within their safe range of 50 ± 0.1 Hz, as shown in Figure 8.13(a). To counteract these frequency deviations, opposite changes in power generations G1 and G2 are obtained in the corresponding networks AC 1 and AC 2, e.g., the negative change in f_1 produces a positive change in P_{g1}. The opposite applies to f_2 and P_{g2}, see Figure 8.13(b) and (c). The output powers of G1 and G2 follow the changes imposed by the VSC-HVDC link new setting, as shown in Figure 8.13(d). In the same figure, the DC link voltage is given, where E_{2dc} remains constant, and E_{1dc} drops slightly to reflect the positive increase in P_{g1}. It ought to be said that these kinds of small DC voltage changes are expected in the operation of VSC-HVDC links when changing the power flow transmission.

The ALFC studies presented above have used contrived two-area power networks, but nonetheless, they carry practical interest and yield great insight into ALFC. The dynamic performance of the VSC-HVDC link used above corresponds to a typical assessment carried out by TSOs where such installations exist. These installations have been used since the early 1960s when the HVDC technology was first introduced. Although such schemes were few at the time, they are now proliferating owing to the marked improvements that have taken place in the HVDC technology. This has progressed from CSC-HVDC links using mercury arc valves and then thyristor valves to the current VSC-HVDC links using IGBT valves. These HVDC installations may be point-to-point or back-to-back. Chapter 2 gives examples of these installations around the globe.

The above ALFC exercise elucidated that VSC-HVDC links are less susceptible to external contingencies, exhibiting reduced power flow oscillations. This reduces the risks of separating the interconnected AC systems during contingencies. This action is often provoked by under-frequency or protection relays sensitive to power oscillations. It follows that VSC-HVDC links installed at critical points of the AC power grid reduce instances of cascading failures, which are frequent in control areas coupled by classic AC transmission lines. Equally important is that TSOs' pre-established energy transactions are more likely to be guaranteed, even during transient events characterised by ample time response windows, such as ALFC-related contingencies, which can persist for several minutes before the interconnected systems reach a new steady-state regime.

8.4 Multiterminal VSC-HVDC power systems

The natural progression of the point-to-point VSC-HVDC link is the multiterminal VSC-HVDC grid, where the number of VSC stations (or terminals) is three or more, depending on the complexity of the DC grid. This is illustrated in schematic form in Figure 8.14(a) for the case of ρ VSC stations.

These multiterminal VSC-HVDC systems would enable the simultaneous interaction of several AC power grids exhibiting different physical and operating

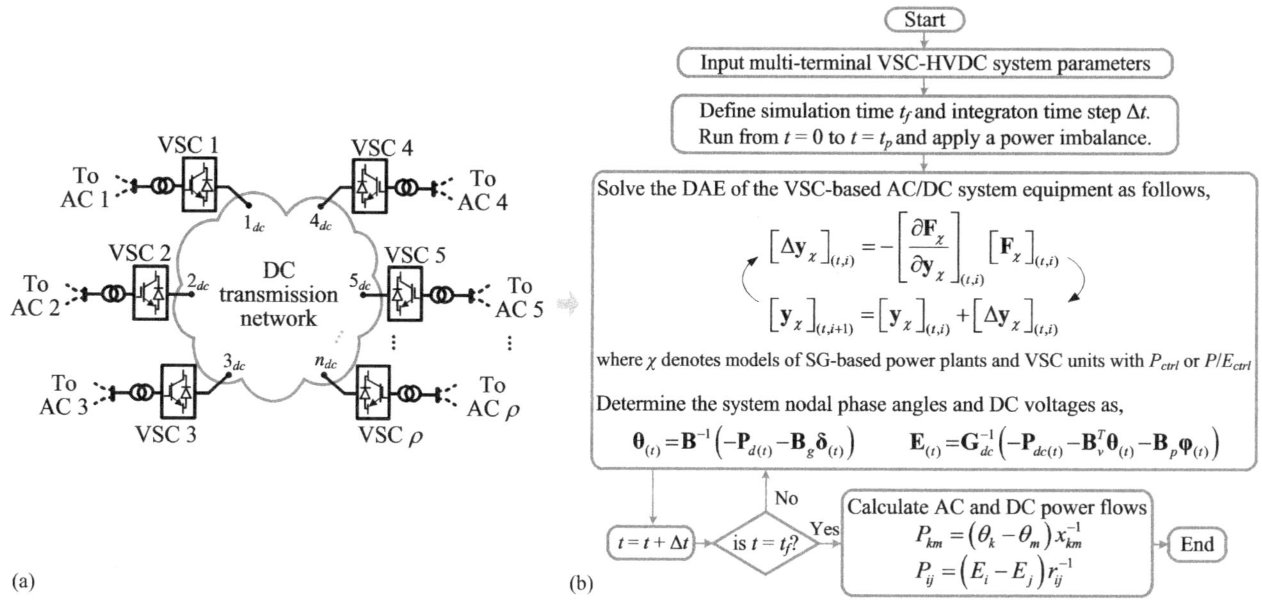

Figure 8.14 (a) Schematic representation of multiterminal VSC-HVDC power systems. (b) Dynamic modelling framework for ALFC studies.

characteristics, such as power grids operating at 50 Hz and 60 Hz, power grids with strong and weak inertial responses, power grids dominated by conventional SGs, and power grids with a high content of wind and solar power generation. However, it ought to be said that no such AC/DC power networks exist at present. As presented in Chapter 2, in the survey of multiterminal HVDC installations, such DC power grids are at their preliminary stage. Still, accelerated progress may be around the corner owing to the many operational advantages that VSC-HVDC power grids have over AC power grids. An aspect of this is presented below.

Similarly to two-terminal VSC-HVDC links, the great technical advantage that multiterminal VSC-HVDC power grids would bring about is the fact that electrical energy exchanges between the interconnected AC systems would be conducted in a highly controlled manner, owing to the rapid power conditioning carried out by the VSC stations and their controls. This applies to both steady-state and dynamic operating regimes. Such power flow control would not be possible with purely AC transmission networks.

Such operating advantages of multiterminal VSC-HVDC systems would allow power systems operation to be more flexible and resilient. For instance, in the case of ALFC events, TSOs would channel their efforts to their local system contingencies rather than the interconnected power systems' external perturbations and operating issues.

However, from a system operating security standpoint, other kinds of challenges would emerge in multiterminal VSC-based AC/DC systems. The DC grid itself may be subject to contingencies. The VSC stations themselves may malfunction and trip off. The DC CB (circuit breaker) technology is still under development, and it may take considerable time before it reaches the level of reliability of AC CBs. An internal contingency in the DC power grid would certainly cause a critical power imbalance in the interconnected AC power systems.

If the DC power grid does not have an effective means for clearing the fault internally, its steady-state and dynamic operation will be affected. Large overvoltages and severe voltage drops will occur, impacting the DC bus voltages of the nearest VSC stations. Following the event, the VSC stations should act cooperatively to carry out an effective power-sharing control to mitigate the initial power imbalance.

The ability to perform power-sharing assignments between VSC converters is exactly what VSC stations with grid-forming control schemes, P/E_{ctrl}, are designed to do during contingencies; see the control loop of VSC 1 of Figure 8.8 and Table 8.2. Its operation is relatively simple. If the VSC experiences a voltage drop at its DC bus, $\Delta E_i = E_{refi} - E_i > 0$, the converter injects extra power to the DC grid by increasing its reference power by $P_{vsc} = P_{ref} + K_{ep}(E_{refi} - E_i)$. In contrast, when the VSC station undergoes a DC voltage rise, $\Delta E_i = E_{refi} - E_i < 0$, the converter reduces its DC power injection, P_{vsc}, using its power/voltage droop control scheme. This converter operation differs from that carried out by VSC stations with fixed-power control scheme P_{ctrl}, as previously discussed for the two-area system with one VSC-HVDC link shown in Figure 8.11. It should be clarified, however, that power converters neither generate nor consume power when performing power-sharing assignments. Rather, they adjust their reference power to increase or reduce its power flow, according to requirements. This affects the power balance on both sides of the converter. It should be noted that the DC grid voltage support by VSC stations with the grid-forming control scheme, P/E_{ctrl}, is like that conducted by SG power plants when providing frequency support to AC systems. In both cases, the aim is to improve the dynamic performance of the AC and DC systems during critical power imbalances that could represent a risk to their operating security.

The flow chart of Figure 8.14(b) shows the dynamic modelling framework for studying ALFC in multi-area AC systems interconnected by multiterminal VSC-HVDC transmission systems [1]. The solution algorithm is like that employed for the case of two-area systems with a two-terminal HVDC link discussed in the previous section. Only straightforward extensions are needed in the vector and matrix terms of (8.18) and (8.19) to make them applicable to n_{ac}-bus AC systems with γ SG power plants and a multiterminal HVDC system with ρ VSC stations and forming an n_{dc}-bus DC power grid. The ensuing dynamic simulation framework is suitable for studying slow electromechanical transients of the ALFC type. It is worth recalling that the DC voltage of VSC stations with grid-following control scheme E_{ctrl}, remains constant throughout the ALFC event. The dynamics of the VSC-HVDC system are dictated by the differential equations (8.10) and (8.11), which apply to each one of the VSC stations having either a grid-forming control scheme, P/E_{ctrl}, or a fixed-power control scheme, P_{ctrl}.

8.4.1 Frequency control in multiterminal VSC-HVDC systems

To exemplify some of the dynamic interactions that are likely to appear in a multiterminal power system due to ALFC-related events, the three-area, hybrid AC/DC power grid shown in Figure 8.15 is employed. The three VSCs, making up the three-terminal VC-HVDC system, are all linked up on their DC sides by four DC transmission lines. There is a DC node with a DC load point. Thermoelectric turbines drive the synchronous generators G1 and G2, and G3 is driven by a hydroelectric turbine. Table 8.4 provides the parameters of the generating power plants and the AC transmission lines, while Table 8.5 provides the parameters of the VSC stations and VSC-HVDC transmission link.

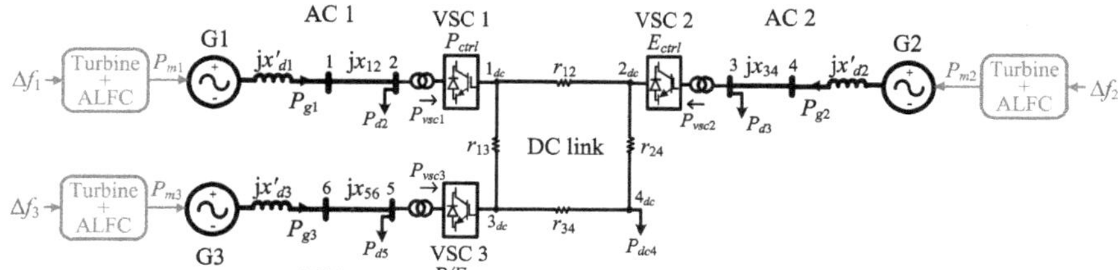

Figure 8.15 Three-area AC/DC system joined by a three-terminal VSC-HVDC system

Table 8.4 Parameters of the three-area VSC-HVDC power system

Power systems AC 1, AC 2, AC 3 100-MW base power		**G1, G2 driven by thermoelectric turbines G3 driven by hydroelectric turbine Generators' P_{nom} base power**
Loads	P_{d2} = 10 pu P_{d3} = 7.5 pu P_{d5} = 5.0 pu	Nominal powers of G 1, G 2 and G3: P_{nom1} = 1200 MW, P_{nom2} = 1000 MW, P_{nom3} = 750 MW. SR_{p1} = 5 %, SR_{s1} = 6%; SR_{p2} = 5 %, SR_{s2} = 10%; SR_{p3} = 5 %, SR_{s3} = 10%;
Generators	P_{g1} = 10.75 pu P_{g2} = 8.25 pu P_{g3} = 4.0 pu	$x'_{d1} = x'_{d2}$ = 0.314 pu; x'_{d3} = 0.333 pu; $D_1 = D_2 = D_3$ = 0 s/rad; $H_1 = H_2$ = 5 s; H_3 = 4 s; $R_1 = R_2 = R_3$ = 0.08 pu; $T_{g1} = T_{g2}$ = 3 s; T_h = 3 s;
Transmission Lines	x_{12} = 0.08 pu x_{34} = 0.07 pu x_{56} = 0.09 pu	$K_{s1} = K_{s2} = K_{s3}$ = 0.15; $B_1 = B_2 = B_3$ = 7.5; $T_{ch1} = T_{ch2}$ = 0.3 s; $T_{rh1} = T_{rh2}$ = 7 s; $T_{co1} = T_{co2}$ = 0.2 s; $K_{hp1} = K_{hp2}$ = 0.3; $K_{ip1} = K_{ip2}$ = 0.4; $K_{lp1} = K_{lp2}$ = 0.3; T_w = 0.5 s;

Table 8.5 Parameters of the three-terminal VSC-HVDC link

Multiterminal VSC-HVDC link100-MW base power	
VSC power rating	P_{nom} = 200 MW
DC lines	r_{12} = 0.0209 pu; r_{13} = 0.0278 pu; r_{24} = 0.0348 pu; r_{34} = 0.0417 pu
DC load	P_{dc4} = 0.5 pu
Phase reactors	$x_{ph1} = x_{ph2} = x_{ph3}$ = 0.075 pu
Transformers	$x_{tr1} = x_{tr2} = x_{tr3}$ = 0.075 pu
VSC 1 – P_{ctrl}	k_p = 1e-4; k_i = 50e-4; K_{ep} = 0
VSC 2 – E_{ctrl}	E_{ref} = 2.0 pu
VSC 3 – P/E_{ctrl}	k_p = 1e-4; k_i = 50e-4; K_{ep} = 150

Concerning the DC power system, VSC 1 is taken to follow a fixed-power control scheme with a power setpoint of P_{ref} = 0.75 pu, VSC 2 is taken to be a grid-following converter with a voltage setpoint of E_{ref} = 2 pu, VSC 3 is taken to be a grid-forming converter with $P_{ref} = -1.0$ pu. The DC power grid contains four nodes, with the first three nodes associated with the DC ports of the three VSC stations. The fourth DC node contains a load point P_{dc4} = 0.5 pu.

For these operating conditions and considering the initial powers of the three SG power plants – see the Loads and Generators entries in Table 8.4, the steady-state power flows through the VSC stations are P_{vsc1} = 75 MW, P_{vsc2} = 75 MW and $P_{vsc3} = -100$ MW. Hence, AC 1 and AC 2 are exporter systems, whereas AC 3 is an importer system.

This study assumes that the DC load $P_{dc4} = 50$ MW is abruptly lost at $t_p = 3$ s of the simulation run. The DC transmission grid experiences a momentary excess of energy, with a positive energy balance of the same value as the disconnected load. The ALFC-related frequency and power dynamics are shown in Figure 8.16. The full simulation run is $t_f = 150$ s.

By close examination of Figure 8.16(a), attention is drawn to the following points: the frequency of AC 1 remains unaffected by the DC contingency, while those of AC 2 and AC 3 are affected in different degrees by the power imbalance.

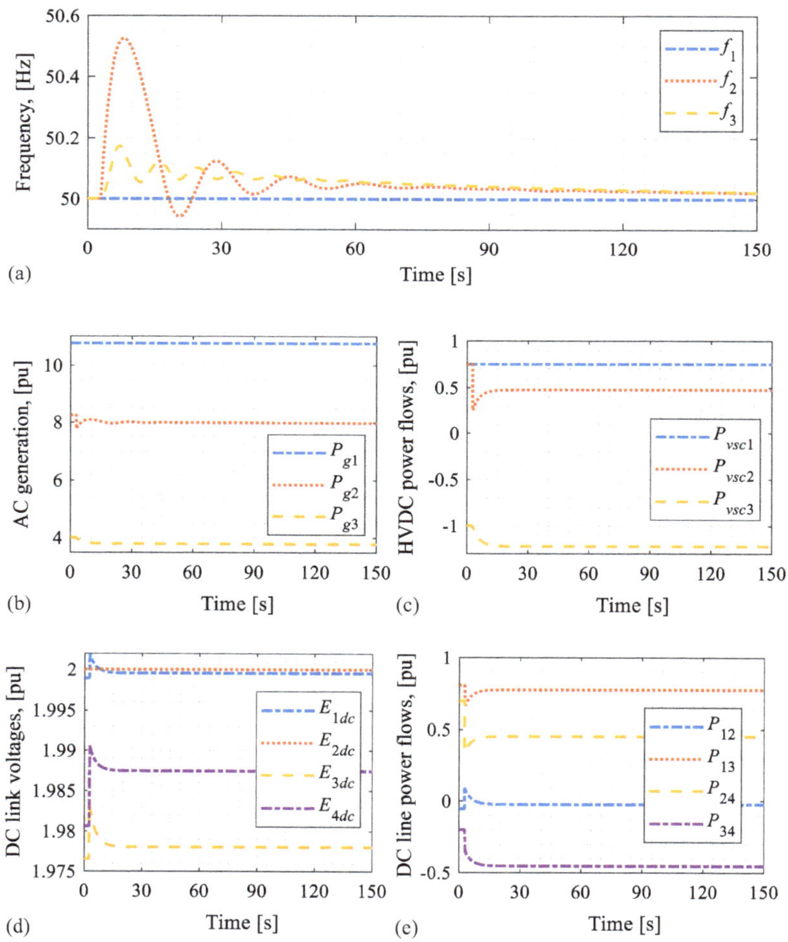

Figure 8.16 Dynamic performance of the three-terminal VSC-HVDC system following a DC load disconnection: (a) generators' frequencies, (b) electrical powers from G1–G3, (c) power flows from VSC 1–VSC 3, (d) DC link voltages, and (e) DC link power flows

The frequency f_2 in AC 2 is the most affected, reaching a peak value of f_2 = 50.52 Hz at about t = 7.5 s, whereas that of AC 3 is always below 50.20 Hz. These frequency dynamics agree with the dynamic behaviours of the electrical powers of the SG power plants shown in Figure 8.16(b). More specifically, when the load decrease occurs in the power network, the AC/DC system presents a transient energy surplus, with the SG power plants in AC 2 and AC 3 accelerating because they are gaining kinetic energy. As expected, the inertial response of the SG power plants makes them reduce their output power, thus opposing the initial disturbance. Seconds later, the ALFC loops of the power plants take over to mitigate the frequency deviations even further.

The dynamic behaviour of the three equivalent areas is directly related to the control schemes of the VSC stations through which they are coupled to the DC network: (i) VSC 1 electrically decouples AC 1 from the rest of the HVDC system since it follows a fixed-power control strategy; (ii) VSC 2, which is in charge of controlling the voltage in the DC network, acts as a 'slack generator' to reestablish, as quickly as possible, the energy balance in the DC grid by taking the required power from its associated system, namely, AC 2; (iii) converter VSC 3 helps to counteract the DC grid unbalance due to its grid-forming control scheme, but only up to a certain level according to both nodal voltage deviations and droop constant.

This assessment is complemented by examining Figure 8.16(c), which shows the dynamic performance of the three VSC stations making up the HVDC system. From these results, it is corroborated that the power through VSC 1 remains unchanged at P_{vsc1} = 75 MW, while VSC 2 and VSC 3 modify their power flows from P_{vsc2} = 75 MW and P_{vsc3} = −100 MW, respectively, during pre-contingency conditions, to P_{vsc2} = 47.425 MW and P_{vsc3} = −122.425 MW, respectively, following the post-contingency conditions, at t = 150 s. In other words, both converters reduce their power flows by ΔP_{vsc2} = −27.575 MW and ΔP_{vsc3} = −22.425 MW, respectively. The sum of both power flows adds up to the 50 MW lost by the load disconnection of P_{dc4}.

The power-sharing characteristic of the multiterminal VSC-HVDC system in this dynamic study illustrates their great potential in ALFC-related applications. Figure 8.16(d) and (e) presents results for the dynamic behaviours of the DC link voltages and power flows. Notice that as soon as the DC load P_{dc4} is disconnected from the network, the DC nodal voltages start increasing, except E_{2dc}, due to the excess of energy in the DC link. The DC nodal voltage that shows more of this effect is the one at node 4, which is where the load disconnection took place.

A second worthwhile case to be assessed is the sudden loss of a VSC station, which physically decouples the associated AC power system from the rest of the AC/DC power grid. To illustrate this operating scenario, it is assumed that there is a sudden converter outage of VSC 1 from the AC/DC system shown in Figure 8.15. This implies the loss of P_{vsc1} = 75 MW in the original system. The results presented in Figure 8.17 show more pronounced variations in the electrical frequencies of AC 1 and AC 2, which reach f_1 = 51.12 Hz and f_2 = 49.01 Hz, respectively. These are expected results since when VSC 1 is lost, setting P_{vsc1} = 0 MW, AC 1 is left with excess energy, while AC 2 and AC 3 are left with energy deficits to different

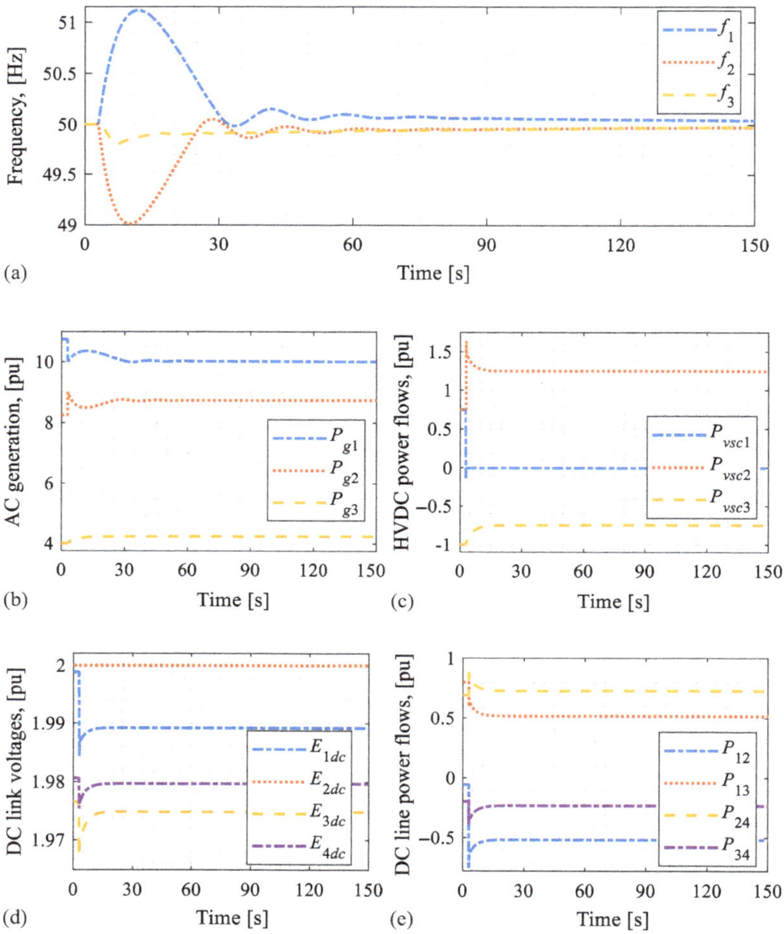

Figure 8.17 Dynamic performance of the three-terminal VSC-HVDC system following the sudden outage of converter VSC 1: (a) generators' frequencies, (b) electrical powers from G1–G3, (c) power flows from VSC 1–VSC 3, (d) DC link voltages, and (e) DC link power flows

extents. All three AC systems mitigate their own frequency deviations aided by the power production of the SG power plants P_{g1}, P_{g2}, and P_{g3}, commanded by their corresponding ALFCs. Likewise, the VSC-HVDC system undergoes marked changes in VSC power flows, P_{vsc2} and P_{vsc3}, DC links voltages E_{1dc}, E_{3dc}, and E_{4dc}, and DC line power flows P_{12}, P_{13}, P_{24}. and P_{34}. In this operating scenario, the DC nodal voltages decrease because the DC transmission network abruptly stopped receiving 75 MW from AC 1 through VSC 1, and, in connection with it, its associated nodal voltage E_{1dc} is the one undergoing the largest deviation. The DC power flows P_{12} and P_{13} also experienced large changes because of the converter outage. The power flow in VSC 3 changed from $P_{vsc3} = -100$ MW to $P_{vsc3} = -74.77$ MW, a drop of about 25 MW less power being drawn from the DC network.

In these simulation studies, the nodal voltages and power flows of the VSC-HVDC system reached new steady-state values in about 5 s following the contingency. This time response is dependent on the integral and control gains of the PI regulators of the VSCs. It may be observed from the set of results presented that the nodal voltages and power flows are rather smooth, with their associated oscillations disappearing relatively faster than those observed in AC systems.

8.5 Concluding remarks

Today's electrical power grids carry larger power throughputs than in the past. This is achieved by using improved power transmission infrastructure, more complex steady-state, and dynamic control strategies. The fundamental task of planners and operators in a control centre is to balance power generation and customer demand. This requires accurate and swift actions. In recent years, such decision-making has been impacted positively by new technologies based on power electronics and the use of more sophisticated controls and powerful computer algorithms residing in industry-grade computer packages. In this context, the use of VSC-HVDC links finds a great area of opportunity. The operation of AC power systems is greatly enhanced when such equipment is strategically placed within the AC power grid. One key aspect is to increase the energy import-export between power grids with a finer power flow control than with classical AC tie-lines, which have no regulating capabilities. However, to assess the cost-benefit of new VSC-HVDC installations, comprehensive dynamic analyses must be carried out to determine its technical impact and the control strategies to use for best performance. This chapter introduced one simple yet effective way of modelling VSC-HVDC links for interconnecting AC power systems subjected to ALFC operations.

The ALFC response of AC/DC networks when subjected to load/generation imbalances was examined. Smoother power oscillations are observed compared to those in classical AC transmission systems. The reason is that the VSC-HVDC link imposes tight control over the voltage and power flow in the DC link. This damps out the power and frequency fluctuations in the interconnected power networks, preventing power oscillations in the DC link. Nonetheless, for the AC area systems to experience smaller oscillations, each AC control area must be able to satisfy its own demand requirements following a disturbance. They must be prepared with enough primary and secondary operating reserves for the ALFC of the entire AC/DC network to operate successfully. Another important aspect to highlight is that some of the functions of the secondary frequency control are achieved locally by the VSC-HVDC link. This contrasts with the AC tie line bias control of the secondary control of the SG power plants, which relies on additional communication systems, data processing, and control systems. Improved dynamic performances were observed with the use of the VSC-HVDC link in terms of time responses and power/frequency oscillations, thanks to the rapid response of the VSC units.

In recent years, the controls of VSC stations have improved greatly, making VSC-HVDC installations and the AC power network more flexible. Besides performing DC voltage and DC power control, VSC stations may also be fitted with voltage/power droop controls that allow them to modify their power setpoint concerning changes in their DC voltages. This key point was discussed within the context of multiterminal VSC-HVDC systems. It was shown that VSC units with voltage/power droop controls help to mitigate the impact of power imbalances on the DC grid and the frequency deviations of the VSC-upgraded AC power systems. This becomes more important when power imbalances occur in the DC power grid.

As evidenced by the comprehensive survey of HVDC installations presented in Chapter 2, the number and capacity of VSC-HVDC systems is increasing globally. The technical advantages of VSC-HVDC with respect to classical AC transmission lines benefit the dynamic and transient operation of modern AC power grids. By judiciously applying their power, voltage, and power/voltage droop control capabilities they can contribute to the controllability of the whole network, just as synchronous generators do.

References

[1] Castro LM. "Simulation framework for automatic load frequency control studies of VSC-based AC/DC power grids". *International Journal of Electrical Power and Energy Systems*, vol. 141, 2022, 108187.

[2] Elgerd OI. *Electric Energy Systems: An Introduction*. New York: McGraw-Hill, 2nd edn, 1982.

[3] Wood AJ and Wollenberg BF. *Power Generation, Operation, and Control*. New York: Wiley, 1996.

[4] Sauer PW and Pai MA. *Power System Dynamics and Stability*. Englewood Cliffs, NJ: Prentice-Hall, 1998.

[5] Stagg GW and El-Abiad AH. *Computer Methods in Power System Analysis*. New York: McGraw-Hill, 1968.

[6] Dommel HW and Sato N. "Fast transient stability solutions". *IEEE Transactions on Power Apparatus and Systems*, vol. PAS-91, pp. 1643–1650, 1972.

[7] Rafian M, Sterling MJH, and Irving MR. "Real-time power system simulation". *IEE Proceedings of the Generation, Transmission and Distribution*, vol, 134, pp. 206–223, 1987.

[8] Acha E, Roncero-Sanchez P, De La Villa-Jaén A, Castro LM, and Kazemtabrizi B. *VSC-FACTS-HVDC: Analysis, Modelling and Simulation in Power Grids*. New York: Wiley, 2019.

[9] Yazdani A and Iravani R. *Voltage-Source Converters in Power Systems – Modeling, Control, and Applications*. New York: Wiley, 2010.

Chapter 9

The role of FACTS and HVDC in wind power generation

Uriel Fernando Sandoval Pérez[1]

9.1 Introduction

Renewable power generation is the process of generating electricity from renewable resources that are replenished naturally and sustainably such as solar, wind, geothermal and biomass energy. These energy sources are becoming increasingly important in transitioning to a low-carbon economy and reducing greenhouse gas emissions to mitigate the effects of climate change.

This electricity can feed residential, commercial and industrial customers with *green energy* or zero-emissions energy. The specific power generation methods vary depending on the type of renewable energy source being used.

Integrating and operating intermittent renewable power sources into the electrical power grid represents a significant challenge due to their power output variability. This is where flexible AC transmission systems (FACTS), HVDC and energy storage systems have a significant role to play. The power electronics converters employed in these technologies improve the stability and reliability of the electrical power grid with renewable power sources. Note that renewable energy is a term used by electrical power engineers to signify electrical energy produced with a primary energy resource that is self-replenishing. These topics follow basic concepts that have been introduced in Chapter 7.

This chapter focuses on the integration of renewable power generation into the grid with the aid of power electronics, with reference to FACTS and HVDC links. It specifically looks at wind power technologies. It discusses the benefits and challenges of integrating these renewable power sources with FACTS and HVDC links. At the closure, the chapter provides future directions for renewable power generation with power electronics equipment, including technological advancements and the role of FACTS and HVDC in future renewable power generation systems.

[1]GE Vernova, Cambridge, MA, USA

9.2 The importance of renewable power in the energy mix and mitigation of climate change

The urgency to address climate change has never been greater, and the so-called renewable power plays a pivotal role in this global endeavour. Fossil fuels, such as coal, oil and gas, are the most significant contributors to global climate change, accounting for over 60% of global greenhouse gas emissions [1]. To avoid the worst impacts of climate change, emissions ought to be reduced by half by 2030 and to reach net zero by 2050.

In this pathway to net zero emissions, almost 90% of the global electricity generation in 2050 should come from renewable energy sources, with solar PV and wind together accounting for 70% [2]. Along with hydroelectric power, these energy sources are readily available, emitting little or no greenhouse gases at their operating stage and being more cost-effective than fossil fuels. As a matter of fact, wind and solar resources incur no cost – so, the 'fuel' is free. To reinforce the point, the cost of electricity from solar PV power fell by 85% between 2010 and 2020. Moreover, the transition from fossil fuels-driven to renewable power has significant economic and health benefits, including the creation of new high-quality jobs. The global renewable power workforce is expected to rise very considerably over the next three decades [3].

In conclusion, the growing ascendency of renewable power in the energy mix is relentless. It is not only a sustainable solution for our energy needs but also a crucial strategy for mitigating climate change and its devastating impact on our ecosystem and way of life.

9.3 Grid code requirements

As discussed in Chapter 7, grid codes govern the incorporation of renewable power sources and the integration of any new energy source into a power grid. The guidelines provided by grid codes dictate the necessary technical specifications, regulations and operational procedures for all components connected to the power system, including power generators, adjustable loads, energy storage systems and any other relevant units.

Grid codes are developed and revised by working groups comprising power transmission and distribution operators and regulators in collaboration with other essential stakeholders. These codes facilitate the interaction and energy exchange between system operators, generator owners and large electricity clients, intending to maintain the grid in a stable and secure operating state.

With the Net-Zero target by 2050 in sight, consolidating clear guidelines and technical specifications becomes even more crucial. These grid codes are pivotal in facilitating the organised integration of renewable power sources into power grids, ensuring a smooth transition without disrupting their routine operations. Many countries have revisited and expanded these regulations to account for renewable power sources and their connections to the grid.

Although the drafting of grid codes varies from nation to nation, they all draw from international experience and technical standards drafted by professional bodies such as IEEE and IEC. This global perspective ensures that the codes are tailored to meet the unique needs of local electricity systems and future development scenarios while also aligning with international best practices. Furthermore, establishing a consensus regarding minimum requirements compels manufacturers to enhance their products in a way suitable for deployment across the globe with little change.

Examples and a general discussion on grid code legislation in various countries were addressed in Section 7.2.1. As pointed out there, grid codes exhibit variations and are subject to adjustments even within the same country such as in the CAISO and ERCOT grid codes in the USA. In the section below, we focus on the common minimum requirements that all grid codes should fulfil when integrating renewable power sources in conjunction with FACTS devices, into the power grid.

9.3.1 Voltage and frequency operating ranges

Defining a minimum operating region for all generators in the power grid is the most basic and mandatory technical requirement. By defining this region, we ensure that all actors are operating at the same harmonious point. Indeed, these limits are defined not by the generators but by the network operators. This requirement avoids having fewer energy sources when needed the most. By way of example, imagine an under-frequency scenario when the operator needs to draw in more synchronised generation to enable a faster frequency recovery.

Generators are expected to operate at their rated active power within the grid's standard operating ranges all the time. They are also expected to stay online within an expanded voltage-frequency range for a short period. For instance, in China, the standard operating voltage varies from 0.9 to 1.1 per unit. If the voltage rises to between 1.1 and 1.2 per unit, the power sources must stay connected and supply active power for a minimum of 10 seconds. If the voltage further increases to between 1.2 and 1.3 per unit, they are required to stay connected and supply power for at least 0.5 s [4].

9.3.2 Zero voltage ride through, low voltage ride through and high voltage ride through requirements

Various kind of faults, like L-G fault, L-L fault and L-L-L faults, can compromise the stability of the entire power system. A voltage swell, or an over-voltage, can occur due to the switching of a capacitor bank. Hence, the fault ride-through (FRT) capability in the power grid sets the requirements for all power generators, including renewable power sources (RES), to remain connected during such transient events. The behaviour of the FRT capability is dependent on the time variation with respect to voltage or frequency. Therefore, it can be represented as a curve in terms of voltage versus time and frequency versus time, respectively [5]. This requirement is crucial for the coordinated operation of protection devices by providing optimised settings. The requirements established for voltage ride-through

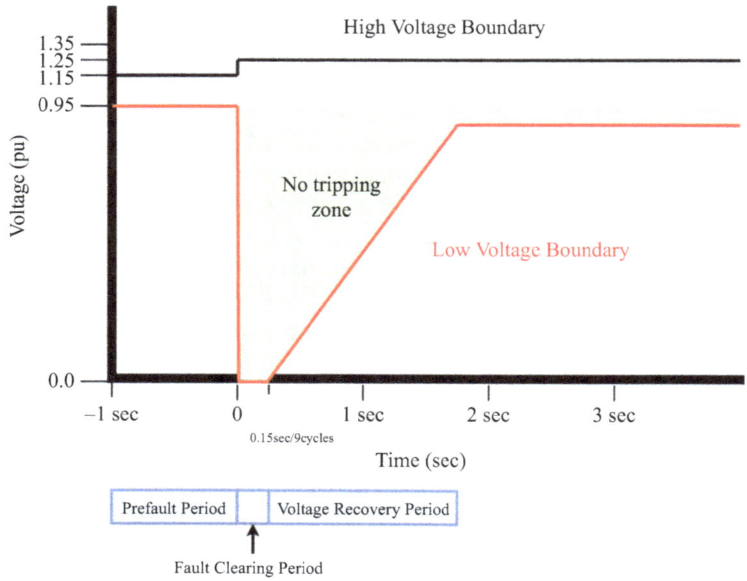

Figure 9.1 Voltage ride-through boundaries

are depicted in Figure 9.1. It is divided into the following three zones: low-voltage, zero voltage and high-voltage ride-through. Section 7.2.3 outlines the history and origin of this requirement.

9.3.2.1 Low-voltage ride through (LVRT)

It refers to the section of the voltage *vs* time curve where the renewable power generator must remain connected despite undergoing a low-voltage condition. During this time, the voltage value may be as low as 0.9 p.u. In most grid codes, it is also established that local reactive power support must participate during such events.

9.3.2.2 Zero-voltage ride through (ZVRT)

The ZVRT is a crucial case of the LVRT, in which the voltage may reach zero at the time of the fault.

9.3.2.3 High-voltage ride through (HVRT)

Similarly to the LVRT, the HVRT curve represents the curve where the renewable power sources remain connected due to temporary high voltage conditions. Reactive power support is also commonly requested by system operators to mitigate the effect on other elements in the power network.

As stated above, these requirements are adapted to each network by considering the following characteristics: short circuit fault and its amplitude, the network design, power control procedures, system protection, type of generator type and short circuit removal period [5]. All these characteristics are considered,

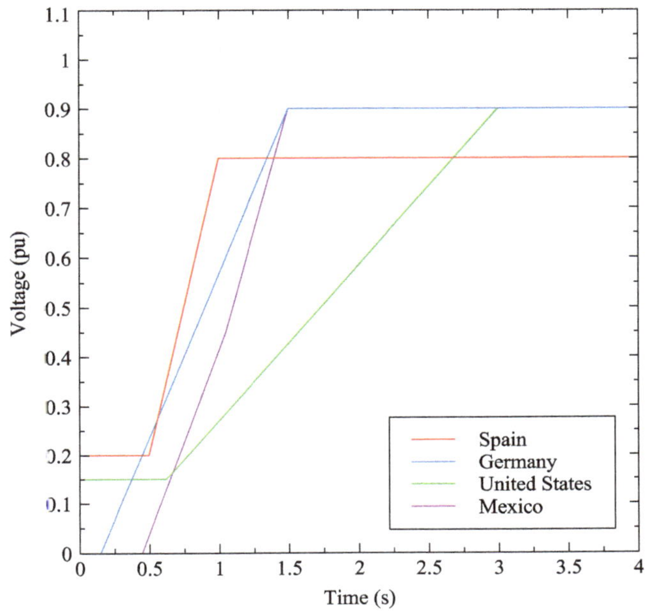

Figure 9.2 LVRT grid code requirements for different countries

together with specific FRT curves that exist in each specific country [6], as exemplified in Figure 9.2.

9.3.3 Grid-forming capabilities

The global trend towards renewable power systems has increased the demand for technologies and operating methods that would enable the operation of AC power grids with lesser reliance on large synchronous generators. VSC-type inverters used in most grid-connected power electronic applications, including renewable power sources and HVDC links, are typically used as current sources and require an external voltage source to provide a frequency reference [4]. While these inverters can be configured to provide some level of voltage and frequency support, such as droops (e.g., P(f), Q(U)) or FFR (fast frequency response), they cannot operate as standalone units. These inverters are commonly called grid-following (GFL) inverters, although other terms, such as grid-supporting or grid-feeding inverters (GFI), are sometimes used in the literature [4,7].

GFL inverters can respond rapidly to control signals or variations in voltage and frequency, typically within a few milliseconds. However, their response is not instantaneous, as it depends on measuring the voltage angle and frequency using a phase-locked loop (PLL) [7,8]. Moreover, GFL inverters do not inherently stabilise the grid during events such as load steps or short circuit faults. Grid-supporting functions must be explicitly programmed into their controls to become grid-forming (GFM) inverters. For instance, a comparative analysis of GFM inverters and GFL inverters responses to both loss of load and generation is presented in [9]; the results indicate that GFM technology demonstrates superior frequency-power control capabilities, characterised by reduced stabilisation times and diminished overshoots.

GFM inverters are designed to operate without an external voltage reference. The active power injection provided by these inverters is now commonly referred to as virtual inertia, as this functionality is distinct from synthetic inertia provided by grid-following wind turbines or FFR provided by other grid-following renewable power sources [4]. Note that virtual inertia does not require frequency measurements, mimicking more closely the inherent inertial response of large synchronous machines (e.g., synchronous generators and synchronous condensers). As the share of renewables continues to rise, other inverter control designs will become necessary to address operational challenges posed by a higher penetration of GFL inverters. Multiple technical concepts are available to implement the relevant inverter controls [10]; some of them are summarised in Figure 9.3, which highlights the functional control diagrams for GFL and GFM inverters. At a high level, GFL and GFM controls can be differentiated by the attributes listed in Table 9.1 [10].

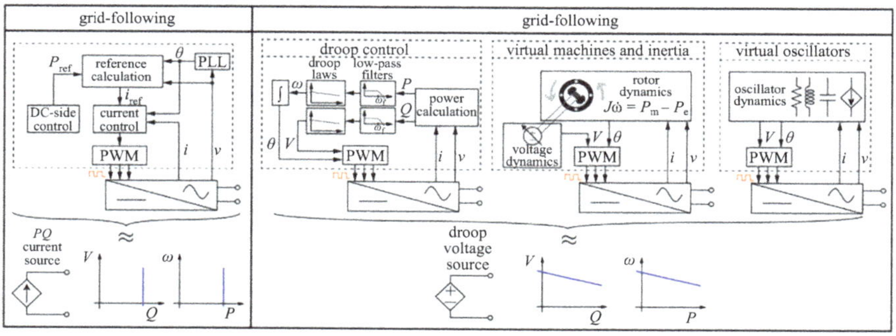

Figure 9.3 Functional diagrams of grid-following and grid-forming inverters. Taken from [10].

Table 9.1 Comparison of GFL and GFM controls parameters [4]

GFL	GFM
Assumes the grid is already formed under normal operation.	It is responsible for forming and maintaining a healthy grid.
Direct-quadrature (dq) vector control of current injected into the grid.	Control of voltage magnitude and frequency/phase.
Decoupled control of P and Q.	Slight coupling between P and Q.
Needs voltage at the point of common coupling to deliver P and Q.	Has black-start capabilities.
It cannot operate at 100% GFL inverter penetration. Instabilities thresholds (tipping points) exist.	Can theoretically operate at 100% power electronics penetration. It coexists with a high penetration of GFL but more than one GFM would be needed depending on the complexity and size of the grid.

However, notice that GFM inverters are not yet widely deployed due to several reasons [4]:

- The need for the virtual inertia functionality is not so evident at low VRE penetration levels.
- The number of resources required to attain stable operation with high penetration levels of instantaneous renewable sources has not yet been determined.
- The quick response of each grid-forming unit requires a specific amount of fast-releasing energy storage, which is seldom available in wind generation systems, and it is only starting to be employed in solar PV generation systems.
- To accommodate rapid load surges, the power rating must be increased by a certain margin, which adds to the overall expense of the system.
- An increased power rating may also be necessary to provide short circuit current during faults and other events requiring overcurrent capacity beyond the rated current.
- There is not yet a universally accepted definition of grid-forming or its expected functionality. This area is still under active research, with only a few original equipment manufacturers offering commercial-grade grid-forming inverters.

9.3.4 Black-start capability

The ability of a power generator to start operating without an external power source and to restore a portion or the whole of the power grid following a blackout is known as black-start capability. Not all generators within the system have this capability, and it is rarely specified as a requirement in grid codes.

Historically, black-start procedures have relied on fast-acting, large synchronous generators like hydro or gas turbine power plants to start up larger coal or gas-fired units. In coal-dominated power systems, black-start capability is often provided by power plants that can operate independently, also known as islanded operation, which can provide its own auxiliary services consumption for several hours following a grid outage. However, with the increasing share of renewable power, the number of black-start, synchronous generator plants being decommissioned and not replaced is increasing. This transition is compelling grid operators to revise their black start strategies and renewable power manufacturers to design and produce black start-capable converters.

The black-start capability is finding additional applications in isolated renewable power facilities, such as in the energy harvesting of green hydrogen in remote projects, where the electricity network is often unavailable or insufficiently robust to synchronise a large renewable power source. Moreover, it is interesting to note that such installations may not be required to transmit their output power, as this energy would be converted into chemical energy. An example of this renewable configuration is the Haru Oni project, commissioned in 2022 in the Magallanes region of Chile [11]. The project comprises a 3.4 MW wind turbine connected to a 1.2 MW electrolyser. It is anticipated that its capacity will increase in the future, compelling the project developer to either enhance the existing grid interconnection or implement a black-start capability in the converters of the wind turbines.

Also, grid codes specify requirements pertaining to frequency operation range, primary and secondary frequency regulation and power quality. However, these requirements will not be addressed in this chapter. For further information, readers are directed to Sections 7.2.4 and 7.2.5, which provide details on frequency regulation and power quality requirements applicable to renewable energy sources.

9.4 Role of FACTS in the integration of renewable power into the power grid

Grid code requirements are becoming increasingly stringent as more and larger renewable power plants are incorporated into power networks. Hence, power utilities are forced to own and operate power generation facilities that comply with local grid codes. To accomplish this, utilities have two options: either redesign and contract customised solutions from generator/technology manufacturers or reinforce their generation site at the point of common coupling (PCC) with the power grid using FACTS equipment.

While the request for a customised solution is physically feasible, it can significantly increase the overall project cost. In general, the most economical solution is to integrate FACTS supporting equipment into the generation facility – for instance, a STATCOM at PCC. When considering this option, it is crucial to keep in mind certain key points when integrating this technology into new or existing renewable generation projects. This will ensure that the integration process is a smooth one, and the many benefits afforded by the FACTS equipment are fully realised, as discussed below.

The optimal location of a FACTS device may not be at PCC between the wind farm and the power transmission grid. However, this may pose a conflicting commercial interest when the wind farm is not owned by the power transmission owner. The former would have its own specific operational and investment worries to attend. In such circumstances, the power grid owner would have different priorities from those of the wind farm owner. Placing a FACTS device at different locations in the network results in different values of network loadability, loss reduction, static voltage enhancement, generation costs and congestion management. This has been an active research topic, with several methods reported in the literature to address these allocation issues. To a large extent, optimisation techniques are used to assess the overall power system operation constrained by operational and physical limits, and grid code constraints. Such an approach suits well cases when the power grid owner and the wind farm owner are the same. Nonetheless, this opens the door for future business opportunities, when the ancillary services market develops more.

9.4.1 Wind energy

Wind power generation is the process of converting wind kinetic energy into electrical energy using wind turbines. It is a clean, renewable power source that has the potential to provide a significant portion of the world's electricity needs. However, the intermittency and variability of the wind represent an operational

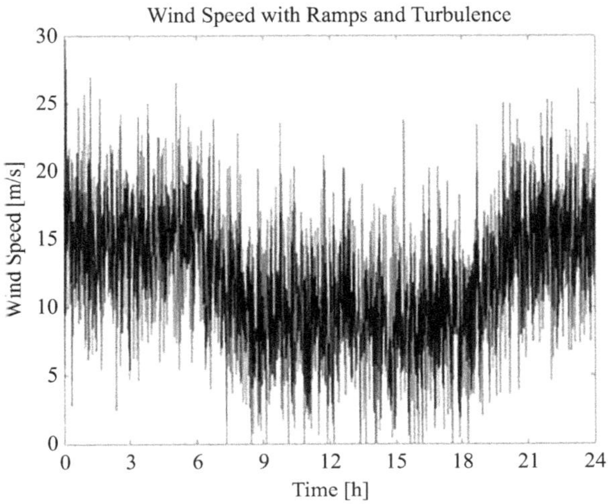

Figure 9.4 A typical wind speed profile with average speeds between 10 m/s and 15 m/s

challenge when integrating these power sources into the grid. By way of example, Figure 9.4 shows a 24-hour wind speed profile with ramps and turbulence for average wind speeds of 15 m/s and 10 m/s. Note that turbulence is a random phenomenon.

In broad terms, the wind energy industry has been divided into two main categories [12]: onshore and offshore.

9.4.1.1 Onshore wind energy

This involves installing wind turbines on land in regions of high winds (e.g., on top of mountain ridges or coastal zones). Despite these being remote zones where typically no transmission facilities existed previously to evacuate the generated power, it has been a cost-effective solution for new projects. However, in some countries, the installation of wind farms encounters natural and social barriers: it requires substantial land extensions and faces opposition from ecological and social groups, which espouse legitimate concerns. The UK is one of such countries, where most of the on-land sites with premium wind resources have already been populated with large wind farms. This has been one of the drivers for wind farm developers going offshore.

9.4.1.2 Offshore wind energy

Offshore wind power is concerned with installing wind turbines on sea waters, either nearshore or deep offshore. These projects typically boast superior production numbers and capacity factors because of the more regular wind speeds on the sea than on land. However, for deep offshore installations, they come with significantly higher installation and maintenance costs than for nearshore and onshore

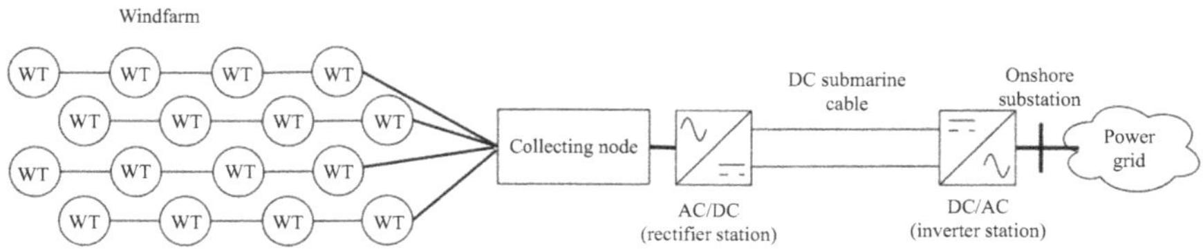

Figure 9.5 Schematic layout of an offshore windfarm interconnection

wind power installations. The reason is that deep offshore installations may require floating platforms. For distances longer than 70 km, submarine power transmission must be carried out using DC power instead of AC power. The many offshore wind farms in the North Sea are prime examples of the use of this technology [13].

However, note that most wind farms installed in the North Sea are not of the floating type, even though they may lie 100 km or more offshore. They lay on shallow waters, taking advantage of natural formations of submerged banks east of London and north of the German coast. However, these areas are already fully populated with wind turbines. Hence, an accelerated push for developing floating wind turbines is ongoing. For instance, a full-scale 3,000-tone floating prototype lays on Aberdeen Bay in North Scotland. This prototype captures enough wind to electrify 35,000 Scottish homes. These developments at the FloWave facility are key to the UK's ambitions to add 34 GW of offshore wind power by 2030. This requires moving to the next frontier on offshore wind farms, namely, MW-level, floating wind power turbines [14]. These floating structures have evolved from the decades-long wealth of expertise developed in that area of Scotland from the oil industry in North Sea waters. Incidentally, these developments may be used in the USA where there are plans to install 30 GW of floating offshore wind capacity by 2035 [15]. Recent research indicates that the Pacific coast could support 100 GW by 2050.

Nonetheless, connecting these types of projects to the grid is a complex undertaking, as it requires a specialised infrastructure to collect electricity from each turbine and transport the entire project's power output to a ground-based interconnection point, as illustrated in Figure 9.5.

Concerning modern power generation technology, two main options for harvesting wind energy are currently available: the so-called DFIG and the PMSG-FRC. Both of them are variable-speed wind turbines. This feature enables them to make optimal use of the wind resources available at any point in time.

9.4.1.3 Doubly fed induction generator (DFIG)

The DFIG wind generator uses a wound-rotor induction machine. The stator of this induction machine is directly connected to the AC grid via a step-up transformer. The rotor winding is connected asynchronously to the AC grid using slip rings. The asynchronous connection to the rotor is carried out through the machine-side converter, as shown in Figure 9.6. The normal electromagnetic coupling between the stator and rotor winding via the air-gap machine stands as normal induction machines. In this DFIG design, the generated power is injected into the AC high-voltage bus directly from the stator via the interfacing transformer and through the back-to-back HVDC link when the DFIG's rotor runs at super-synchronous speed. Conversely, when the DFIG's rotor runs at sub-synchronous speed, power is injected from the stator into the interfacing transformer, but a portion of that power returns to the DFIG's rotor via the back-to-back HVDC link. The rotor converter is ¼ the size of the stator-side converter, which is fully rated.

Although not shown, the rotor circuit contains a thyristor-controlled mechanism called crossbar or crowbar, activated during grid faults. This mechanism protects the rotor-side converter from the large currents generated by the fault, allowing it to ride through the fault. Essentially, this mechanism transforms the DFIG into a squirrel-cage induction generator, enabling it to keep operating but with no variable speed features for the duration of the fault.

9.4.1.4 Permanent magnet synchronous generator-fully rated-converter (PMSG-FRC)

In this generator, all the energy from the rotor is transferred onto the stator in electromagnetic form and onto the step-up transformer through a back-to-back VSC-HVDC link, as illustrated in Figure 9.7. The generator typically uses a multi-pole, permanent magnet synchronous machine since the rotor rotates at the speed of the blades [13]. These machines are axially short but radially quite large – several meters in diameter for multi-MW machines. This type of wind turbine generator is very flexible regarding connection to the grid. The two VSC converters employed for interconnection provide a decoupled control of active power and voltage control at the high-voltage side of the step-up transformer. Furthermore, it boasts a wide range of low-voltage ride-through capability, as reported in [16], with three-phase faults lasting more than 600 ms still being controllable. It features a reactive power capability curve that rivals conventional synchronous generators, underlining its adaptability.

Note that in this design although the generator's rotor runs at synchronous speed, the frequency of its rotating magnetic field is very low, compared to the grid frequency, owing to the lack of a gearbox. A crucial role of the back-to-back HVDC link is to act as a frequency converter to produce a compatible frequency with the AC grid. Moreover, the stator-side converter bestows variable speed

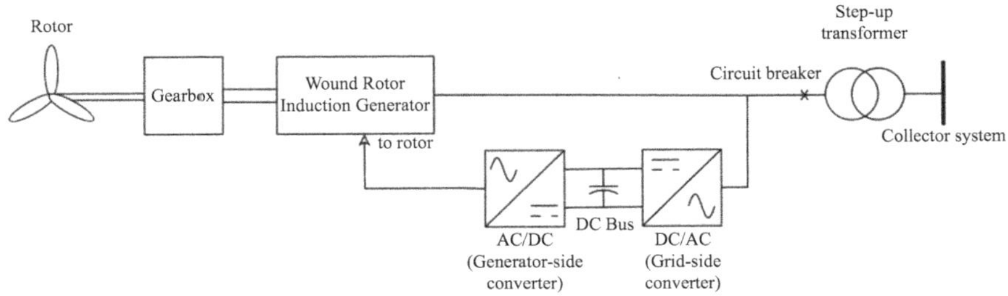

Figure 9.6 Schematic representation of DFIG wind turbine

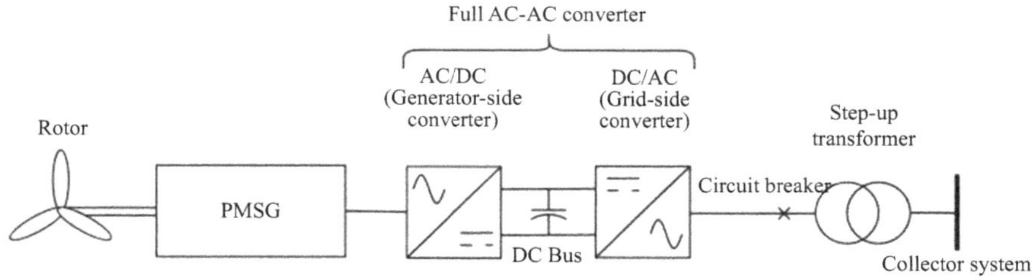

Figure 9.7 Schematic representation of a PMSG-FRC wind turbine

characteristics on the generator by modulating the generator's stator currents, whereas the grid-side converter acts like a STATCOM. Also, note that contrary to conventional synchronous generators, the PMSG does not possess an automatic voltage regulation (AVR) system. It is fluxed at a constant rate by permanent magnets embedded in the fabric of the generator at the construction stage. The stator-side converter acts as its AVR system.

9.4.2 Reactive power improvement using FACTS

The variable nature of wind power, influenced by weather conditions and the specific turbine design characteristics, may cause power output fluctuations that can impact the grid voltage levels and create instability. This is a more critical issue in wind farms that use fixed-speed wind turbines, as opposed to DFIG or PMSG-FRC. As mentioned above, wind farms are frequently situated in remote locations and use radial AC transmission lines of relatively low voltage to evacuate their power outputs, causing power losses and voltage regulation issues. This problem becomes more acute as the number and size of wind farms grow in a specific area of the power grid.

Broadly, the variable speed capability of the newer generation of wind power turbines – afforded by power electronics – has reigned over the rapid variations of the wind but not the longer-term variations. For these, we need ancillary equipment: energy storage equipment to provide active power support and FACTS equipment to provide reactive power support. The former may involve BESS and SMES, whereas the latter may be SVCs and STATCOMs.

By strategically installing these devices together with TCSCs and UPFCs at critical points of the power grid, transmission system operators could manage more effectively the power fluctuations, regulate voltage levels better and improve grid stability.

STATCOMs provide precise and rapid reactive power compensation, aiding in stabilising grid voltage and improving power factor. By dynamically injecting or absorbing reactive power, according to requirements, STATCOMs mitigate the impact of variable wind power on grid voltage, ensuring smooth and stable operation of the electrical system. Similarly, SVCs offer dynamic voltage support and reactive power control, enhancing grid stability and power quality

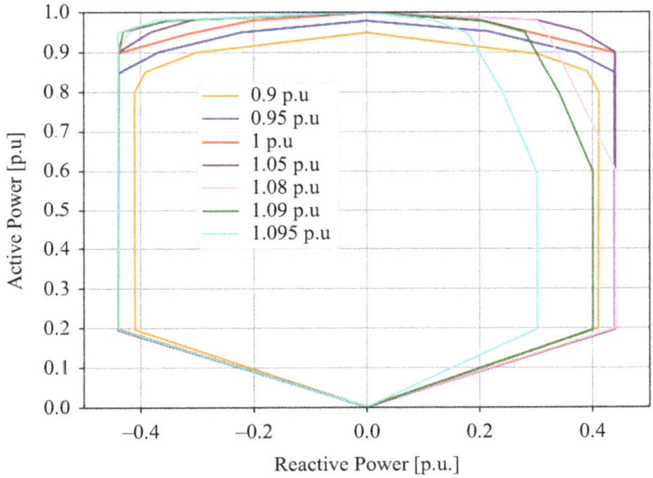

Figure 9.8 Reactive power capability curve of a 2 MW DFIG taken from [17]

even when faced with rapid changes in wind power outputs. Note that the STATCOM performs the same function as the SVC but exhibits an all-round better performance.

By way of example, a typical capability curve of a DFIG with reactive power limits is shown in Figure 9.8 [17]. These curves represent the reactive power output at different levels of active power and voltage magnitude of a 2 MW DFIG, measured at the primary side of the interfacing transformer. Note that each colour curve corresponds to a given voltage magnitude, as indicated in the figure, ranging from 0.9 p.u. to 1.095 p.u.

The reactive power capability varies across different levels of voltage and active power, due to the number of control systems and physical limitations. The production of reactive power is typically constrained by the rotor current limits. Conversely, the absorption of reactive power is constrained by the stator current limits [18].

Grid code requirements establish the minimum capability curve at which the wind farm must supply reactive power at different levels of active power. Typically, this requirement is measured at PCC, which may be located some kilometres away from the wind farm. In some instances, to satisfy this grid code requirement, a STATCOM of suitable dimensions is used to improve the wind farm's reactive power output at PCC. Figure 9.9 shows the PQ curve required by the Egyptian grid code, together with the reactive power available with and with no STATCOM [17].

Figure 9.9 shows that, in some cases, the reactive power capability of the wind turbines is insufficient to satisfy the reactive power requirements at PCC. This reactive power requirement is represented by the curve in the green line, and the

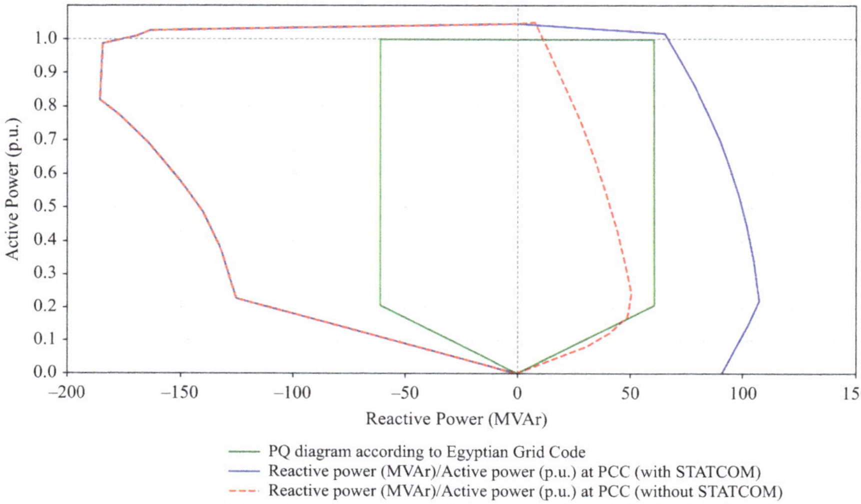

Figure 9.9 PQ diagram for voltage at PCC = 1 per unit taken from [17]

overall reactive power capability of the wind farm is represented by the curve in the red line. When using a STATCOM at PCC, the wind farm can fulfil the grid code's reactive power requirement, as represented by the curve in blue line.

9.4.3 LVRT and HVRT improvement using FACTS equipment

Similarly to the generator's capability curves, LVRT and HVRT curves are improved when installing shunt FACTS equipment (e.g., STATCOM) at PCC or near PCC. These devices provide dynamic reactive power support, enabling wind turbines to withstand voltage sags and swell, maintaining a more stable operation. By bolstering the fault tolerance of wind farms, FACTS devices enhance grid resilience and reliability, ensuring a smoother integration of wind energy into the power system.

The test system shown in Figure 9.10 illustrates the benefits of employing FACTS equipment in power grids with wind farms, concerning LVRT and HVRT improvements.*

The test system corresponds to a large utility power grid with a significant, localised component of fixed-speed wind power generators. The test system represents in detail the southeast region of the large, interconnected Mexican power grid, which concentrates the dominant part of the country's hydro and wind resources. The high-voltage transmission system of the Mexican power grid is rated at 400 kV. As seen from the large substation JUI-400, it may seem to comprise two

*Test case prepared by Norberto Garcia.

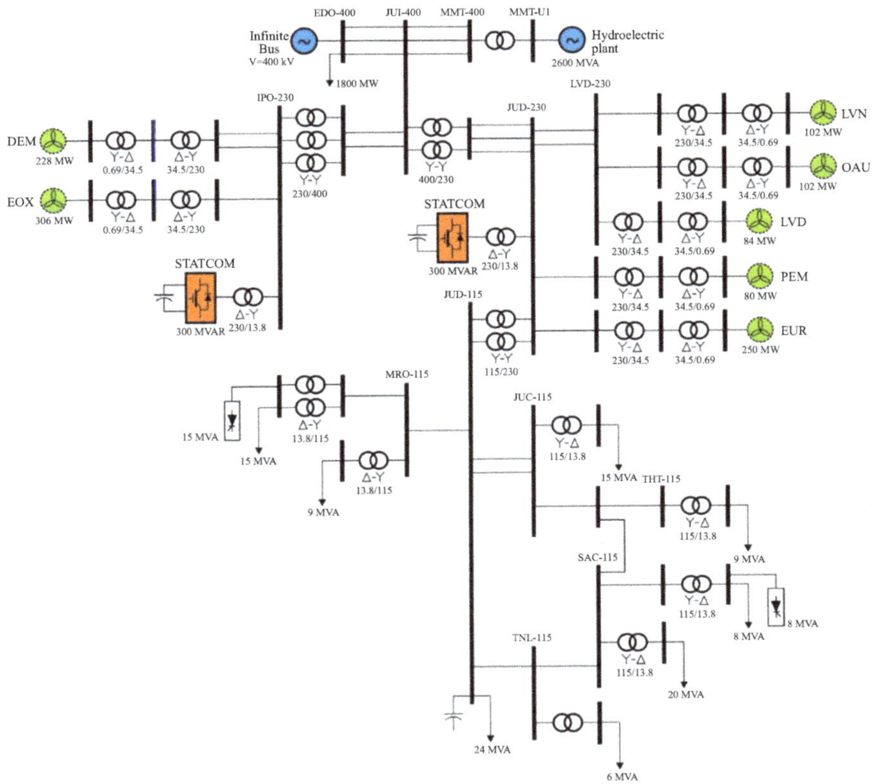

Figure 9.10 A portion of a large utility power grid system with seven wind farms [19]

equivalent systems and a detailed representation of the power grid containing the wind farms under study. One equivalent corresponds to the hydropower plants complex with an installed capacity of 2,600 MVA. In this case study, this hydropower system is represented by one equivalent hydro synchronous generator delivering 800 MW. The other equivalent corresponds to the rest of the interconnected Mexican power grid comprising a grid of 400 kV, 230 kV and 115 kV transmission circuits, large synchronous generators and bulk load points. The load is taken to be 1,929 MW and concentrated in the substations EDO-400 and JUD-115. The voltage and frequency at EDO-400 are assumed to be constant owing to the large size of the rest of the interconnected power system.

Seven wind farms with a total power rating of 1,152 MW are located at various points of the power grid in that region of Mexico, which experiences one of the best wind regimes in the country. The seven wind farms are taken to deliver 1,000 MW. The individual wind turbines are fixed-speed and rated at 2 MW.

To investigate the benefits of reinforcing the power grid with FACTS equipment, owing to the relatively large amount of fixed-speed wind power penetration in the area, two STATCOMs rated each at ±300 MVAR are assumed to be connected at substations IPO-230 and JUD-230.

The STATCOMs are two-level VSCs switching at 1,980 Hz (i.e., $m_f = 33$). The system power frequency is 60 Hz. One STATCOM is connected at JUD-230 using a 230/13.8 kV step-up transformer, providing a nearby source of dynamic reactive power for 618 MW of rapidly fluctuating wind power. Likewise, the second STATCOM is connected at IPO-230 using a 230/13.8 kV step-up transformer and is aimed at ameliorating the power output variations of two wind farms having a combined output of 534 MW. Notice that each STATCOM is rated at around 50% of the maximum power output of the wind farms they are compensating.

The parameters of the induction generators and coupling transformers used in the wind farms are given in Tables 9.2 and 9.3, respectively. For the rest of the parameters of the test power grid in Figure 9.10, refer to [19].

To assess the LVRT and HVRT capabilities of the fixed-speed wind farms in this utility-level power system, with and with no FACTS equipment connected at neighbouring locations, a three-phase-to-ground short-circuit fault was applied at MRO-115, which is the feeding substation of an industrial complex. The electromagnetic transient simulation studies are carried out with the de facto industry standard PSCAD simulation package. The fault takes place at 0.1s and lasts for 0.1667 s (ten fundamental-frequency cycles). The total simulation time is 2s. Figures 9.11–9.14 present the ensuing nodal voltages at the most critical nodes, namely, JUI-400, IPO-230, JUD-230 and MRO-115, respectively.

Table 9.2 Induction generator parameters

Parameter	Value
Rated power	2 MW
Rated voltage	0.690 kV
Number of poles	4
Stator resistance	0.01 p.u.
Rotor resistance	0.01 p.u.
Stator leakage inductance	0.12 p.u.
Rotor leakage inductance	0.096 p.u.
Magnetising inductance	3.6 p.u.
Inertia	4.32 s
Compensating capacitor	0.5 p.u.

Table 9.3 Coupling transformer parameters

Parameter	Value
Rated power	2.7 MVA
Winding voltages	0.69 kV/34.5 kV
Leakage reactance	0.06 p.u.
Stator resistance	0.01 p.u.

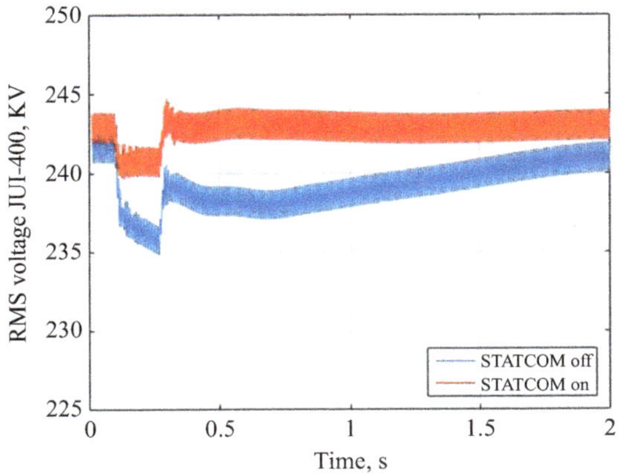

Figure 9.11 Nodal voltage responses at JUI-400

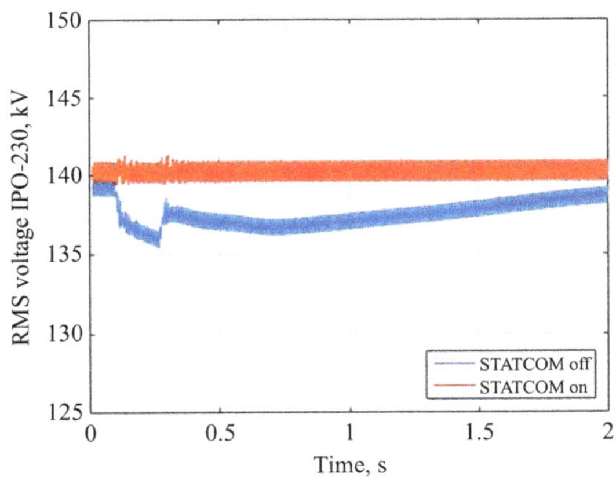

Figure 9.12 Nodal voltage responses at IPO-230

As expected, during the duration of the fault, the voltage drops to zero at MRO-115; it recovers at 2s of the simulation, following the clearance of the fault. In contrast, it recovers almost instantaneously when the STATCOMs are connected (see Figure 9.14). If no dynamic reactive compensation is applied, voltages at nodes JUD-230, IPO-230 and JUI-400, drop 24%, 3% and 3%, as observed in Figures 9.13, 9.12 and 9.11, respectively. The most severe voltage drop occurs at JUD-230 because it is electrically closer to where the fault occurred. When the STATCOMs are connected at nodes IPO-230 and JUD-230 the voltage drops at JUI-400, IPO-230 and JUD-230 are much improved. They hold the following values: 1%, 0 and 11%, respectively. With the STATCOMs in place, the voltages recover almost immediately following the liberation of the fault. Note that the voltages are displayed as RMS, line-to-neutral voltages at nodes JUI-400, IPO-230, JUD-230 and MRO-115.

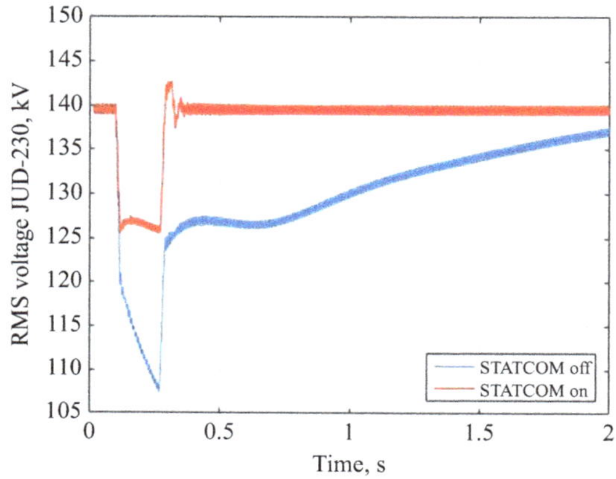

Figure 9.13 Nodal voltage responses at JUD-230

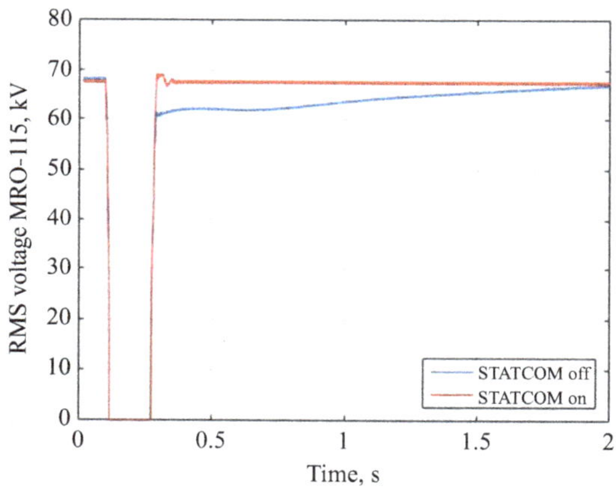

Figure 9.14 Nodal voltage responses at MRO-115

This set of electromagnetic-type simulations carried out with PSCAD illustrates quite well the positive impact of using one or more STATCOMs in the vicinity of wind farms to bestow LVRT and HVRT capabilities on the wind farms themselves and the whole power system area. This is particularly important if the wind farms are of the fixed speed type, but it may also apply to wind farms with variable speed wind turbines, as illustrated in Section 9.4.2.

9.4.4 *Interconnection of wind energy parks using HVDC transmission systems*

Today's wind farms, also known as wind energy parks, are large-capacity power plants reaching several hundred MWs. They are in remote on-land areas or offshore areas, where wind resources are plentiful [12]. Incorporating these parks into existing electrical power grids presents many challenges such as transporting electricity over long distances, under the sea or under the sea and land. Two options are available for transporting electrical power on land, using AC transmission lines or DC transmission lines. Moreover, these transmission lines may be overhead wires or underground cables. Concerning submarine power transmission, only the AC or DC cable option is practical. On technical/economic grounds, the former option applies to distances shorter than 70 km, whereas the latter nowadays covers distances of up to 740 km, as exemplified by the North Sea Link running between Kvilldal, Norway and Blyth, UK, and the Viking link connecting Revsing, Denmark and Bicker Fern, UK. These distances are expected to increase as the cable technology improves further. Refer to Chapter 2 to get a fuller picture of these developments.

It should be noted that the long-held concept of 'break-even distance' associated with long-distance AC vs DC power transmission is largely outdated, and it could be quite misleading. It refers to the point at which the cost savings from reduced transmission line expenses and reduced power losses equal the costs incurred by the power converter stations required by an HVDC link. A bipolar HVDC line uses two sets of conductors rather than the three employed by AC transmission lines. Hence, DC transmission corridors require narrower rights-of-way, more compact transmission towers and incur lower power transmission losses, assuming AC and DC lines of similar capacities. A rough estimate of the savings in line construction is 30% [20]. While the break-even distance is influenced by the expenses of right-of-way and line construction, it pays little attention to operational issues. Take the case of a typical 500 km long AC transmission line. This AC transmission line requires a shunt and series reactive power compensation at the mid-point. Likewise, a 1,000 km long AC transmission line requires two compensation substations located at 1/3 and 2/3 along the length of the line, respectively. Also, special towers would be required to carry out conductors' transpositions. The inclusion of these ancillary equipment increases quite substantially the costs for AC power transmission.

Nowadays, most point-to-point power transmission applications either on land or offshore favour the DC option over the AC option, when controllability and power grid integrity represent a major concern. Inter-area connections are being carried out more and more using back-to-back DC links in favour of the classic AC tie-lines. This issue is discussed in depth in Chapter 8.

VSC-HVDC power transmission systems, which only appeared in the commercial arena in around 2,000, have rapidly become the key solution to enable the efficient and dependable tapping of wind energy parks, particularly those laying on deep seas. Additionally, this technology improves grid stability during power outages since the rectifier/inverter stations are, in fact, fully fledged STATCOMs [21,22]. The current DC technology is not yet competitive with the highly developed AC technology in forming arbitrarily meshed power grids. For one, LCC-HVDC exhibits fundamental limitations that forbid them to form meshed circuits. Such limitations do not apply to VSC-HVDC, which, in principle at least, can be used to make up multiterminal HVDC systems of arbitrary configurations, owing to their inherent voltage source principles. However, as elaborated in Section 9.3.3, further theoretical developments are required concerning *grid-forming inverters*. Perhaps more importantly, there is a lack of field experience with MW-level DC switching breakers. It is reported that all the major manufacturers of HVDC equipment are actively working on DC CB development and that Hitachi Energy has verified DC CBs of up to 350 kV, and a further 525 kV unit will undergo verification tests this year. However, NR in China has developed DC CB rated at 500 kV and three kA, with 16 units employed in the first ring-type, four-terminal VSC-HVDC system in the world. So far, there seems to be no independent verification of the operational performance of the DC CB or the multiterminal VSC-HVDC system. Refer to Section 2.2.3 for additional information.

Economic factors always play a crucial role when deciding on the kind of technology to use to connect an offshore wind park. The entry-level question is AC or DC. However, drawing on the vast experience of the North Sea wind energy installations, VSC-HVDC is the preferred option if the wind farm lies more than 70 km from shore. The AC submarine power cable option would be left for connecting near-shore wind farms, up to about 50 km [23]. As stated above, HVDC links are highly reliable in transporting large amounts of electrical power over long and extra-long distances with the utmost controllability.

In cases of offshore wind farm interconnections using a DC link, this would be done with the VSC-HVDC option using power converters mounted on marine platforms. In principle, an LCC-HVDC link could be considered for such a task, but this carries a major operational handicap in the rectifier plant – the dreaded commutation failure phenomenon arising from not having a solid commutation source. Wind power generators are unsuitable for commutation sources – their sizes are relatively small, and their designs are not amenable to this task. Synchronous condensers in this marine application would be ruled out based on size, weight and cost. Today, this leaves us with the natural option to use point-to-point VSC-HVDC links for this task.

The use of the VSC-HVDC technology for evacuating offshore wind power is on the increase, encompassing increasingly longer distances as the submarine cable technology continues to improve. This is a pressing issue as floating platforms suitable for supporting MW-level wind turbines in deep sea waters are being built, as we write, in readiness for fulfilling rather ambitious quotas of wind power that several European countries with coastal access to the North Sea are committing themselves to fulfil by 2030. For instance, the UK has plans to add 34 GW of offshore wind power by 2030, *en route* to decarbonisation [14]. Such accelerated developments of deep offshore wind power installations call for a multiterminal VSC-HVDC system in the North Sea, building on the dozens of point-to-point VSC-HVDC links that have been installed there over the last two decades for various purposes: evacuating wind power, feeding in oil and gas rigs, interconnecting countries across the North Sea, e.g., UK and Norway [24] and UK and Denmark [25].

The recently commissioned three-terminal Shetlands-Caithness-Moray link in the north of Scotland gives a taste of what may be about to come in VSC-HVDC link interconnections to form multiterminal VSC-HVDC systems [26]. The 260 km submarine cable from Kergord, Shetland, rated at 600 MW, takes wind power to Staxigoe, in DC power form. At that point, it joins with an existing submarine VSC-HVDC link commissioned in 2018 that goes from Spittal, Caithness, to Blackhillock, Moray, rated at 1,200 MW [27]. Together, they form a Y-connected, three-terminal VSC-HVDC link. Incidentally, this is the first multiterminal system of this kind in Europe [28]. This development should be read together with what is taking place in China, where the first meshed HVDC system in the world came into operation in 2021. This is the four-terminal Kangbao-Zhangbei-Fengning-Changping DC system near Beijing [29]. For more detailed information concerning multiterminal HVDC installations, refer to Chapter 2. The topic of multiterminal VSC-HVDC systems has been presented in depth in Chapter 5, from the vantage of steady-state modelling, simulation and control using power flow theory.

9.4.5 *Test case of a multiterminal VSC-HVDC system with DFIG-based wind farms*

The AC/DC power network in Figure 9.15 is a DC hexagon with six VSC stations and two 200-MW wind farms, each comprising twenty 10-MW DFIG turbines. VSC_1 is a GFM inverter and is taken to be the reference (VSC_{Slack}). VSC_2 is a GFL inverter that feeds into a passive network drawing 100 MW. VSC_3 and VSC_5 are GFL converters. VSC_{Psch}, set to draw each P_{sch} = 125 MW from the DC hexagon.

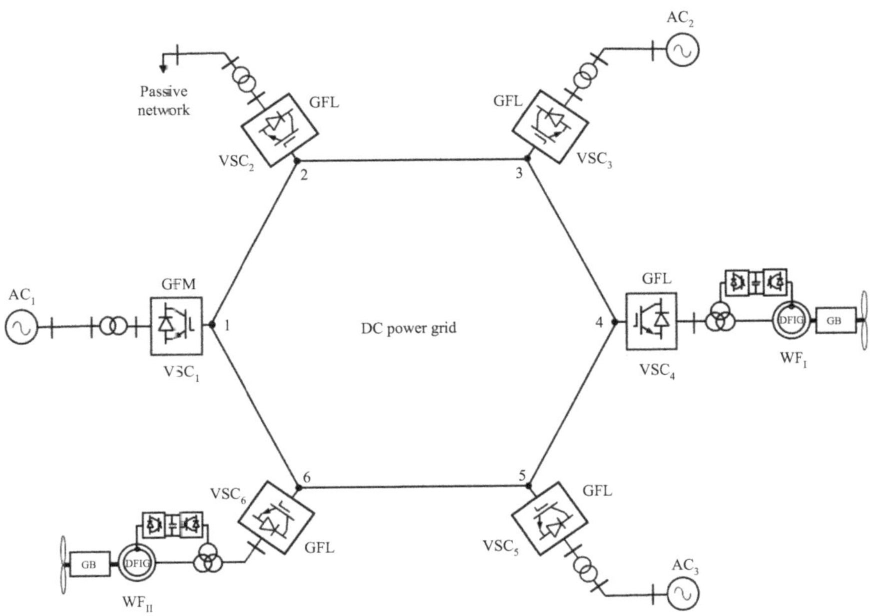

Figure 9.15 A hybrid AC/DC power grid with DFIG-based wind farms

The two wind farms, WF_I and WF_{II}, are connected to the DC grid through the GFL inverters VSC_4 and VSC_6, respectively. The three AC networks, AC_1, AC_2 and AC_3, which are coupled to VSC_1, VSC_3 and VSC_5, respectively, are represented by the Thevenin equivalents in Figure 9.15.

Nowadays, there are very powerful simulation tools that enable comprehensive assessments of these kinds of future power grids. Perhaps the family of EMTP packages, widely available in industry and academia, is the best-placed tool for such a task. For instance, EMTP-RV, PSCAD and Simscape Electrical from Simulink are very popular packages. To a greater or lesser extent, all these software packages show similar model capabilities and achieve the same accuracy of results. The latter software has been used to solve the hybrid AC/DC power grid shown in Figure 9.15. All the basic parameters of the DC power grid, the Thevenin equivalents, wind farms, loads and VSCs are available in [30].

Two events are programmed to observe the dynamic responses of the VSCs and the AC/DC power grid: (i) at $t = 0.5$ s, the scheduled power, P_{sch}, of VSC_3 abruptly changes from 125 MW to 200 MW; (ii) at $t = 3$ s, the wind speed, V_w, of wind farm I, connected to VSC_4, increases from 8 m/s to 10 m/s with a rate of change of $\Delta V_w/\Delta t = 1\ \text{m·s}^{-1}$/s.

Table 9.4 gives the steady-state operating condition of the power grid with the two wind farms connected. These *steady-state* values are the simulation values exhibited by Simulink at the end of 50 s of simulation.

The steady-state information provided for the six VSCs is DC voltages (E_{DC}), amplitude modulation ratios (m_a) and the powers drawn by each VSC at their AC sides (P_{VSC}). The DC powers at each DC line of the hexagon grid (P_{DC}) are also given. Concerning the wind farms, WF_I and WF_{II}, the following parameters are given: the terminal voltage magnitude (V_k), rotational speed (ω_m), mechanical torque (T_m) and generated power (P_g).

Table 9.4 Steady-state solution of the hybrid AC/DC power system in Figure 9.15

			EMT simulation (Simscape Electrical) @ t = 50 s			
Network and equipment			**VSC stations**			**DC lines**
Grid	**No.**	**VSC type**	E_{DC} **[p.u]**	m_a	P_{VSC} **[MW]**	P_{DC} **[MW]**
AC 1	1	VSC_{Slack}	2.0001	0.878	146.815	$P_{12} = 159.60$
PN	2	VSC_{Pass}	1.9965	0.862	−100.0	$P_{23} = 53.82$
AC 2	3	VSC_{Psch}	1.9942	0.832	−125.0	$P_{34} = -78.17$
WF I	4	VSC_{WF}	1.9960	0.792	60.823	$P_{45} = -20.71$
AC 3	5	VSC_{Psch}	1.9965	0.824	−125.0	$P_{56} = -153.24$
WF II	6	VSC_{WF}	2.0010	0.856	186.261	$P_{61} = 22.11$
			VSC-connected DFIG-based wind farms			
Wind farm			V_k [p.u]	ω_m [p.u]	T_m [p.u]	Total P_g [MW]
I – WG 1-3			1.0021	0.9567	−0.3155	59.765
II – WG 4-6			1.0013	1.2463	−0.7155	183.985

Having established the steady-state operation, the dynamic study starts by rescheduling the power in VSC_3 to reflect an increase in imported power by AC_2. According to the planned scenario, P_{sch} will increase from 125 MW to 200 MW. The transient response is shown in Figure 9.16. The AC powers at the

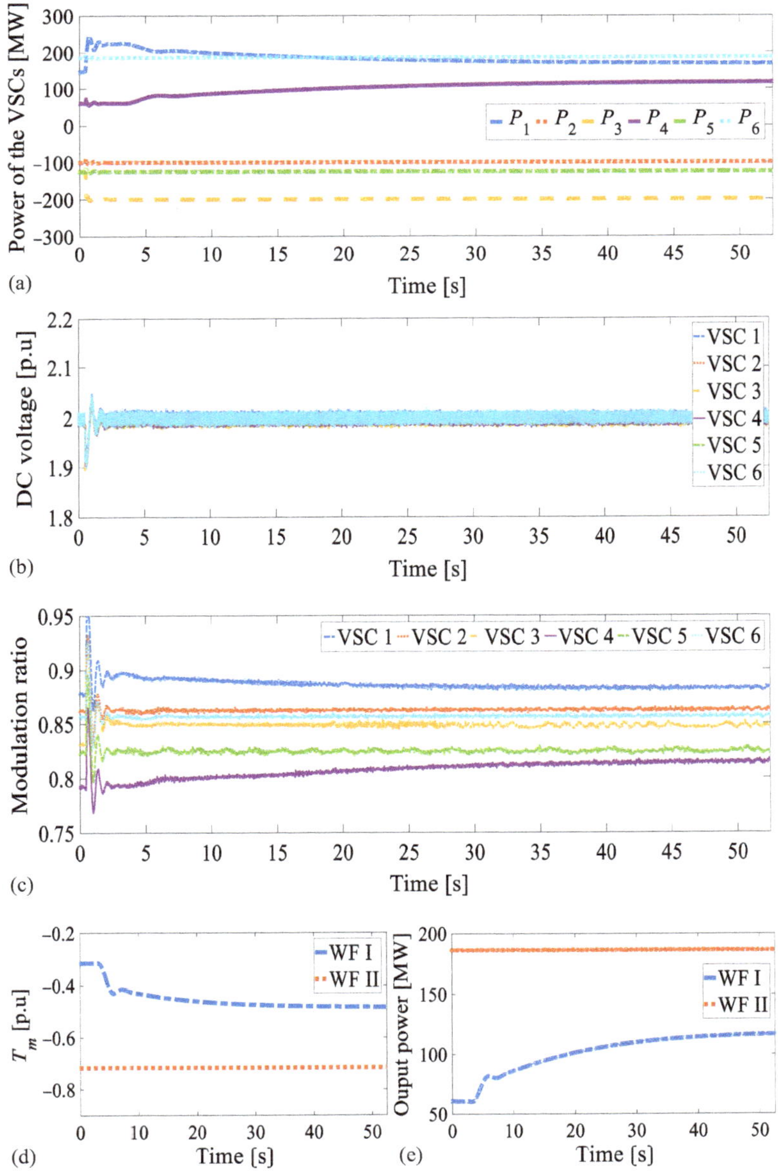

Figure 9.16 Transient simulation of the hybrid AC/DC power systems in Figure 9.15. (a) Active power at terminals of the VSCs, (b) DC voltage at terminals of the VSCs, (c) modulation ratios of the VSCs, (d) mechanical torque of WF1 and WF2, and (e) active power at terminals of WF1 and WF2.

VSCs' terminals and the DC voltages are shown in Figure 9.16(a) and (b), respectively. The increased power demand causes a transient discharge of the energy stored in the DC capacitors, leading to nodal voltage drops in the DC network. The almost instantaneous action of VSC_1, which is a GFM inverter, is to inject power into the DC network – as seen by the profile of P_1 in Figure 9.16(a). This has the effect of restoring the DC grid voltage very rapidly, i.e., the oscillations die out at 2 s. Notice that in the time range of 0.5–3 s, the perturbation mainly involves VSC_1 and VSC_3 and produces almost no impact on the other VSC converters' powers.

At $t = 3$ s, the power output of WF_I starts increasing owing to the wind speed going up from 8 m/s to 10 m/s. It reaches a mechanical torque of $T_m = -0.458$ p.u. and a power generation $P_g = 113.49$ MW at $t = 40$ s, approximately when the new steady state is attained. This is observed in Figure 9.16(d) and (e), respectively. Notice that owing to the rate of change of $\Delta V_w/\Delta t = 1\ \text{m}\cdot\text{s}^{-1}/\text{s}$, the wind speed is 9 m/s at 4 s and 10 m/s at 5 s.

The additional wind speed in WF_1 results in an increased power entering VSC_4 and into the DC network, as seen in Figure 9.15(a). The additional power in the DC network is counteracted by the GFM inverter (VSC_1), which reduces its power contribution, restoring the power balance in the whole AC/DC power grid, attaining a new steady state at the end of the simulation period, i.e., 53 s.

Figure 9.16(c) shows the dynamic performances of the modulation ratios of the six VSCs. These are dynamically adjusted to comply with the events taking place in the AC/DC power grid. The VSC modulation ratios experience more significant oscillations following the abrupt change in P_{sch}. The change in wind speed in WF_1 induces a gradual increase in the modulation ratio of VSC_4 and a gradual decrease in the modulation ratio of VSC_1. These largely follow the upward and downward trends of powers P_1 and P_4.

Notice that the variation of V_w in WF_I only impacts the DC network and AC_1, which is the AC system coupled to the GFM inverter (VSC_{Slack}). The other power grids experience minor variations. This is the critical characteristic of an HVDC system, which blocks the electrical interactions between AC power networks. However, it may be observed that when power changes occur at any of the VSC-connected networks, the AC power grids coupled to GFM inverters do experience power deviations. This situation is expected to occur in the context of multiterminal HVDC links.

HVDC transmission systems offer substantial advantages over conventional AC systems in point-to-point applications, particularly for connecting offshore wind farms to mainland grids when distances longer than 70 km are involved. HVDC provides enhanced grid stability and control, with VSC-based HVDC enabling precise power flow management, thereby mitigating the risk of electrical instability. Also, HVDC links decouple neighbouring AC grids, ensuring that disturbances in one grid do not directly affect other grids, a critical feature for offshore installations subjected to fluctuating power conditions.

9.5 Conclusions

The global integration of wind power generation into the power grid is relentless. According to the WWEA Annual Report 2023, the cumulative wind power capacity has surpassed 1,000 GW. This contrasts with the 320 GW that existed in 2013 – a trebling installed wind power capacity in the last decade. Wind power technology has improved vastly. Nowadays, the old, fixed-speed wind turbines are seldom used. Instead, all new wind farms use variable-speed wind turbines to enable optimum use of the available wind energy resource at any point in time. These wind power turbines are the so-called DFIG and PMSG-FRC. Both turbines use power electronics circuits to achieve their highly developed functionality. The former is used in onshore and offshore wind power installations, whereas the latter is suitable for large, offshore wind power installations owing to their high ratings, exceeding 10 MW and nearing 15 MW. Gross estimates indicate that the current ratio of installed onshore wind power to offshore wind power is 90% in favour of the former. To evacuate the power output of wind farms, except for the very small capacity installations, the use of power electronics converters becomes necessary, either in the form of FACTS equipment or HVDC systems.

In general, STATCOMs or SVCs are placed at the PCC between onshore wind farms and the power grid to ensure that the connection meets today's stringent grid codes' requirements. For transporting extra-large blocks of wind power generated onshore over extra-large distances, the point-to-point CSC-HVDC option is unassailable. This is well exemplified by the Gansu Wind Farm in northern China, being developed incrementally to produce up to 20 GW of wind power and its 2383 km long LCC-HVDC link. Note that being an onshore wind farm, installing a commutation source at the rectifier substation is not a problem.

Concerning offshore wind farms, near-shore wind farms are connected to the power grid using AC cables and depending on the size of the wind farm and the strength of the power grid at PCC, FACTS equipment may or may not be necessary. Drawing on the vast experience gained from dozens of offshore wind farms in the North Sea, for distances longer than 70 km from shore, the evacuation of wind power is carried out using point-to-point, VSC-HVDC links. In this application, neither the AC cable option nor LCC-HVDC links are suitable. Among other things, the former is ruled out because of the excessive capacitive reactance exhibited by a long cable and the latter owing to the lack of a commutation source at the rectifier end of the DC link. The VSC-HVDC transmission option does not exhibit limitations since it does not require a commutation source to commutate, and no charging currents are generated in the cable at 0 Hz. Furthermore, looking forward to the future, VSC-HVDC links are suitable for forming DC meshed grids and AC power grids dominated by converter-interfaced power generation sources, owing to their *grid-forming* capabilities.

References

[1] Intergovernmental Panel on Climate Change: *Climate Change 2014: Mitigation of Climate Change: Working Group III Contribution to the IPCC Fifth Assessment Report* (Cambridge: Cambridge University Press, 2015).

[2] International Energy Agency: *Net Zero by 2050 – A Roadmap for the Global Energy Sector* (Paris: IEA, 2021).

[3] International Renewable Energy Agency: *Global Energy Transformation: A Roadmap to 2050* (Abu Dhabi: IRENA, 2019).

[4] International Renewable Energy Agency: *Grid Codes for Renewable Powered Systems* (Abu Dhabi: IRENA, 2022).

[5] Tarafdar Hagh, M., and Khalili, T.: 'A review of fault ride through of PV and wind renewable energies in grid codes'. *Int. J. Energy Res.*, 2019, 43 (4), pp. 1342–1356.

[6] Yadav, M., Pal, N., and Saini, D.K.: 'Low voltage ride through capability for resilient electrical distribution system integrated with renewable energy resources'. *Energy Rep.*, 2023, 9, pp. 833–858.

[7] Zhang, M., Yuan, X., and Hu, J.: 'Inertia and primary frequency provisions of PLL-synchronized VSC HVDC when attached to islanded AC system'. *IEEE Trans. Power Syst.*, 2018, 33 (4), pp. 4179–4188.

[8] Alatrash, H., Mensah, A., Mark, E., Haddad, G., and Enslin, J.: 'Generator emulation controls for photovoltaic inverters'. *IEEE Trans. Smart Grid*, 2012, 3 (2), pp. 996–1011.

[9] Elkhatib, M.E., Du, W., and Lasseter, R.H.: 'Evaluation of inverter-based grid frequency support using frequency-watt and grid-forming PV inverters', in *2018 IEEE Power & Energy Society General Meeting (PESGM)* (Piscataway, NJ: IEEE, 2018), pp. 1–5.

[10] Lin, Y., Eto, J.H., Johnson, B.B., *et al.*: *Research Roadmap on Grid-Forming Inverters* (Golden, CO: National Renewable Energy Lab. (NREL), 2020).

[11] Cacciuttolo, C., Navarrete, M., and Atencio, E.: 'Renewable wind energy implementation in South America: A comprehensive review and sustainable prospects'. *Sustainability*, 2024, 16 (14), p. 6082.

[12] Rahman, S., Khan, I., Alkhammash, H.I., and Nadeem, M.F.: 'A comparison review on transmission mode for onshore integration of offshore wind farms: HVDC or HVAC'. *Electronics*, 2021, 10 (12), p. 1489.

[13] Ackermann, T. (ed.): *Wind Power in Power Systems* (New York: Wiley, 2012, 2nd edn.).

[14] Fairley, P.: 'Buoyant behemoths: The global race is on to tap potent winds far offshore'. *IEEE Spectr.*, 2023, 60 (11), pp. 30–37.

[15] Hansen, T.A., Wilson, E.J., Fitts, J.P., *et al.*: 'Five grand challenges of offshore wind financing in the United States'. *Energy Res. Soc. Sci.*, 2024, 107, p. 103329.

[16] Xiang, D., Turu, J.C., Muratel, S.M., and Wang, T.: 'On-site LVRT testing method for full-power converter wind turbines'. *IEEE Trans. Sustain. Energy*, 2017, 8 (1), pp. 395–403.

[17] Hassan, M.H., Helmi, D., Elshahed, M., and Abd-Elkhalek, H.: 'Improving the capability curves of a grid-connected wind farm: Gabel El-Zeit, Egypt', in *2017 Nineteenth International Middle East Power Systems Conference (MEPCON)* (Piscataway, NJ: IEEE, 2017), pp. 300–307.

[18] Lund, T., Sørensen, P., and Eek, J.: 'Reactive power capability of a wind turbine with doubly fed induction generator'. *Wind Energy*, 2007, 10 (4), pp. 379–394.

[19] Garcia, N., and Acha, E.: 'On the efficient calculation of the periodic steady-state response of grid-connected wind parks—Part I'. *IEEE Trans. Sustain. Energy*, 2017, 8 (2), pp. 458–467.

[20] Bahrman, M.P., and Johnson, B.K.: 'The ABCs of HVDC transmission technologies'. *IEEE Power Energy Mag.*, 2007, 5 (2), pp. 32–44.

[21] Korompili, A., Wu, Q., and Zhao, H.: 'Review of VSC HVDC connection for offshore wind power integration'. *Renew. Sustain. Energy Rev.*, 2016, 59, pp. 1405–1414.

[22] Renaudin, F.: 'Integration and stability of a large offshore wind farm with HVDC transmission in the Norwegian power system' *Master thesis* (Trondheim: Institutt for elkraftteknikk, 2009).

[23] Lazaridis, L.: *Economic Comparison of HVAC and HVDC Solutions for Large Offshore Wind Farms Under Special Consideration of Reliability* (Stockholm: KTH, Electric Power Systems, 2005).

[24] Hook, L.: 'UK and Norway complete world's longest subsea electricity cable'. *Financial Times*, 2021. Available: https://www.ft.com/content/399c1c37-3f7a-4770-af13-66741df01135

[25] 'National Grid's Viking Link interconnector recognised by GUINNESS WORLD RECORDS™, *National Grid Group*', https://www.nationalgrid.com/media-centre/press-releases/viking-link-interconnector-guinness-world-records (accessed September 2024)

[26] Tracey: 'UK wind power reaches "historic" 30 GW milestone', https://www.modernpowersystems.com/news/uk-wind-power-reaches-historic-30-gw-milestone/ (accessed September 2024).

[27] Caithness – Moray HVDC link', https://www.ssen-transmission.co.uk/link/5222e9ad35ea421cba4381355b37231a.aspx (accessed September 2024).

[28] Fairley, P.: 'High-voltage DC power roars ashore in Europe: Scotland's next-generation transmission network foreshadows a global grid transformation'. *IEEE Spectr.*, 2024, 61 (1), pp. 28–31.

[29] NR Electric Co., Ltd.: 'Case study: NR's VSC-HVDC solution 500 kV Zhangbei DC grid for boosting large scale hybrid renewables penetration', 2019. Available: https://www.nrec.com/en/web/upload/2019/05/14/15578229444108hgcuh.pdf.

[30] Castro, L.M., and Acha, E.: 'On the dynamic modeling of marine VSC-HVDC power grids including offshore wind farms'. *IEEE Trans. Sustain. Energy*, 2020, 11 (4), pp. 2889–2900.

Chapter 10

Protection units and systems in AC and DC power networks

Ernesto Vázquez Martínez[1]

10.1 Hybrid AC–DC power systems

AC–DC power systems are the new face of the modern electrical power infrastructure, combining the best attributes of alternating current (AC) and direct current (DC) technologies. These hybrid systems offer enhanced efficiency, flexibility and reliability, making them suitable for a wide range of applications, ranging from renewable energy integration to industrial processes. AC–DC power systems are characterised by their ability to reliably convert energy between AC and DC forms. Today, these conversion processes are achieved through solid-state power rectifiers, which convert AC to DC, and inverters, which convert DC to AC. The efficiency of these conversions is crucial for minimising energy losses and maximising the overall performance of the power system.

One of the standout features of AC–DC power systems is the HVDC power transmission option, which is unassailable in long-distance power transmission applications. HVDC can transport large amounts of electrical power over vast distances with reduced losses and voltage drops. It is very popular in undersea and underground power transmission where the AC option would be less effective or unyielding. This capability is particularly beneficial for connecting remote offshore renewable energy sources to the main grid.

Many renewable energy sources, such as solar PV panels and fuel cells, generate DC power. AC–DC power systems enable the integration of these sources into the power grid by converting the generated DC power to AC for grid compatibility. Wind turbines generate AC power with significant variations owing to the variability of the incoming winds. Energy storage systems, such as fluid batteries, when fully developed, may become instrumental in firming the power output of wind power generators These systems can also extract excess energy from the AC power grid and store the energy within battery systems, which can then be converted back into AC power on demand. In industrial and commercial settings, AC–DC power systems provide a versatile power supply to meet diverse energy demands. Many

[1]Faculty of Mechanical and Electrical Engineering, University of Nuevo Leon, México

industrial processes, such as electrolysis, require DC power, while others operate on AC. Hybrid AC–DC systems ensure that all machinery and processes receive the appropriate type of power, improving operational efficiency.

Energy storage systems, such as battery energy storage systems (BESS), use DC power to store and discharge electricity, which can be crucial for balancing supply and demand. These systems can store energy generated from renewable sources or during off-peak hours and then convert it to AC power when needed, ensuring a stable and reliable AC power supply. This capability is essential for maintaining grid stability and enhancing energy security. Smart charging solutions and vehicle-to-grid (V2G) technologies that allow electric vehicles (EVs) to feed stored energy back into the grid are key components of modern AC–DC power systems.

High voltage direct current (HVDC) transmission is a critical technology for efficiently transferring electricity over long distances and integrating renewable energy sources. Its robustness and efficiency are well documented, but understanding its behaviour under fault conditions, both AC and DC, is crucial for ensuring the reliability and stability of power systems. The widespread incorporation of power electronics equipment in the electrical power network at the generation, transmission and distribution systems levels has a marked impact on the theory and practice of power systems protection. Incorporating power generation sources at the power network's low and very low voltage points has called for a revision of traditional power system protection practices due to the real possibility of reverse power flows and because of the coexistence of AC and DC power systems and equipment at the generation, transmission and distribution levels will become the norm.

Since 1954, the protection of classic HVDC (CSC-HVDC) links has not presented significant changes because of the natural response of this technology when undergoing fault conditions, due to the fast action of the thyristors [1]. However, modern HVDC or VSC-HVDC links do not handle faults well of faults. This is due to the limited power capacity of the IGBT valves and converters during a fault and the control strategies implemented, reaching current magnitudes of up to 2 p.u., and a fast voltage decrease compared to CSC-HVDC. Therefore, a modern HVDC system requires a new solution in relay protection.

Several CIGRÉ working groups, such as B4-56, B4/B5-59, JWG A3-B4.34 and CELENEC TC8x WG6, are dedicated to HVDC protection. Other related standards are IEC 61869, IEC 60255, IEC 61850 and IEC 60834. According to these groups, HVDC protection aims are similar to AC systems operating philosophy: security, dependability and selectivity. Aspects such as redundancy and backup protection are considered with the objective of increasing reliability and are presented in [2]. These aspects are key to enhancing multi-vendor strategies in DC grids [3]. The links that make up a DC grid must have fully selective protections to enable only the outage of the failed DC conductor (underground or overhead) instead of the entire DC grid, as illustrated in Figure 10.1.

For completeness and as a quick reference when reading the chapter, Appendix A presents a comprehensive list of relays methods and principles available for AC power circuits.

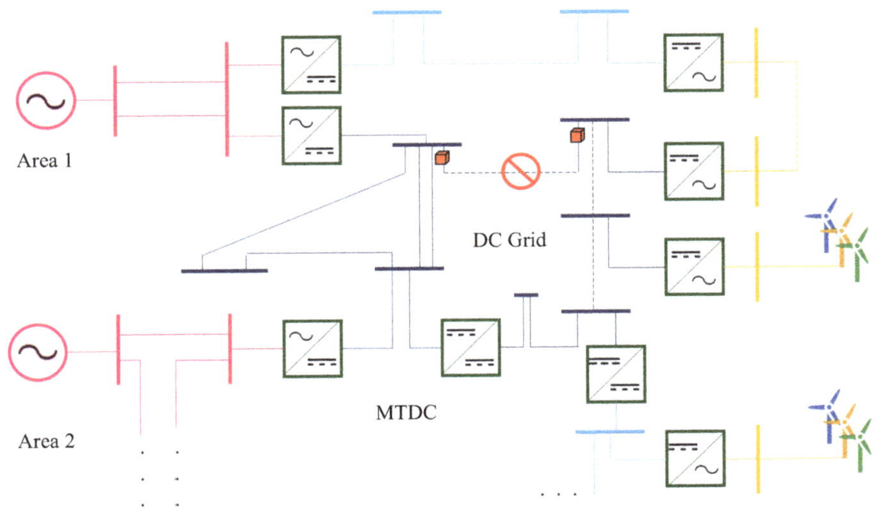

Figure 10.1 Example of selective protection in a DC grid, adapted from [4,5]

Due to the operating characteristics of DC signals, the full-scheme protection of a DC link comprises several elements acting as a single unit protection [4] (differential travelling wave, current differential) and backup protection [5] (over-current and undervoltage relays).

Based on Figure 10.1, the DC link protection scheme considers all the possible faults inside and outside the DC link. Internal faults or 'DC faults' take place along the length of the DC conductor while faults at the converter station are excluded. However, external faults or 'AC faults' occur only between the AC grid and the converter station. External faults produce a deviation of the nominal power in the DC link and are observed by the DC protection as a change of the current measured, producing false trip signals. If the HVDC protection is based on overcurrent units, any deviation that exceeds the pick-up current can produce false trips. Therefore, using overcurrent relays is not recommended in DC links as a unit protection, restricting its use only for backup protection in DC links and DC grids [6]. For this reason, the capacity of discrimination of a relay is an important point to be covered by the protection algorithms programmed in the protection unit. This requirement must be present in each converter station.

10.2 HVDC fault studies

A point-to-point HVDC system converts AC power to DC power at the sending end (i.e., rectifier), transmits it over long distances and then converts it back to AC power at the receiving end (i.e., inverter). This technology offers several advantages over traditional AC transmission, including reduced losses, the ability to connect asynchronous grids and improved stability and controllability. As discussed in Chapters 2 and 4, HVDC systems can be classified into two main types [7]:

- Line commutated converters (LCC-HVDC). These use thyristors and are dependent on the AC system strength for commutation and consume reactive power.
- Voltage source converters (VSC-HVDC). These use IGBTs or other fully controllable semiconductors and can independently control active and reactive power.

AC faults, such as short circuits or line-to-ground faults, can significantly impact HVDC systems. The main effects include:

- Impact on LCC-HVDC systems. AC faults can lead to commutation failures in LCC-HVDC systems. This occurs when the AC voltage drops below a critical level, preventing the thyristors from turning off properly. Multiple commutation failures can cause a temporary loss of power transmission capability and require immediate control actions to stabilise the system.
- Impact on VSC-HVDC systems. VSC-HVDC systems are more resilient to AC faults due to their ability to continue operation with reduced AC voltage. However, severe AC faults can still cause significant voltage drops and potential overcurrents. VSC systems typically employ fast control strategies to ride through faults, including dynamic voltage support and current limiting.

DC faults, such as pole-to-pole or pole-to-ground faults, pose unique challenges to HVDC systems:

- Pole-to-pole faults: These faults cause a direct short circuit across the DC bipolar link, resulting in high fault currents. In LCC-HVDC systems, the fault current is primarily limited by the inherent impedance of the DC line, smoothing reactors and the converters. Protection schemes, such as DC circuit-breakers, are required to quickly isolate the faulted section rapidly.
- Pole-to-ground faults: These faults can lead to high ground currents and potential damage to the insulation of the DC line. Effective grounding and protection systems are essential to quickly detect and clear such faults. VSC-HVDC systems can manage these faults by temporarily reducing the DC voltage and using fast protection mechanisms.

Another key issue in DC faults is the localisation of the fault point on the DC conductor. Although the location of the fault is an important functionality, it is in the second order of priority because fault detection is the main function of a protection device. Contrary to AC signals, a DC signal contains no information of unbalance, phase or operating frequency, but VSC-HVDC protection has incorporated new forms to discriminate between a faulted condition and a nominal operating condition. In this case, the estimation of the distance to the fault can be performed by new algorithms based on waveform information.

10.3 Transient behaviour of AC–DC systems under fault conditions

The lack of zero crossing, phase, sequence components and nominal frequency has represented a major challenge in DC protection algorithms. This contrasts with the wide range of applications that exist in conventional AC protection methodologies – refer to Appendix A for a comprehensive list of relays methods and principles available for AC power circuits. In DC links, the DC component of the fault signal is higher than in AC signals. Normally, the frequency present in a DC voltage is several kHz, being the main component of the commutation frequency in VSC-HVDC links. The DC component of the fault signal acts as an offset, which does not allow the zero crossing. Therefore, the fault current interruption is a serious inconvenience [8]. These factors require the use of new techniques in DC protection.

During fault conditions, AC–DC systems present some recurrent characteristics, such as currents directions, the waveforms of the signals and the variations of the measured magnitudes with respect to their nominal values. For each case, this aspect presents the same 'signature' or behaviour. In other words, although a fault condition is a severe situation in electrical circuits, it is possible to characterise some signal peculiarities.

10.3.1 AC faults characterisation

As stated above, AC–DC systems present both internal and external faults. In the case of external faults, there are factors that can impact the behaviour of the fault current like the fault resistance, the location of the AC fault and the short circuit ratio [9]. The converter transformer, for example, acts as an inductive element that limits the current, making this action valid only for the first instant after the fault and remaining in a nominal value once the controls act. The converters must control the fault current to be below 2 p.u. to avoid the tripping of the IGBT valves by overcurrent protection. Since the fault current control depends on control actions, a second element that avoids the tripping of the DC breaker is needed. Figure 10.2 presents a typical DC response to an AC fault.

Under fault conditions on the AC side, the voltage decreases to zero, and the current reaches magnitudes of up to 20 p.u. The dynamics on the DC side present an oscillatory component, both in voltage and current. This transient increase can affect the response of conventional relays that use thresholds to operate. Selecting an ideal threshold is hard work due to the arbitrary structure of the connecting AC grids. However, several solutions can be developed, including intrinsic factors of the DC voltage and current waveforms as opposed to only their magnitudes. During an AC fault situation, both the voltage and the current in a point-to-point DC link present similar waveform responses. This is exemplified in Figure 10.3. Such behaviour can be explained because the current direction remains, given that the fault occurs outside the AC–DC system. Therefore, converters do not contribute to the short circuit current. That is, the change in the original DC current owing to the fault, remains, given that the direction of the currents does not change.

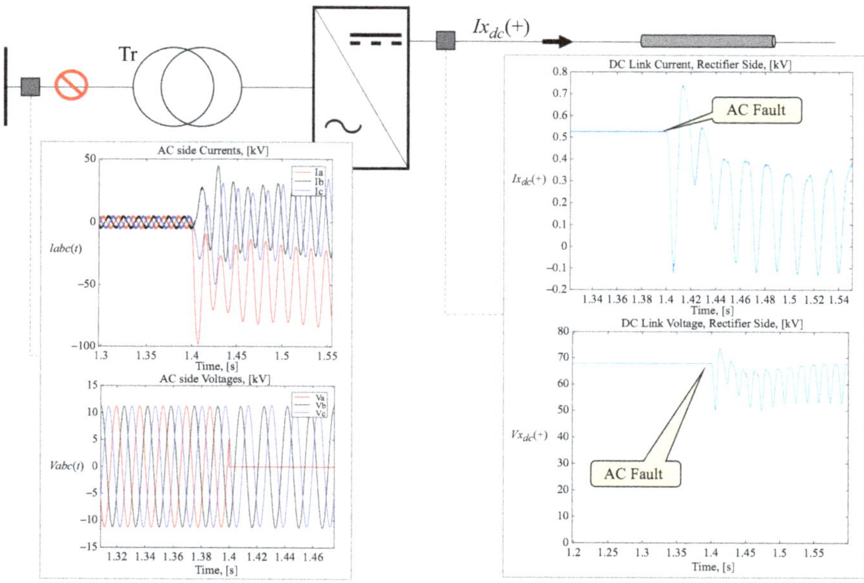

Figure 10.2 Signals measured in AC side and DC side under AC fault (phase A with ground (AG))

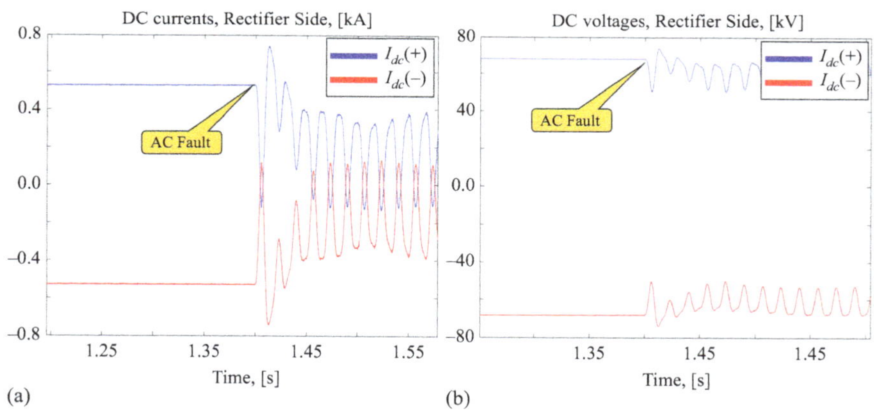

Figure 10.3 DC voltages and currents in the rectifier station output under phase-ground fault. (a) DC currents at rectifier side, (b) DC voltages at rectifier side

The recurrent feature in this case is the direction of the currents and the pattern similarity of the DC voltage and current waveforms. This pattern is an attribute specific to this type of fault. This property could be used as an identifier or discriminator in AC–DC systems as part of a global protection algorithm.

AC system faults can be categorised into symmetrical and asymmetrical faults, each having distinct characteristics and impacting the system differently.

- Symmetrical faults affect all three phases equally and are the most severe but not the most common. The main type of symmetrical fault is the three-phase fault, where all three phases are short-circuited together. Symmetrical faults yield high fault currents and significant system disturbances.
- Asymmetrical faults are more common than symmetrical faults, affecting the system asymmetrically. They include:
  - Single line-to-ground fault.
  - Line-to-line fault.
  - Double line-to-ground fault.

Faults in AC systems may arise from various sources, including natural causes, such as lightning strikes, storms, high winds and falling trees, which can damage transmission lines. Also, a potential source of faults is equipment failure, owing to ageing infrastructure, manufacturing defects, poor quality installation or lack of maintenance. Substation equipment typically includes transformers, circuit breakers, bushings and protection & control equipment. Human error is never far away when operating such a complex system as it is the power grid: operational mistakes, such as incorrect relay settings and installation errors, can introduce faults. Animals, such as birds or squirrels, and overgrown vegetation can cause short circuits and ground faults by coming into contact with power lines. The occurrence of faults in AC systems has very significant and wide-ranging effects:

- Voltage drops: Faults cause sudden and severe voltage drops, with the potential to destabilise the power system and lead to brownouts or blackouts.
- High fault currents: Faults generate high currents that can damage equipment, such as transformers and generators, and pose serious safety hazards.
- System instability: Faults can induce oscillations and instabilities in the power system, affecting the quality of the power supply and potentially leading to widespread disturbances.
- Power outages: Severe faults can result in extended power outages, disrupting services and causing exorbitant economic losses and social hardship.

10.3.2 DC faults characterisation

DC faults can be assessed by means of the following two approaches: Transient response (Laplace) and travelling wave response (Telegrapher's equations). Both methods enable a way to assess the dynamics of the HVDC signals undergoing faulted conditions, in the frequency and time domains, respectively. The time response of the HVDC system (seconds and milliseconds) is larger than the travelling wave response (microseconds), since it includes the dynamics of the voltage in the DC capacitors, the current flowing through the DC conductor and the various stages of freewheeling when the AC side feeds into the fault through the power converters. It is necessary to distinguish between two types of DC faults in HVDC systems: asymmetrical and symmetrical faults. The first type involves the contact

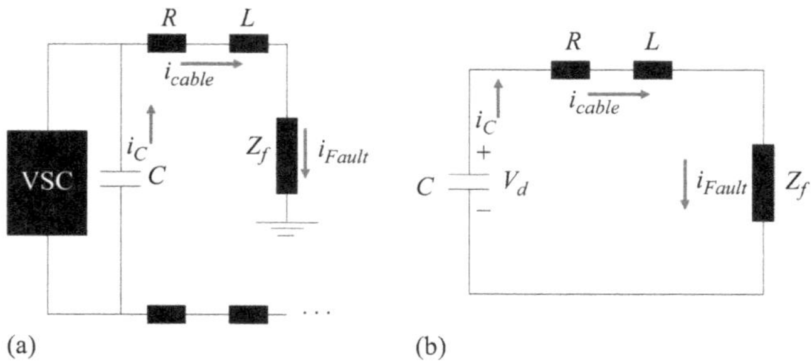

Figure 10.4 (a) Simplified circuit of the rectifier side undergoing an asymmetrical DC fault, (b) Equivalent circuit

between one pole (positive or negative) and ground, linked by the fault impedance Z_f. Normally, this parameter is in the range of 1–10Ω, and it is usually present when the conductor touches the structure of the tower establishing an electric arc in a foggy environment surrounding the conductors. A simplified representation of this case is illustrated in Figure 10.4. Another important issue to be considered is the neutral's grounding of the converter transformer and the DC grounding midpoint, due to the unbalanced conditions arising. If the rectifier transformer's winding is delta-connected, in the asymmetrical fault condition, the healthy pole will undergo an overvoltage of nearly 2 p.u. (twice the nominal value) [10]. However, this issue can be resolved by way of an adequate insulation coordination design.

During an asymmetrical fault, since only one pole has fails, the voltage in the DC link capacitor decreases to near zero, so the IGBT remains connected to the converter. In this situation, while the voltage decreases, the current increases fast. Considering the HVDC system equivalent representation shown in Figure 10.4(a), and its associated RLC circuit (Figure 10.4(b)), then the fault current is expressed by (10.1).

$$i_{cable}(t) = C\frac{d(v_d(t))}{dt} = \gamma_1\rho_1 e^{\rho_1 t} + \gamma_2\rho_2 e^{\rho_2 t} \tag{10.1}$$

$$\gamma_{1,2} = \frac{1}{\rho_2 - \rho_1}\left(\pm\rho_2 V_0 + \frac{I_0}{C}\right)$$

$$\rho_{1,2} = -\left(\frac{Z_f + R}{L}\right) \pm \sqrt{\left(\frac{R + Z_f}{L}\right)^2 - \frac{1}{LC}}$$

During this time, the system response is governed by the DC link capacitor, the conductor inductance and the AC grid. At this point, the voltage of the DC link increases due to the energy contribution from the AC side. This is characterised by oscillations in the current and voltage signals that present lower magnitudes than in cases of pole-to-pole faults [11,12].

For symmetrical faults due to the DC conductors coming into contact with each other, as illustrated in contrived form by the equivalent circuit in Figure 10.5(a), the circuit's voltage falls to zero, as inferred from the equivalent circuit in Figure 10.5(b).

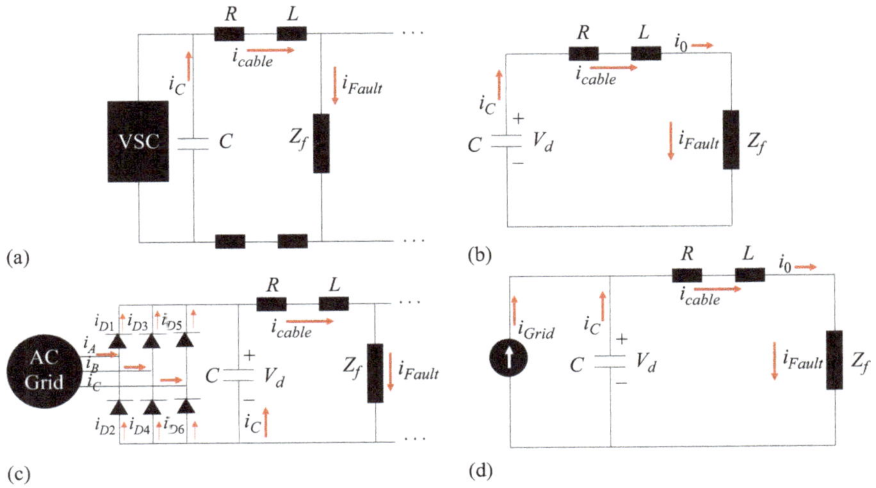

Figure 10.5 (a) Simplified circuit of the rectifier side undergoing a symmetrical DC fault, (b) equivalent circuit, (c) non-controllable diode bridge, (d) AC grid feeding

Therefore, the IGBT becomes blocked for low voltage self-protection, leaving the reverse diodes exposed to overcurrent [23]. The overcurrent produced in this stage is harmful to the diodes since each converter leg carries one-third of the fault current. This is illustrated in Figure 10.5(c) where the non-controlled diode bridge remains a part of the equivalent circuit, but not the IGBTs that have been blocked. In the last stage, the DC link circuit is fed by an equivalent current source, as depicted in Figure 10.5(d). In HVDC links with modular multilevel converters (MMC) with one DC capacitor for both conductors, the DC link is replaced by a low-value cascade capacitor. For this reason, MMC-HVDC can be tolerant to DC faults, due to the transient response being governed by the slow response of the DC capacitor.

An expression to calculate the fault current flowing through the cable is given by (10,2). Considering $V_c(t) = V_0$ and $i_{cable}(t_0) = I_0$, the fault current may be expressed as follows:

$$i_{cable}(t) = C\frac{d(v_c(t))}{dt} = -\left(\frac{I_0\omega_0}{\omega}\right)e^{-\delta t}Sin(\omega t - \beta) + \left(\frac{V_0}{\omega L}\right)e^{-\delta t}Sin(\omega t) \tag{10.2}$$

$$\delta = \frac{R}{2L}$$

$$\omega^2 = \frac{1}{LC} - \left(\frac{R}{2L}\right)^2$$

$$\omega_0 = \sqrt{\delta^2 + \omega^2}$$

$$\beta = \arctan\left(\frac{\omega}{\delta}\right)$$

The typical DC system faults are:

- A pole-to-ground fault occurs when one of the conductors, either the positive or the negative pole, makes direct contact with the ground. The direct connection to the ground can result in a large fault current flowing through the system. This has the potential to damage conductors, insulation and connected equipment. Moreover, it presents serious safety risks to personnel, potentially leading to electrical shocks, fires and explosions. A pole-to-ground fault can create a voltage imbalance within the system, affecting the performance and stability of the entire network.
- A pole-to-pole fault occurs when the positive and negative conductors come into contact with each other. Due to the low impedance between the two conductors, the fault current can be extremely high. This rapid current flow generates significant heat, which can cause substantial damage. It can result in a severe voltage drop, leading to a loss of power in the system and affecting all the connected loads. The intense energy released during a pole-to-pole fault can damage and destroy conductors, connectors and other components, requiring extensive repairs and replacements.
- Ground faults on both poles occur when both the positive and negative conductors experience ground faults simultaneously. In these kinds of faults, the current flows through multiple paths, including the ground. This introduces a rather complex scenario, which makes it difficult to analyse and mitigate the fault current effectively, causing severe voltage fluctuations and instability within the system, potentially leading to equipment malfunction and even total failure. Detecting and isolating ground faults on both poles require advanced protection schemes and equipment to ensure a quick and accurate fault identification.

Arc faults occur when an electrical arc is formed between conductors or between a conductor and the ground. These faults can generate significant heat and potentially lead to fires. Arc fault detection devices are used to identify and mitigate these hazards promptly.

Overcurrent conditions can arise from excessive system loading or short circuits, while overvoltage conditions may result from external factors such as lightning strikes or switching surges. Protective devices like fuses, circuit breakers and protective relays are essential for managing these conditions and protecting the system from damage.

10.4 Faults in AC–DC rectifiers and inverters

AC–DC rectifiers and DC–AC inverters are essential components in electrical power systems which enable the conversion and control between AC and DC powers [13]. Understanding the types of faults that can occur in these power electronics devices is crucial to maintaining system reliability and performance.

10.4.1 AC faults

Fluctuations and variations in the AC input voltage can affect the operation of AC–DC rectifiers. Voltage sags, swells and interruptions can lead to instability in the rectifier's output DC voltage, affecting the connected loads and equipment.

Phase imbalance in the AC supply can occur due to unequal load distribution between the phases or faults in the distribution network. This imbalance can result in uneven current sharing among rectifier phases, causing overheating, reduced efficiency and potential damage to rectifier components.

Harmonics in the AC supply, typically caused by non-linear loads or power electronics equipment, can affect the performance of AC–DC rectifiers. High harmonic content can lead to increased losses, overheating of rectifier components and interference with other sensitive equipment connected to the same power system.

Overcurrent protection devices in AC–DC rectifiers are designed to trip and disconnect the rectifier from the AC supply in case of excessive current flow. Overcurrent faults can occur due to short circuits, overloads or faults in the rectifier circuitry, requiring troubleshooting and resetting procedures to restore normal operation.

Cooling system failures, such as fan malfunctions or inadequate heat dissipation, can lead to overheating of the rectifier components. This can reduce the rectifier's efficiency and reliability, potentially causing thermal damage and requiring maintenance or replacement of cooling components. This issue has been outlined in Chapter 3.

10.4.2 DC faults

DC link capacitors in DC–AC inverters are critical for filtering and stabilising the DC voltage before inversion to AC. Capacitor failures due to ageing, overvoltage conditions and thermal stress can lead to increased ripple voltage, instability in the DC bus and potential damage to inverter components.

Overvoltage protection mechanisms in DC–AC inverters are activated when the AC output voltage exceeds safe operating limits. This can occur due to sudden load changes, resonance conditions or faults in the inverter circuitry, requiring corrective actions to prevent damage to the connected loads and equipment.

As detailed in Chapters 3 and 4, DC–AC inverters use power semiconductor valves such as insulated gate bipolar transistors (IGBTs) or metal-oxide-semiconductor field-effect transistors (MOSFETs). These devices are very susceptible to overcurrent, voltage spikes and thermal overstress, disrupting the switching operation and overall performance of the inverter.

Frequency instability in the AC output of DC–AC inverters can occur due to disturbances in the DC link voltage, load variations or control system issues. Also, inverter output frequency deviations can affect the operation of connected equipment, particularly sensitive devices requiring precise AC supply frequencies.

Errors or faults in the control system of DC–AC inverters can lead to improper voltage regulation, frequency deviation or failure to synchronise with the grid. Control system faults may require diagnostic procedures and recalibration to restore normal operation and ensure compatibility with AC grid requirements.

10.5 Protective relays used in AC–DC systems

Protective relays are essential to ensuring the reliability and stability of AC–DC power systems. These schemes detect and isolate faults to prevent damage to equipment and maintain the continuity of service. Figure 10.6 shows the zones arrangement of an HVDC system [14] and the protection schemes used:

- AC busbar (BB) zone protection: differential (87B), breaker failure (50BF) and detector of direct transfer trip for faults on the AC side.
- Converter transformer zone protection: overcurrent (51).
- Converter/pole zone protection: asymmetry protection (fundamental and second harmonic frequencies), pole differential protection (87) to detect ground faults on DC side, AC–DC differential to detect valve short circuit (87), DC–AC differential to detect converter failures (87), DC overcurrent to detect overcurrent in the HVDC link (51), AC overcurrent to detect overcurrent in any of the valve winding connections (51), AC overvoltage (59), AC under-voltage (27), thermal protection (49). Inside this zone, additional protections are required as valve unit protection, valve hall fire detection protection and valve hall HVAC system.
- Neutral zone protection: overvoltage protection (59), current unbalance (46), overcurrent protection (51N).
- DC transmission link zone protection: differential (87), overcurrent (51), DC under-voltage (27), DC over-voltage (59), cable balance (60).

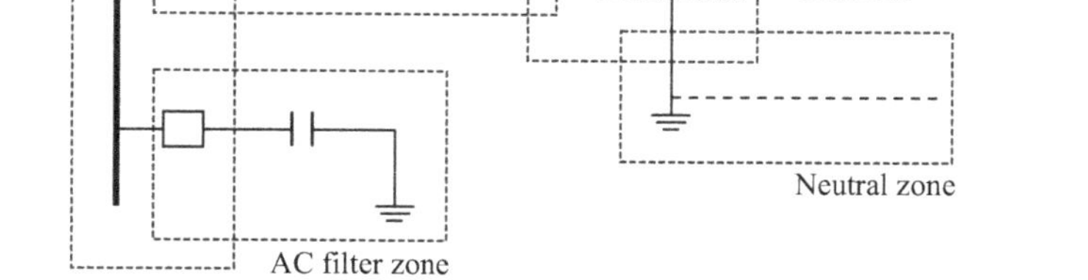

Figure 10.6 Zones arrangement of an HVDC system

- DC filter zone protection: overall protection (51/49), overcurrent (51), unbalanced current (46), thermal (49).

Each protection deals with stochastic signals of voltage and currents at the terminals of the rectifier and inverter stations, with sampling frequencies in the order of 4.8 kHz and up to 96 kHz. This is to measure voltage and current signals at the fundamental frequency or sampling frequencies of about 100 kHz and higher to capture transient events, for cases when the protection uses travelling wave information.

Even though there is a strong drive for standardisation and modularity in power electronics installations and their design and construction, each installation has its own peculiarities and the protection schemes are selected based on component configuration: Figure 10.7 shows a simplified scheme protection for a rectifier

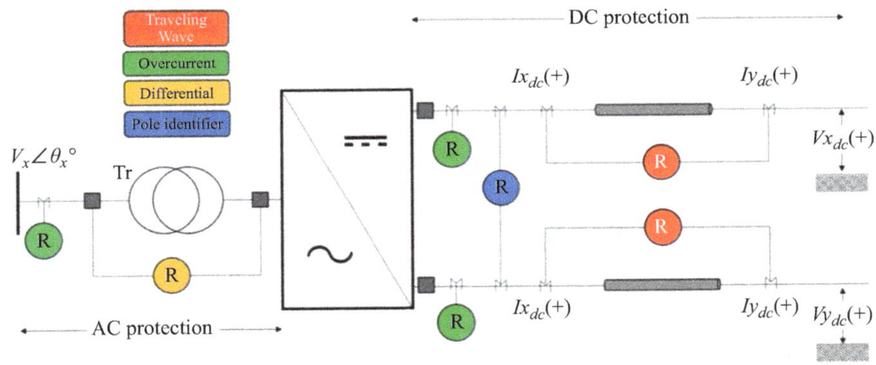

Figure 10.7 Full-scheme protection of an HVDC system

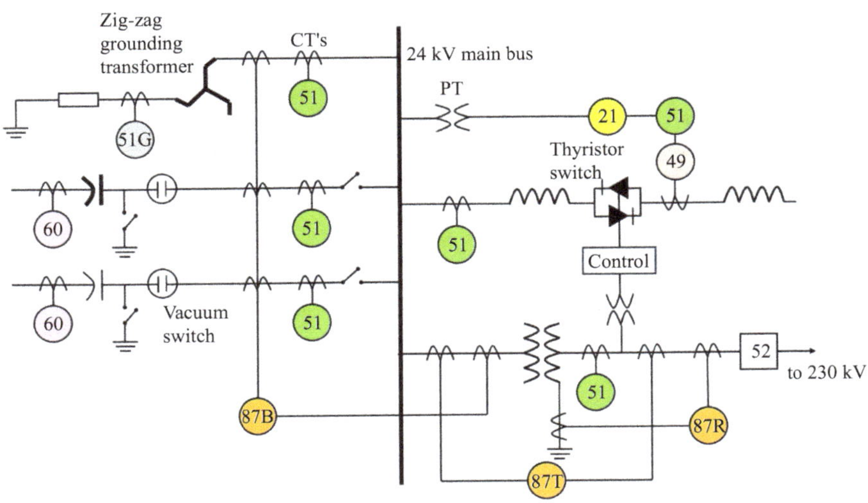

Figure 10.8 One-line diagram for a 230 kV SVC, adapted from [15]

station, whereas Figure 10.8 shows the protection system of a 230 kV static var compensator (SVC) [15]. This protection scheme includes phase and ground protection (51, 51G) for the zig-zag grounding transformer, overcurrent (51) and voltage balance (60) for capacitor branches, differential protection (87B) for the main bus, differential protection (87T), overcurrent protection (51) and ground differential protection (87R) for main transformer and the thyristor switch branches overcurrent protection (51), overload protection (49) and distance protection (21). Note that Figure 10.8 does not include internal protection function related to the operation and control of thyristors.

10.6 Principle of the travelling wave

In ultra-high voltage power systems, high-speed fault clearance is included to improve transient stability. The post-fault voltage and current are initially dominated by electromagnetic travelling waves. Based on the analysis of these transient state signals, the travelling wave-based protective relays can detect and locate the fault within milliseconds after the fault occurrence. It is envisaged that the new generation of fibre optic voltage and current measuring systems based on travelling wave protection will have wide application in DC link protection.

To develop the mathematical foundations of the travelling wave, consider the circuit shown in Figure 10.9. The DC line is charged by the voltage source connected to it. This action produces incremental electric and magnetic fields, expressed in the following manner: $\Delta Q = C\,\Delta x\,v$ and $\Delta\phi = L\,\Delta x\,i$, which are equally distributed over the line segment. These expressions are functions of distance and time. Hence, following an incremental approach, with the segments' line taking to be infinitesimally small (tending to zero), both fields give rise to the voltage and current wave expressions in (10.3) and (10.4) [16].

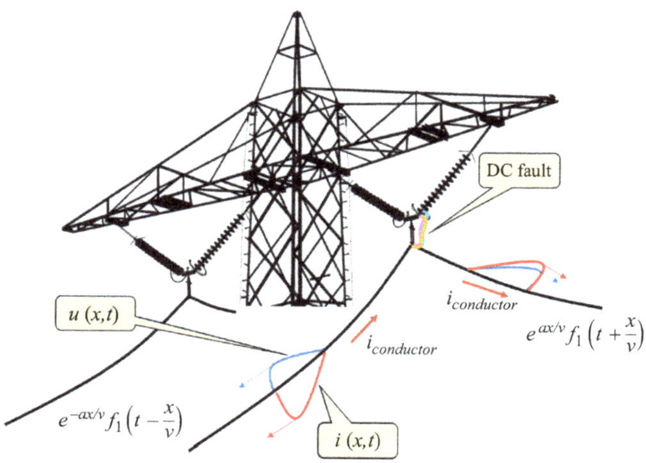

Figure 10.9 Voltage and current travelling wave directions in a DC transmission line

$$v(x,t) = e^{\frac{\alpha x}{v}} f_1\left(t + \frac{x}{v}\right) + e^{-\frac{\alpha x}{v}} f_1\left(t - \frac{x}{v}\right) \tag{10.3}$$

$$i(x,t) = \frac{1}{Z_C}\left[e^{\frac{\alpha x}{v}} f_1\left(t + \frac{x}{v}\right) + e^{-\frac{\alpha x}{v}} f_1\left(t - \frac{x}{v}\right)\right] \tag{10.4}$$

These equations represent the transient phenomena of the travelling wave in a DC transmission line. The negative sign in the argument of the second term in (10.3) and (10.4) is for the backward travelling wave due to its displacement in the negative-x direction. This satisfies the direction of the currents and the polarity of the voltage, as in conventional electric circuit theory (Figure 10.9). These waves travel at near-light-velocity in overhead conductors and approximately at one-half of this velocity in underground cables owing to their higher attenuation.

10.7 DC fault protection methods in HVDC systems

The long-enduring conventional AC protection schemes, such as distance, overcurrent and differential protection, run into difficulties when applied to HVDC systems. Refer to Appendix A for a comprehensive list and summary of AC protection schemes. In overcurrent protection, the tripping action takes place when the current reaches the threshold value. However, the same relay may protect adjacent lines as backup protection. Differential relays operate if the difference between the input and output currents in a transmission line element exceeds a threshold. Applying this procedure to a short-distance transmission line and busbar protection is straightforward, but for long transmission lines, the operation time becomes a restriction. Distance relays estimate the apparent impedance by measuring the voltage and current at the relay location. This can detect the fault as primary or backup protection for AC transmission lines. However, the 'impedance' at the fundamental frequency is different in an HVDC link. Therefore, conventional relays aimed at AC protection applications are not suitable for DC circuits protection, and developing suitable protection schemes for HVDC systems is an ongoing challenge. Several options, such as the derivative of the signal and the wavelet transform (WT)-based protection schemes, have been suggested because of their short response time.

Another problem is the operation time. In an AC grid, a Fourier Transform (FT) is applied to extract voltage and current phasor information for the relay computations. However, the concept of phasor does not exist in DC circuits. Accordingly, the need has arisen to develop protection algorithms based on transient responses rather than the fundamental frequency component.

Several methods have been put forward for addressing such problems in DC grids: some based on the voltage derivative, travelling waves, currents differential and DC voltage-level protection schemes. Some methods detect the fault using the time difference between the two reflected waves using different correlation methods. The algorithms based on the voltage derivative method calculate the derivatives of different DC voltages and compare them with a threshold: this is defined as a heuristic process relaying on running many simulation scenarios. However, these methods are vulnerable to fault loop impedance and long-distance faults. Another method uses the concept of travelling waves (typically current signals). In this case, the fault is detected by resorting to a waveform pattern recognition process that discriminates between steady state and faulted conditions. These classes of DC protection algorithms use detection methods such as wavelets and transient-based protection schemes.

10.8 State of the art in HVDC protection algorithms based on the travelling wave principle

Several protection algorithms have been put forward for VSC-HVDC links ever since the first installation in Hellsjön, Sweden, in 1997 [17]. As discussed above, due to the nature of the DC voltages and currents, these protection strategies have been based on travelling wave information. This is a challenging area of research, and there is still much to do to have reliable and efficient protection algorithms to meet the same level of dependency as the algorithms employed in AC power circuits.

Figure 10.10 shows a simplified representation of a DC link. When a fault occurs in a DC conductor, the circuit in Figure 10.10(a) can be represented by the equivalent circuit shown in Figure 10.10(b). L is the total length of the conductor, $Ix_{dc}(+)$ and $Iy_{dc}(+)$ are the input and the output currents, holding the following relationship $Ix_{dc}(+) = -Iy_{dc}(+)$, in steady-state operation. $Vx_{dc}(+)$ and $Vy_{dc}(+)$ are the voltages at the sending and receiving ends, between the conductor and ground, with $Vx_{dc}(+) > Vy_{dc}(+)$ to enable power transfer between terminals.

According to [18], the point in fault generates components of voltage and current: ΔV_{Pdc} and ΔI_{Pdc}, where $p = x, y$. These signals are superimposed

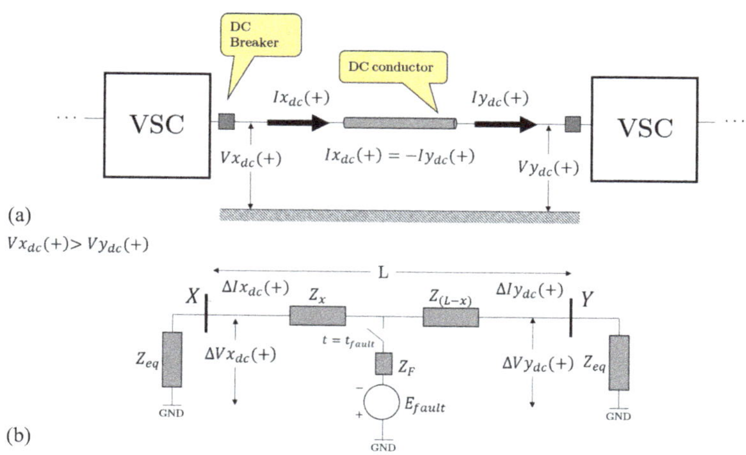

Figure 10.10 (a) Electrical variables of a DC link, in (b) equivalent circuit

quantities on the measured quantities, which can be detected as incremental changes, enabling to identify the location of the point in fault using (10.5). The time difference is the main indicator of the distance to the point in fault when one of the terminals is used as the reference. For example, if terminal X is taken to be the reference, it implies that the time t_a is the time in which this terminal captures the first wavefront and t_b is the time used by the opposite terminal. Based on Figure 10.10, when two waves are produced and travel to their respective end points at a specific velocity, depending on whether the conductor is an overhead or an underground conductor, the distance to the fault is given by (10.5).

$$d = \frac{L - v(t_b - t_a)}{2} \tag{10.5}$$

The algorithms presented in [19–23] and [24] employ the travelling wave principle to locate faults in point-to-point HVDC links. They represent a fair selection of the state of the art in this challenging area of protection, having advanced significantly the task in hand. Nonetheless, they all have drawbacks that are worth discussing.

In [20–22], the algorithms use the travelling wave concept and carry out the fault location identification with a second algorithm, namely, S-transform [20], FastICA [21] and voltage gradient [23]. These algorithms provide a complete way to achieve fault detection. The main drawback of these methodologies is that they have a strong dependency on external factors such as the sampling frequency (S-transform) or high computational burdens (optimisation equations in FastICA). In [23], the computation of the voltage gradient, based on the travelling wave principle, is carried out to detect and determine the type of fault (internal or external). A further disadvantage of these algorithms is the use of empirical thresholds (set with no proper methodology) to determine whether the faults are external or internal. The algorithms put forward in [22] and [24] show similar weaknesses. The determination of thresholds in protection algorithms and, in general, the setting of the relay characteristics, is an issue of paramount importance in protection equipment. Methodologies supported by proper mathematical foundations are required in a relay setting to ensure the algorithms perform reliably, even under challenging circumstances, such as cases of high impedance faults or external faults.

The capacity to discriminate between internal and external faults using a reduced number of settings has been addressed satisfactorily in recent publications relating to DC link protection. However, there are additional aspects that need to be covered to count with reliable, efficient and computationally light algorithms. The number of pre-processing, processing and post-processing stages is the key factor to determine whether or not a protection algorithm is feasible for implementation in real time, embedded in a chip. Several hybrid algorithms have been put forward in [25–28]. In [25], several methods of pattern extraction are addressed: Hilbert-

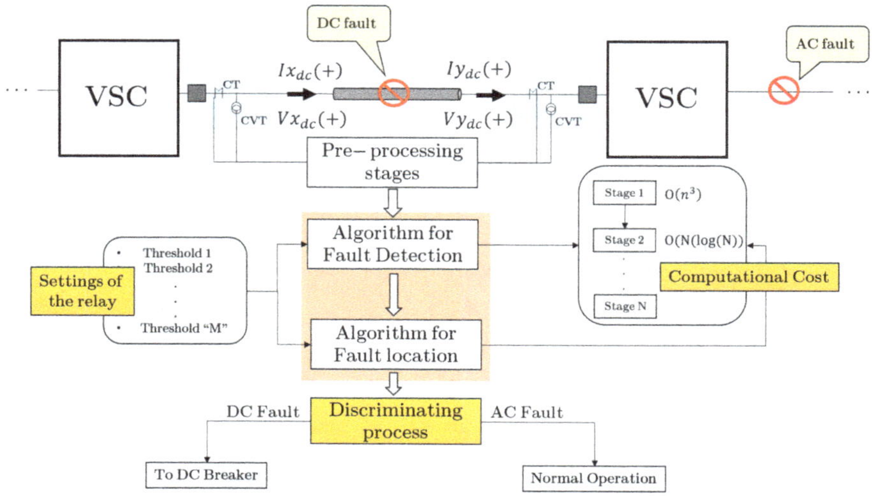

Figure 10.11 Stages of a feasible protection algorithm applied in a DC link

Huang transform, support vector regression and an optimisation technique termed Bat Algorithm. They are combined and applied to the task of estimating the point of fault on a DC conductor. It is reported that this hybrid algorithm is dependent on the selection of the sampling frequency. A similar weakness occurs with the algorithms presented in [26] and [27].

The work in [28] presents a method to detect and locate faults in DC grids using pattern recognition and optimisation techniques based on the WT. The objective is to obtain coefficients to make up a Pattern Vector comprising the main components to classify the type of fault by resorting to fuzzy logic techniques. The setting of each stage is carried out by genetic algorithms. Pattern recognition techniques offer great potential for DC protection once training has been carried out. However, the high number of stages complicates its implementation.

It may be concluded that although the mathematical foundations of the latest techniques that have been put forward are sound, their real-time implementation is yet outstanding. As indicated in Figure 10.11, the pre-processing stages modify the input signals with filters. Once they are clean, two algorithms (fault detection and fault location) are executed concurrently. As a final stage, the discriminating process stage consists of a classification algorithm to detect internal and external failures. Altogether, this a complex algorithm that demands a fair share of computing power, and its real-time implementation remains a big challenge.

Appendix A

A.1 Overcurrent protection [29]

An electrical fault represents a hazardous situation in an electrical power system, putting personnel and equipment at risk. The effects of faults must be limited or minimised and, when possible, avoided. Perhaps the most frequent fault is the short circuit. Since the early years of the electrical power industry, short circuits have been a great problem for power system operators. Short circuits are characterised by a sudden increase in the current magnitude, reaching several times the rated current. Efforts to develop control mechanisms to limit the damage caused by short-circuit events and to enhance reliability have been a topic of great importance since the last quarter of the 19th century, when the power industry first started. Protection against excessive currents (i.e., overcurrents) in the system was the first type of system protection to be designed.

Early studies on the effects of short circuit currents in electrical power system equipment showed the potential damage that can be caused, as well as thermal and mechanical stress damages. The thermal damage is caused by heat generated by the large short circuit currents through the equipment. A measurement for that heat-related stress is the so-called I^2t, which is typically used as energy measurement [30]. It is generally expressed in ampere-square-second units.

Overcurrent protection devices, such as relays, fuses and circuit breakers, require specific settings to operate effectively and protect electrical power systems from damage due to excessive current. In general, the parameters for adjusting overcurrent protection are the following:

A.1.1 Current rating (I_n)

The current rating, also known as the nominal or rated current, is the maximum continuous current that a protective device can handle without tripping. This parameter is critical because it defines the operational capacity of the device under normal conditions. The current rating should be chosen based on the expected load of the circuit and the capacity of the conductors. See (A.1). Setting the current rating correctly ensures that the device does not trip during normal operation while providing adequate protection against overloading conditions.

$$I_n = kI_{\max LOAD} \tag{A.1}$$

A.1.2 Long-time pickup setting (LTPS)

The long-time pickup setting determines the current level at which the protective device starts timing for an overload condition. This setting allows the system to manage temporary overcurrent situations, such as inrush currents for large induction motor startings, without tripping. The LTPS should be set slightly above the normal operating current to avoid unnecessary interruptions while still providing protection against sustained overloads that could damage equipment and conductors.

A.1.3 Long-time delay setting (LTDS)

The long-time delay setting specifies the duration that the protective device will wait before tripping once the long-time pickup current is exceeded. This delay helps to prevent the device from tripping due to short-duration overloads such as those caused by temporary surges. The LTDS should be configured based on the load characteristics, ensuring that the system remains operational during minor fluctuations but trips to protect the system during prolonged overloads.

A.1.4 Short-time pickup setting (STPS)

The short-time pickup setting defines the current level at which the protective device responds to short-duration overcurrent. This setting is essential for protecting the system against short circuits and high fault currents. The STPS should be set to a value higher than the long-time pickup setting but lower than the maximum fault current to ensure a quick response to significant overcurrent events, preventing extensive damage.

A.1.5 Short-time delay setting (STDS)

The short-time delay setting determines the brief delay before the protective device trips in response to a short-time pickup condition. This delay allows the device to differentiate between temporary surges and actual fault conditions, enhancing selectivity. Typically, the STDS is set to a short duration to ensure rapid tripping during high fault currents while avoiding unnecessary trips for brief surges.

A.1.6 Instantaneous pickup setting (IPS)

The instantaneous pickup setting specifies the current level at which the protective device will trip immediately without any intentional delay. This setting is crucial for providing immediate protection against extremely high fault currents such as those caused by direct short circuits. The IPS should be set significantly above the highest expected operating current but below the maximum fault current to ensure immediate tripping during severe faults.

A.1.7 Ground fault protection (GFP)

Ground fault protection settings detect and respond to ground faults, where current flows directly to the ground, bypassing the load. This protection is vital for preventing significant damage and safety hazards associated with ground faults. The GFP should be configured based on system requirements and safety considerations, typically set lower than phase fault settings to ensure sensitive detection and response to ground faults.

Figure A.1 describes the typical curves associated with each parameter for adjusting overcurrent protection. These parameters can be classified into three groups: instantaneous, definite time and inverse time. In the first two parameters, relays operate instantaneously when the current reaches a predetermined value. The inverse time parameter has a trip time that is inversely proportional to the magnitude of the fault current. In other words, the greater the fault current magnitude, the faster the relay will operate. Inverse-time relays are generally classified in accordance with their characteristic curve that indicates the speed of operation; based on this, they are

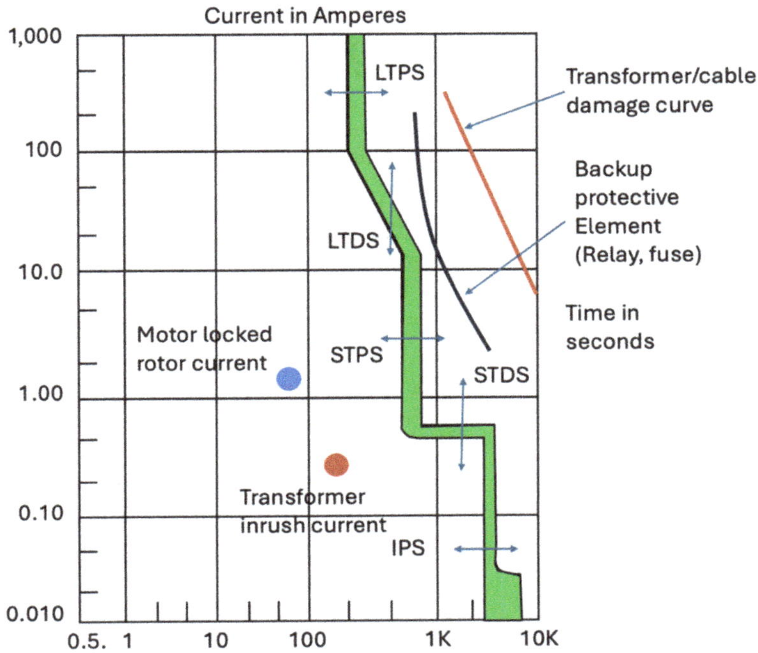

Figure A.1 Typical overcurrent time-current characteristics

commonly defined as being inverse, very inverse or extremely inverse. Inverse-time relays are also called inverse definite minimum time (IDMT) overcurrent relays. As a consequence of their performance, relays with instantaneous and definite time parameters cannot be coordinated with inverse time parameter relays.

A.2 Distance protection [29]

Distance protection, also known as impedance protection, is a protection scheme used to detect and isolate faults in high-voltage transmission lines. This scheme measures the impedance between the relay's location and the fault's point. Because impedance is proportional to the distance between these points, distance protection effectively determines the fault location and responds accordingly.

The principle of operation of distance protection is based on the relationship between voltage, current and impedance. The impedance (Z) is calculated using the equation $Z = V/I$, where V represents voltage and I represents current. Under normal operating conditions, a transmission line's impedance remains relatively stable. However, when a fault occurs, there is a significant drop in voltage and a corresponding rise in current, causing a sudden change in impedance.

Distance relays are designed to detect these changes. They continuously measure the voltage and current in the line to calculate the impedance. When the

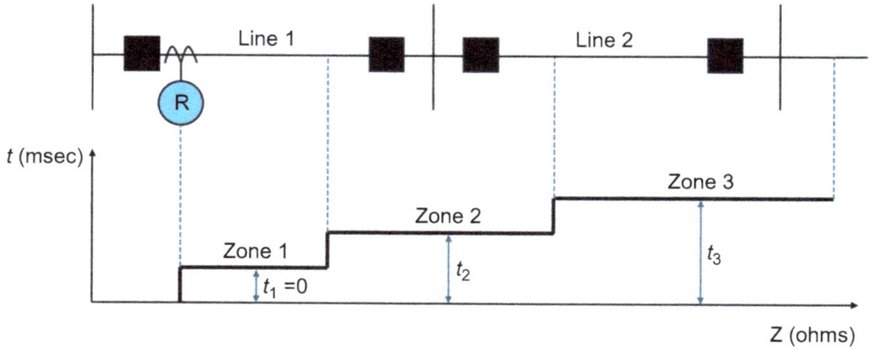

Figure A.2 Protection zone with distance relays

calculated impedance falls within a predetermined range, indicative of a fault, the relay initiates a trip signal to isolate the affected section of the line.

To ensure precise fault detection and isolation, distance protection relays are configured with multiple zones of protection (Figure A.2):

- Zone 1: This zone provides instantaneous protection and covers 80–90% of the transmission line length. It operates without intentional delay for a rapid fault clearance within its designated distance coverage.
- Zone 2: Extending beyond Zone 1, this zone includes the remaining 10–20% of the line and a portion of the adjacent line. Zone 2 has a slight time delay to allow for coordination with Zone 1 and neighbouring protection devices, ensuring selective fault isolation.
- Zone 3: This zone operates as a backup, covering the entire protected line and extending into the next line section. With a longer time delay, Zone 3 ensures fault clearance if Zone 1 and Zone 2 protections fail.

Distance protection offers several advantages over other protection methods. One significant advantage is its relative immunity to changes in load and system conditions, providing consistent and reliable operation. The multi-zone configuration enhances selectivity, ensuring that only the faulty section is isolated, thereby minimising the impact on the overall system.

Furthermore, distance protection is effective over long distances, making it ideal for high-voltage transmission lines. Its ability to quickly and accurately identify faults helps maintain the stability and reliability of the power grid.

In practical applications, distance protection is often integrated with other protective schemes, such as overcurrent and differential protection, to provide comprehensive coverage. Properly coordinating these schemes is crucial to ensuring timely and selective fault isolation. By carefully setting the parameters for each protection zone and coordinating with adjacent relays, engineers can optimise the protection strategy for maximum effectiveness.

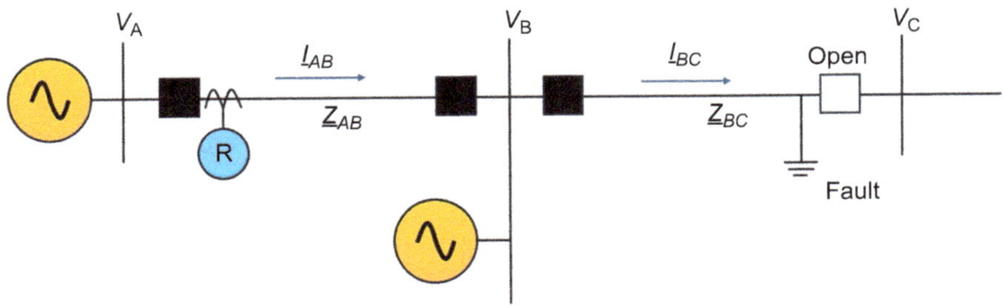

Figure A.3 Impact of infeed effect in distance protections

When a fault current source is within the operating range of a distance relay, the relay's range decreases depending on the magnitude of the intermediate source current. This effect, known as positive intermediate source (or infeed), is illustrated in Figure A.3. In this figure, there is a current path towards the fault from the source connected to bus B, with no current circulating through the protection at bus A. When a fault occurs at bus C, the voltage measured by the protection at bus A equals the voltage drop across the AB line plus the voltage drop across the BC line. Then, the impedance measured by protection A ($\underline{Z}_A$) is given by (A.2) and (A.3):

$$\underline{Z}_A = \frac{\underline{V}_A}{\underline{I}_{AB}} = \frac{\underline{I}_{AB}\underline{Z}_{AB} + \underline{I}_{BC}\underline{Z}_{BC}}{\underline{Z}_{AB}} = \underline{Z}_{AB} + k_i\underline{Z}_{BC} \tag{A.2}$$

where k_i is the current distribution factor, defined as:

$$k_i = \frac{\underline{I}_{BC}}{\underline{I}_{AB}} \tag{A.3}$$

With no generation source in bus B, $\underline{I}_{AB} = \underline{I}_{BC}$, so the impedance measured by protection A is: $\underline{Z}_{AB} + \underline{Z}_{BC}$.

From the analysis, when there is no intermediate source effect, the impedance measured by protection in bus A is the actual impedance between the protection and the fault point. As the intermediate source effect increases, making $\underline{I}_{AB} < \underline{I}_{BC}$, the term corresponding to the impedance $\underline{Z}_{BC}$ is increased by the factor k_i. As the impedance measured by the protection is greater than the impedance that exists up to the point of failure, the relay range will decrease. If the relay is set to reach exactly up to bus C with no intermediate source (i.e., $\underline{Z}_{AB} + \underline{Z}_{BC}$) when this is present, its range is reduced and it stops covering part of the BC line. That is, the intermediate source effect causes under-range in distance protection.

However, if the protection of bus A is adjusted to reach up to bus C with the intermediate source present, when this does not exist, the relay over-reach the busbar C and enters the adjacent line. If the current flowing through the protection is greater than the total fault current, an intermediate source effect is also present, but negative (or outfeed).

A.3 Differential protection [29]

Differential protection is the main protection used to safeguard electrical equipment, such as transformers, generators and busbars, by detecting faults within a specific zone. This protection scheme operates on the principle of comparing the current entering and leaving the protected zone. If the difference between these currents exceeds a predetermined threshold, the relay is activated and isolates the faulted section, ensuring minimal damage and system disruption.

The fundamental concept behind differential protection is Kirchhoff's Current Law, which states that the sum of currents entering a node (or zone) must equal the sum of currents leaving that node. Under normal operating conditions, the current entering the protected zone should be nearly equal to the current exiting it, accounting for minor differences due to measurement errors and load variations.

To implement differential protection, current transformers (CTs) are installed at the boundaries of the protected zone (e.g., the primary and secondary sides of a transformer). These CTs measure the current and feed the values to a differential relay. The relay calculates the difference (or differential current) between the incoming and outgoing currents. If this differential current exceeds a preset threshold, it indicates a fault within the protected zone (Figure A.4).

When a fault occurs within the protected zone, the current balance is disrupted. For example, in the case of an internal fault in a transformer, the current entering the transformer may not equal the current leaving it, resulting in a significant differential

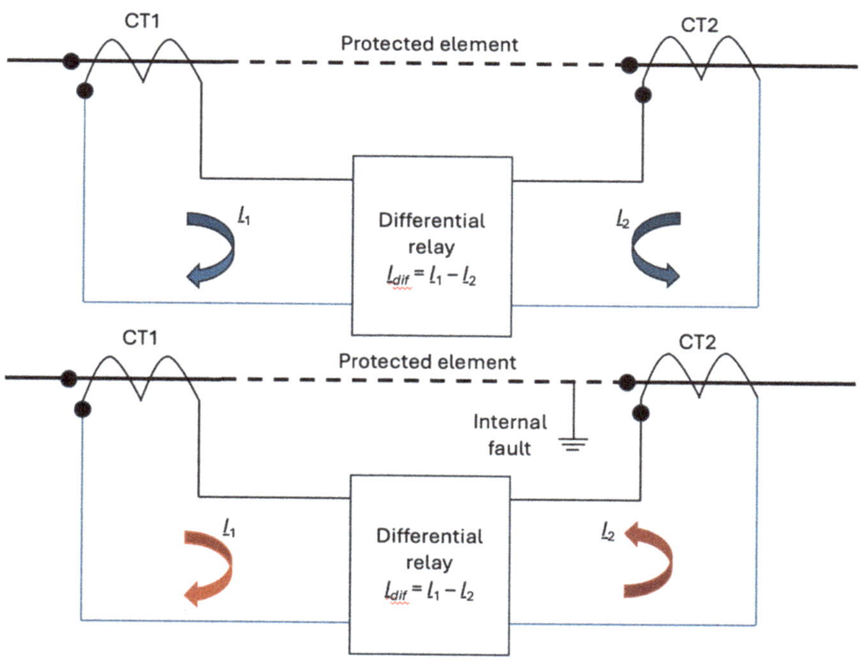

Figure A.4 Principle of differential protection

current. The differential relay detects this imbalance and quickly sends a trip signal to the circuit breakers, isolating the faulty section from the rest of the system. This rapid response helps to prevent equipment damage and maintain system stability.

In contrast, for external faults or normal operating conditions, the currents measured by the CTs will be nearly equal, and the differential current will remain below the threshold. This ensures that the differential relay does not operate unnecessarily, providing selective protection.

Differential protection offers several advantages, making it the preferred choice for critical equipment protection:

- High sensitivity and selectivity: Differential protection is highly sensitive to internal faults and can detect even minor imbalances. Its selectivity ensures that only the faulted section is isolated, minimising the impact on the rest of the system.
- Fast response: The differential relay operates very fast upon detecting a fault, reducing the time fault currents flow through the equipment. This rapid action limits damage and enhances the system's reliability.
- Stability for external faults: Differential protection schemes are stable for faults outside the protected zone, ensuring that the relay does not trip for external disturbances. This stability is crucial for maintaining continuous operation of the unaffected parts of the system.
- Adaptability: Differential protection can be applied to various types of electrical equipment, including transformers, generators, motors and busbars. Its adaptability makes it a versatile protection method.

In practical applications, differential protection schemes are often coordinated with other protection devices to provide comprehensive coverage. For instance, in a transformer protection setup, differential protection may be combined with overcurrent and distance protection to address different fault scenarios. Proper coordination ensures that the differential relay operates only for internal faults, while other relays handle external or system-wide disturbances.

References

[1] G. Ebner, D. Döring, F. Schettler, K. Würflinger and M. Zeller, "Fault handling at hybrid high-voltage AC/DC transmission lines with VSC converters," *IEEE Transactions on Power Delivery*, vol. 33, no. 2, pp. 901–908, 2018, doi:10.1109/TPWRD.2017.2748163.

[2] P. M. Anderson, C. Henville, R. Rifaat, B. Johnson, and S. Meliopoulos, "Voltage source converter HVDC protection," in *Power System Protection*. New York: Wiley; 2021. pp. 1021–1040.

[3] M. Wang, M. Abedrabbo, W. Leterme, *et al.*, "A review on AC and DC protection equipment and technologies: Towards multivendor solution," in *Cigrè Winnipeg 2017 Colloquium*, Cigrè, 2017, pp. 1–11.

[4] J. Zheng, M. Wen, Y. Chen, and X. Shao, "A novel differential protection scheme for HVDC transmission lines," *International Journal of Electrical Power & Energy Systems*, vol. 94, pp. 171–178, 2018.

[5] S. Gautam and S. M. Brahma, "Detection of high impedance fault in power distribution systems using mathematical morphology," *IEEE Transactions on Power Systems*, vol. 28, no. 2, pp. 1226–1234, 2013.
[6] B. Chang, O. Cwikowski, M. Barnes, R. Shuttleworth, A. Beddard, and P. Coventry, "Review of different fault detection methods and their impact on pre-emptive VSC-HVDC DC protection performance," *High Voltage*, vol. 2, pp. 211–219, 2017, https://doi.org/10.1049/hve.2017.0024.
[7] H. Rao, Y. Zhou, C. Zou, *et al.*, "Design aspects of hybrid HVDC system," *CSEE Journal of Power and Energy Systems*, vol. 7, no. 3, pp. 644–653, 2021.
[8] W. Lin, D. Jovcic, S. Nguefeu and H. Saad, "Modelling of high power mechanical DC circuit breaker," in *2015 IEEE PES Asia-Pacific Power and Energy Engineering Conference (APPEEC)*, Brisbane, QLD, Australia, 2015, pp. 1–5.
[9] D. Jovcic, *High Voltage Direct Current Transmission*. University of Aberdeen. New York: Wiley; 2015.
[10] P. Chen, B. Xu and J. Li, "A traveling wave based fault locating system for HVDC transmission lines," in *2006 International Conference on Power System Technology*, Chongqing, China, 2006, pp. 1–4.
[11] J. Yang, J. E. Fletcher and J. O'Reilly, "Short-circuit and ground fault analyses and location in VSC-based DC network cables," *IEEE Transactions on Industrial Electronics*, vol. 59, no. 10, pp. 3827–3837, 2012.
[12] P. Zhao, Q. Chen and L. Xing, "DC fault analysis of VSC-HVDC and DC cable protection principle," in 2015 *IEEE PES Asia-Pacific Power and Energy Engineering Conference (APPEEC)*, Brisbane, QLD, Australia, 2015, pp. 1–5.
[13] S. Liao, D. Gan and Z. Shi, "A study on HVDC overcurrent during commutation failure," *IEEE Transactions on Power Delivery*, vol. 39, no. 3, pp. 1963–1974, 2024.
[14] Technical Specification, Protection & Control for HVDC Systems, National Grid, Internal and Contract Specific, TS3.24.90 (RES) – Issue 1, October 2014.
[15] T. J. E. Miller, *Reactive Power Control in Electric Systems*. New York: Wiley; 1982.
[16] R. C. Bernardino, G. S. Salge, C. M. Martins, P. S. Pereira, G. E. Lourenço and P. S. P. Junior, "Ultra-high speed transmission line protection performance in time domain due to several contingency situations," in *17th International Conference on Developments in Power System Protection (DPSP 2024)*, Manchester, UK, 2024, pp. 115–121.
[17] U. Axelsson, A. Holm, C. Liljegren, K. Eriksson, and L. Weimers, "Gotland HVDC light transmission-world's first commercial small scale DC transmission," in *CIRED Conference*, May 1999, p. 5.
[18] E. Vazquez, J. Castruita, O. L. Chacon, and A. Conde, "A new approach traveling-wave distance protection—Part I: Algorithm," *IEEE Transactions on Power Delivery*, vol. 22, no. 2, pp. 795–800, 2007.

[19] Z. Liu, H. Gao, and S. Luo, "Non-unit protection for long-distance LCC-HVDC transmission lines based on waveform characteristics of voltage travelling waves," *International Journal of Electrical Power & Energy Systems*, vol. 150, pp. 142–154, 2023.

[20] C. Xi, Q. Chen, and L. Wang, "A single-terminal traveling wave fault location method for VSC-HVDC transmission lines based on S-transform," in *2016 IEEE PES Asia-Pacific Power and Energy Engineering Conference (APPEEC)*, Xi'an, 2016, pp. 1008–1012.

[21] L. Yue, "Research on HVDC single-ended fault traveling wave location method based on FastICA," in *2016 International Conference on Audio, Language and Image Processing (ICALIP)*, Shanghai, China, 2016, pp. 509–514.

[22] L. Wang, Q. Chen and C. Xi, "Study on the traveling wave differential protection and the improvement scheme for VSC-HVDC transmission lines," in *2016 IEEE PES Asia-Pacific Power and Energy Engineering Conference (APPEEC)*, Xi'an, 2016, pp. 1943–1947.

[23] S. Rajimol, and K. N. Chandrabose, "Fault analysis and traveling waves based protection in bipolar 500 kV HVDC system," in *2017 International Conference on Intelligent Computing, Instrumentation and Control Technologies (ICICICT)*, Kerala, India, 2017, pp. 478–485.

[24] X. Li, S. Lin, and S. Jiang, "Identification method for internal and external faults based on amplitude feature of backward traveling wave for HVDC transmission line," in *2015 5th International Conference on Electric Utility Deregulation and Restructuring and Power Technologies (DRPT)*, Changsha, China, 2015, pp. 1166–1169.

[25] Y. Hao, Q. Wang, Y. Li, and W. Song, "An intelligent algorithm for fault location on VSC-HVDC system, *International Journal of Electrical Power & Energy Systems*," vol. 94, pp. 116–123, 2018.

[26] B. Bahmani-Firouzi, E. Farjah, T. Niknam, and E. Azad Farsani, "A new hybrid HBMO-SFLA algorithm for multi-objective distribution feeder reconfiguration problem considering distributed generator units," *Iranian Journal of Science and Technology Transaction A Science*, vol. 36, pp. 51–66, 2012.

[27] T. Rajic, and Z. Stojanovic, "An algorithm for longitudinal differential protection of transmission lines," *International Journal of Electrical Power & Energy Systems*, vol. 94, pp. 276–286, 2018, 10.1016/j.ijepes.2017.07.001.

[28] R. Bertho, V. A. Lacerda, R. M. Monaro, J. C. M. Vieira and D. V. Coury, "Selective nonunit protection technique for multiterminal VSC-HVDC grids," *IEEE Transactions on Power Delivery*, vol. 33, no. 5, pp. 2106–2114, 2018.

[29] J. L. Blackburn, and T. J. Domin, *Protective Relaying: Principles and Applications*, 4th edn. Boca Raton, FL: CRC Press; 2014.

[30] A. D. Polykrati, C. G. Karagiannopoulos, and P. D. Bourkas, Thermal effect on electric power network components under short-circuit currents, *Electric Power Systems Research*, vol. 72, no. 3, pp. 261–267, 2004, https://doi.org/10.1016/j.epsr.2004.04.010.

Chapter 11
Conclusions and future trends

Claudio R. Fuerte-Esquivel[1]

The global electric power industry is undergoing significant changes in its planning, operation and control processes, due to the integration of non-synchronous generation sources, new types of loads being supplied from the power grid and major changes in its business model. These non-synchronous generation sources, which to a large extent are environmentally harmonious, have unique characteristics such as variability, uncertainty and modularity. This shift has prompted engineers and researchers to develop solutions aimed at improving the real-time operation and control of electric power systems. The goal is to enhance the security and reliability of such systems, ensuring they can operate under normal conditions and recover rapidly from emergency situations without jeopardising the system's integrity. This is achieved by adding operating flexibility through electronic power controllers based on the concepts of flexible AC transmission systems (FACTS) and high voltage direct current (HVDC) power transmission. This chapter provides an overview of the topics covered in this book and draws on the main conclusions. It outlines future trends aimed at enhancing the secure operation of power systems through the application of FACTS controllers and HVDC links.

11.1 Conclusions

The concepts of FACTS and HVDC transmission aim to make the electrical power transmission network electronically controllable, enabling increased operational flexibility, maximising power transfers and ensuring the stability of the electrical power system in the event of disturbances. Various types of FACTS controllers and DC links have been implemented worldwide, each with their specific designs and operating characteristics, as described in Chapter 2. In this context, the technical requirements associated with each installation are described in detail. This chapter also presents the operational problems that motivated the use of FACTS devices and DC links and emphasises the benefits of using these technologies in *network-wide* applications. The chapter details actual applications of different types of FACTS equipment, ranging from those developed to control voltage magnitude

[1]Faculty of Electrical Engineering, University of Michoacan, México

using electronically modulated reactive power support to those exerting simultaneous control of voltage magnitude and active and reactive power flows through transmission elements using controllers based on voltage source converters. The chapter presents a comprehensive survey of FACTS and HVDC installations, which demonstrates the worth of these technologies, given that each of them was selected to address a specific problem where a potentially equivalent AC solution was either not feasible or at least, not competitive against the respective adopted FACTS or HVDC solution. It may be concluded that the power transfer of clean electrical power encompassing intercontinental distances is now possible, using the DC option. This has been demonstrated by the extra-long-distance schemes in China and Brazil, the long-distance submarine schemes between the UK and the Nordic countries and projected DC transmission schemes between Australia and Southeast Asia, as well as between Northwest Africa and Northern Europe. It may also be concluded that modular multilevel converters (MMC) and VSC-HVDC have made the reliable integration of deep offshore windfarms in the North Sea into the onshore European power grid possible.

Semiconductor materials have facilitated the development of switching elements integrated into power electronic circuits, essential components of the controller devices used to enhance the electrical network's operating and control flexibility. Electronic valves appeared during the first quarter of the 20th century, the century of the avant-garde, and the development of the mercury arc rectifier for traction applications and then for power transmission promoted innovation and competition, leading to the development of power semiconductor devices, from the triode to the ignitron. As described in Chapter 3, the invention of the silicon transistor in the 1950s triggered the beginning of a new era in power electronics that allowed the development of thyristor-based switching elements such as the silicon-controlled rectifier. New proposals of electronic-based switches have continued until the present thanks to advances in semiconductor materials technology, which have given rise to various types of solid-state switching devices such as bipolar junction transistors (BJTs), metal-oxide field-effect transistors (MOSFETs), gate-off switches (GTOs), the MOS-controlled thyristor (MCT) and the insulated-gate bipolar junction transistor (IGBT). Chapter 3 provides a detailed explanation of these switching devices' structure and operating characteristics. Additionally, it presents the relationship between the development of power semiconductor devices and power electronic-based controllers used in electric power systems. It may be concluded that Si as a semiconductor material has served very well the electronics industry in general and the power semiconductor industry in particular. However, as the demand for clean energy continues unabated and this energy is transported over thousands of kilometres using state-of-the-art DC transmission circuits that operate at increasingly higher voltages, the Si valves employed are subjected to higher electric and magnetic field stresses. It is concluded that although Si switching valves will continue to be used in most FACTS and HVDC applications for the foreseeable future, the use of wide bandgap semiconductor materials will come as a natural replacement for Si in due course.

Chapter 4 underscores the importance of understanding the physical structures and control mechanisms of thyristor and isolated-gate bipolar transistor (IGBT)-based converter for voltage/reactive power control, active power control, phase-angle regulation and impedance regulation. The topologies, control modes and operation of current source converters (CSCs) and voltage source converters (VSCs) are explained to visualise their capabilities. Both types of converters are used in high-voltage direct current (HVDC) transmission systems. The primary distinction between these two technologies is that the VSC acts as an AC variable voltage source, eliminating the need for large filters and reactive power support. On the other hand, the CSC operates as a constant current source and requires major AC filtering to eliminate harmonics in the system's AC side, as well as substantial capacitive reactive power support. The IGBT-based VSC offers higher efficiency than the thyristor-based CSC, featuring greater controllability and operational flexibility. The CSC, however, is traditionally more suitable for bulk power transmission over long and extra-long distances owing to the more robust nature of the thyristor valves compared to IGBT valves and their lower manufacturing costs. It may be concluded that both flavours of the HVDC technology will coexist alongside each other for the foreseeable future and that the selection between the VSC and CSC options depends on the specific requirements of the HVDC system and its intended application.

Before installing FACTS controllers and HVDC links, it is essential to quantify their technical and economic benefits in power grids through computational studies. These studies are crucial to understand their impact on the grids' operation. Hence, adapting and updating the computational tools used for planning the electric system's future expansion and quantifying its operation in real time is essential to accomplishing such an objective. From a steady-state operation standpoint, power flow studies and derived formulations are not just essential but of the utmost importance in evaluating the performance of the overall power system with the new equipment, assessing the electric power flows through the network and implementing appropriate control measures to strengthen the power system under normal and disturbance conditions. Consequently, an area of great activity in many research laboratories has been the modelling and implementation of FACTS and HVDC models, computing algorithms and control methods in power flow, optimal power flow and state estimation computer programs. Within this context, Chapter 5 discusses the modelling of selected FACTS equipment and HVDC systems from a steady-state operation context, through which a quantitative and qualitative assessment of the control capabilities of this equipment in electrical networks is carried out. Study cases are presented to analyse their impact and overall performance in a selected portion of a real-life electric power system. Their effectiveness in achieving control of power flows and nodal voltage magnitudes has been assessed. The simulation results indicate that the embedded controllers do improve the network's operating performance without adversely impacting its overall operating cost. It is concluded that the steady-state models of FACTS and HVDC presented in the chapter are effective in assessing the control actions of these equipment on their targeted power network parameters such as nodal voltage, line impedance, phase angle and power flows. It is also concluded that network-wide computation algorithms like power flows and derived formulations play a significant and amenable role in power network expansion planning by incorporating the models of new transmission technologies.

Chapter 6 delves into the evolution of distribution systems from passive to active networks driven by the widespread use of low-capacity power generators and electric vehicles (EVs) as well as battery energy storage systems (BESS). These local generators may fulfil a significant portion of the local power demand, but they can trigger a reverse power flow phenomenon during periods of low demand, causing the distribution system to appear as an equivalent generator capable of providing ancillary services at specific points of interconnection with the transmission system. Power electronic-based devices embedded in the distribution network can also provide these kinds of services through proper coordination from a control centre, including voltage magnitude control, voltage dip regulation, power factor correction and harmonic distortion mitigation, among others. The bidirectionality of power flows in active distribution networks has required a rethink in the design and configuration of protection units and systems, such as using bidirectional relays, which has become necessary to achieve adequate electrical network protection in active power distribution networks. Photovoltaic generation is one of the most used technologies in low-voltage distribution networks. However, this technology involves converter devices that introduce harmonic distortion into the distribution network, which makes the planning, operation and control of this type of network even more complex. One promising solution to cope with these challenges is the development of hybrid AC/DC power distribution grids by integrating VSC-HVDC links. This integration holds significant potential for streamlining power distribution systems, where DC links could be designed from their most basic configuration, involving a multiterminal, back-to-back VSC-HVDC link to a fully fledged multiterminal DC distribution network. Even though these concepts have not been implemented in practice anywhere, the power flow studies numerically illustrate a positive impact on the operation of power distribution systems. The study results show that hybrid networks enable the operation of DC systems at higher voltage levels than AC systems, increasing their transmission capacity. Furthermore, the service quality is enhanced owing to the operating voltage that can be maintained nearly constant on the customer's end. Additionally, as no reactive power flows through the DC systems, voltage and power losses are reduced, allowing the use of smaller cables. It is concluded that the low-voltage power distribution network and the low-end customers are about to receive the tangible benefits that the new technology affords. This may be, but not limited to, business opportunities opening up for small concerns and households, who may become prosumers. It may also be concluded that a faster and more common engineering sense is needed to increase the power-carrying capabilities of the existing power distribution infrastructure by becoming hybrid AC/DC systems. A fast revamp of these networks is needed to accommodate, for instance, the many recharging points for the soon-to-be ubiquitous EV.

Chapter 7 presents wind and solar photovoltaic power generation principles and the way these resources can be predicted for short-term operational planning purposes. It addresses the regulatory requirements for integrating these types of generation into the electrical grid and the role of energy storage devices in firming the power output of these so-called renewable energy sources. It discusses the loss of inertia with increasing shares of renewable energy and how utilities deal with this issue. It also touches upon energy storage technologies to offer operational flexibility and support the large-scale integration of renewable resources in the power system.

Global projections indicate a remarkable growth in the installed capacity of solar and wind power generation, estimated at around 2,000 and 1,000 billion kilowatt-hours, respectively, by 2050. The key to this growth is associated with the significant reduction in the installation and operation costs of these generators, which has been achieved thanks to international efforts to adopt global energy policies aimed at environmental improvements together with government geostrategic interests.

Due to global climate variability, the power output of these resources is heavily influenced by weather conditions such that the expected power generation differs from one location to another. Hence, temporal and spatial forecasting of meteorological condition changes is required to represent climate effects on large-scale energy system models. This forecast will also allow analysts to determine the variability and uncertainty of renewable power in each period of time to allocate a generation dispatch that allows better operation and real-time control of the electrical system.

Based on statistics produced by international energy institutes and government agencies, it is concluded that the rise of solar PV power and wind power is relentless and continues to gather pace. It follows that as the power industry continues to accumulate more and more constructional and operational experience with these newer technologies, and their associated costs decrease even further, it is not difficult to conclude that they will eat inexorably into the fossil fuels' share of electricity production. It is recommended that we continue to prepare ourselves in industry and academia for conforming a much larger body of engineers and energy professionals that understand the new technologies at root.

Chapter 8 provides a qualitative analysis of the impact of renewable power generators' variability and lack of inertia on the dynamic behaviour of the electrical system, highlighting potential challenges. The electrical system's inertia is associated with its ability to oppose instantaneous changes in its operating state in the event of power imbalances. This is achieved through the kinetic energy reserve embodied in the mass of the synchronous generators' rotating parts. Since renewable power generators, such as wind turbines, are interfaced with the transmission system through power converters, this type of interconnection is characterised by a lack of inertia and only exchanges active and reactive power with the grid based on specified setpoints. Hence, the way these generators participate in the system frequency control is by changing their power outputs. This type of control aims to limit the transient frequency excursion by mimicking the synchronous generator natural response associated with its spinning masses, which is why it is called virtual or synthetic inertia.

To maintain the system frequency within a range of values corresponding to normal operation, the control of active power imbalances using the DC links' converters is introduced. To develop the algorithm, the transmission network is taken to be lossless, and VSC-HVDC links are used as tie-lines. Different control schemes for VSC stations, such as grid-forming, grid-following and fixed-power control schemes, are utilised to manage frequency control. The simulation results show that VSC-HVDC links yield a more effective inter-area control than the classical AC tie-lines, helping to mitigate power oscillations owing to the rapid action of the VSC station's controls. A further advantage is that VSC-HVDC tie-lines help to maintain system stability by confining frequency disturbances to the control area where the disturbance occurred. This ensures that this area is solely responsible for adjusting its generation dispatches to restore power balance and maintain frequency within operating limits while minimising corrective actions, like unwelcome load shedding and generation tripping. At the close, the chapter addresses ALFC in multiterminal VSC-HVDC systems, demonstrating the feasibility of control and the increased operational flexibility of hybrid AC/DC electrical power grids.

The overall conclusion drawn from the set of simulation studies presented in this chapter is that the upgrading of a conventional AC tie-line with a VSC-HVDC tie-line would bring about a very positive impact on the dynamic and transient operation of inter-area AC power grids. The improved dynamic performances of the modified test system, in terms of its time responses and power/frequency oscillations, are due to the almost instantaneous responses of the VSC units. It is further concluded that by leveraging their power, voltage and power/voltage droop control capabilities in a strategic manner, VSC-HVDC systems enhance the controllability of the entire network, similarly to the function of synchronous generators, but in a more flexible and economical manner.

Each country establishes a series of minimum technical and operational requirements – framed in a grid code document – that generators integrated into the power grid must comply with to ensure the reliability of the electrical power supply under steady-state and dynamic operating conditions. The renewable generators' intermittent output power needs to be controlled for grid code compliance, and this is achieved locally by using power converters and energy storage systems. From a wide-area context, FACTS devices and HVDC links play an important role in grid code compliance. Chapter 9 details the main requirements in grid codes for the operation of wind power generators in terms of operating voltage and frequency ranges in the event of transmission grid contingencies. It also compares the operating and control characteristics of the grid-forming and grid-following converters used to interconnect these types of generation to the power grid. It is concluded from this comparison that grid-forming converters have superior frequency-power control capabilities in terms of faster stabilisation times together with lower values of frequency excursions. Chapters 5 and 8 illustrate quantitatively the operational flexibility afforded by FACTS controllers. Chapter 9 extends these analyses to show how using these controllers enables renewable power generators to comply with specific grid code requirements. The numerical results show that additional reactive power support is needed to enable the wind generators to ride-out voltage dips, remain connected, and fulfil grid code requirements. These results conclude that the low-voltage ride-through capability of a wind generator improves by using FACTS devices, which also allows the reliable integration of this type of generation into the power network. Regarding the interconnection of wind energy parks using HVDC transmission systems, it is concluded that the best option for evacuating offshore wind power is the VSC-HVDC technology. In addition, the increasing penetration of wind energy in specific geographical areas could be better integrated into the network through multiterminal HVDC links, to enhance the network controllability and decrease undesirable interactions between wind farms and AC power networks.

With the relentless integration of renewable energy resources and the accompanying penetration of power electronic converters into the electrical power system, transmission system operators are facing rapidly changing power systems that they must operate in an economical and secure manner, ensuring their integrity and the continuous supply of demanded energy. Achieving these aims require operation procedures adapted to the new reality, which include a robust protection system that can promptly, reliably, selectively and cost-effectively clear any menacing scenario to the power system. While protection schemes for AC power grids are well established, the experience and equipment in this area may not directly be applied to cope with faults on the DC side of hybrid AC/DC power systems. In this regard, Chapter 10 describes the basic concepts and fundamental differences of protection schemes in AC and DC power networks. When it comes to DC protection schemes, it is important to consider both internal and external faults. Internal faults refer to faults occurring solely on the DC link, while external faults take place between the AC network and the converter station. External faults can lead to a change in the power flowing through the link, which the DC protection scheme may misconstrue as a change in DC current flow, leading to a false line tripping. Therefore, using overcurrent relays as DC link protection units may not be advisable. The overarching conclusion concerning DC protection systems is that this is an ongoing R&D area, where the long-enduring practices of AC power systems may only apply to a limited extent and that a great deal of innovative work is still to come forward from equipment manufacturers and academic circles. For one, on-chip travelling wave protection algorithms would need to be realised. This is an imperative necessity as we move towards fully fledged DC power grids of arbitrary configurations.

11.2 Future developments and trends

Over the past 30 years, or so, the power system industry has been experiencing significant changes in its planning, operation and control processes due to a variety of commercial, technology and climate change issues. On the technology front, these changes are associated with the increasing use of converter-interfaced renewable energy sources (CIRES) and phasor measurement units, together with advances in information and communication technologies. Within this new operating paradigm, electrical power utilities must increase their operational flexibility, reliability and security by strengthening their infrastructure [1]. The choice of the type of technology to achieve these objectives is influenced by various factors, including techno-economic, government policies and social considerations [2]. All these factors allow for defining a wide range of opportunities and future trends in using FACTS controllers and HVDC technologies, some of which will be detailed below in terms of what has been described in the book's chapters.

As the penetration levels of wind and solar PV power generation increase, the predictability of power flows through transmission elements decreases, affecting the control of scheduled power flows through tie-lines. The unpredictable changes in power flows in the entire grid can lead to congestion problems [3], resulting in operational transitions from normal to alert or emergency states [4]. Even though FACTS devices and HVDC links have been demonstrated to be good solutions for ameliorating such operational problem, it is BESS, with their fast speed of response and ramping rate to inject active power into the grid, that is better poised to combat the variations and limited predictability of solar and wind power. Hence, looking to the near future, it is expected that the variability and uncertainty of wind and solar PV powers will be counterbalanced with the integration of suitable energy storage systems, such as fluid batteries, which are sturdier than Li-ion batteries, exhibit higher energy densities, do not catch fire and will potentially be cheaper to manufacture and to maintain. Within this context, delving into future power systems, engineers should be less concerned about intermittence and more concerned with technology and operational practices to make this a worry of the recent past.

In practice, various types of oscillations caused by CIRES have been reported [5]. In the case of inter-area oscillations, power system stabilisers (PSSs) have been used to improve the small-signal stability based on using local signals as feedback inputs. Note, however, that damping out oscillations with these types of signals is limited owing to the lack of global observability of the system's modal characteristics [6]. In such cases, remote signals are employed, taking into account the time delay in data communication [6]. Recently, a methodology has been proposed to calculate participation factors of algebraic variables, which can be used to select a remote signal as a power input to improve modal observability [7]. In addition to the above, PSSs can be coordinated with FACTS devices [6,8] and HVDCs [9,10] to achieve better oscillation damping. In this context, the location of FACTS controllers can be selected based on the algebraic variables with the most significant participation in the oscillation modes to be damped. These algebraic states will be adjusted in real-time to achieve better control, as demonstrated in [6] and [11] for the coordination between PSSs and static VAR compensators. Inter-area power oscillations are an important problem to address in AC-dominated power grids, which could be solved in planning future power systems by interconnecting different parts of the electrical network through multiterminal DC systems. Note, however, that in this operating context, the controllers' gains must be appropriately tuned to avoid the appearance of unwanted oscillatory modes in the AC/DC hybrid power system and DC voltage oscillations.

Coordination between controller devices for wide-area control can be achieved through a dynamic state estimator [6,12]. Even though there are already proposals to integrate FACTS devices and HVDC links into state estimators [13,14], it will be necessary to integrate these types of devices into a wide area dynamic state estimation formulation [15]. This formulation aims to co-estimate the transient evolution of the nodal voltages of the transmission network and the dynamic states associated with the synchronous machines and the controller devices [15]. The estimation will provide situational awareness of the system's operating state and allow for real-time analysis of interactions between different types of controllers within the system. An excellent discussion on the classification of controls and the different approaches to analysing their interactions is presented in [16–18].

The increasing use of inverter-based renewable energy sources reduces the bulk power system's inertia, which concerns operators due to fast ramping frequency gradients in the event of grid contingencies, resulting in lower frequency nadirs and higher frequency zeniths. In an electrical system with increasing or prominent penetration of renewable energies, the risk of transmission line congestion rises considerably, so frequency control needs greater flexibility to balance the power flow fluctuations that occur due to the intermittency of these resources [19]. In [20], it is reported that a 10% integration of photovoltaics would require approximately a 2.5% increase in the reserve capacity to achieve adequate frequency control.

Due to their active power control capability, FACTS and HVDC controllers can assist in frequency control. The use of HVDCs allows the isolation of frequency disturbances between the interconnected systems such that the system or control area where the disturbance occurs is responsible for the generation redispatch to restore the frequency to its nominal value. A disadvantage of this decoupling is that a remedial action scheme is triggered if the affected system does not have enough active power available to maintain reliability. This issue could be avoided by supplying active power through the HVDC link, requiring real-time coordination between the active power reserve of the supplying system and the control specifications of the HVDC link.

In the case of AC links, frequency control is performed through automatic generation control (AGC), considering the area control error signal formed by the frequency bias and the net exchange of active power between systems or control areas. The frequency bias is an adjustable gain with units [MW/Hz] whose value is defined in each control area. A new area injection error (AIE) signal based on active power measurements has been proposed to perform the AGC [21]. This method significantly reduces the need to estimate the bias of each control area accurately to ensure a non-interactive response between areas [21]. The results of implementing this approach to analysing real power systems while preserving the network structure [22] demonstrate an improvement in the response of the speed changer as it presents a decentralised control action. This means that an active power dispatch occurs in the affected area, with minimal impact on neighbouring areas. Based on these results, the AIE-based AGC can be implemented in practice to perform a decentralised frequency control while having an active power reserve from the entire power system in case the disturbed control area cannot achieve the power balance.

References

[1] International Energy Agency. *Power Systems in Transition: Challenges and Opportunities Ahead for Electricity Security*, OECD Publishing, Paris, 2020.
[2] Arcia-Garibaldi G., Cruz-Romero P. and Gómez-Expósito A. Future power transmission: Visions, technologies and challenges. *Renewable and Sustainable Energy Reviews*, vol. 94, 2018, pp. 285–301.
[3] ENTSO-E. European power system b2040. Completing the map: System needs analysis, part of ENTSO-E's 2025, 2030, 2040 Network Development Plan 2018. October 2019. Available online: https://eepublicdownloads.blob.core.windows.net/public-cdn-container/clean-documents/tyndp-documents/TYNDP2018/european_power_system_2040.pdf (accessed 11 December 2024).
[4] Gotham D.J. and Heydt G.T. Power flow control and power flow studies for systems with FACTS devices. *IEEE Transactions on Power Systems*, vol. 13, no 1, 1998, pp. 60–65.
[5] Cardozo C., Prevost T., Huang S.H., *et al.* Promises and challenges of grid forming: Transmission system operator, manufacturer and academic view points. *Electric Power Systems Research*, vol. 245, 2024, pp. 1–30.
[6] Ortega Rivera I.L., Vittal V., Heydt G.T., Fuerte-Esquivel C.R. and Angeles-Camacho C. A dynamic state estimator based control for power system damping. *IEEE Transactions on Power Systems*, vol. 33, no. 6, 2018, pp. 6839–6848.
[7] Tzounas G., Dassios I. and Milano F. Modal participation factors of algebraic variables. *IEEE Transactions on Power Systems*, vol. 35, no. 1, 2020, pp. 742–750.
[8] Pourbeik P. and Gibbard, M.J. Simultaneous coordination of power system stabilizers and FACTS device stabilizers in a multimachine power system for enhancing dynamic performance. *IEEE Transactions on Power Systems*, vol. 13, no. 2, 1998, pp. 473–479.
[9] Smed, T. and Andersson, G. Utilizing HVDC to damp power oscillations. *IEEE Transactions on Power Delivery*, vol. 8, no. 2, 1993, pp. 620–627.
[10] Elizondo M.A., Fan R., Kirkham H., *et al.* Interarea oscillation damping control using high-voltage DC transmission: A survey. *IEEE Transactions on Power Systems*, vol. 33, no. 6, 2018, pp. 6915–6923.
[11] Lara-Hernández E.I. *Generalized Modal Analysis for Damping Oscillations in Power Systems with Renewable Energy Integration* (in Spanish), MSc. thesis, University of Michoacan, Mexico, 2022.

[12] Zhao J., Netto M., Huang Z., *et al.* Roles of dynamic state estimation in power system modeling, monitoring and operation. *IEEE Transactions on Power Systems*, vol. 36, no. 3, 2021, pp. 2462–2472.

[13] Cardenas E.A., Alcaide-Moreno B. and Fuerte-Esquivel C.R. State estimation of flexible AC transmission systems considering synchronized phasor measurements. *Electric Power Systems Research*, vol. 106, 2014, pp. 120–133.

[14] Martinez-Parrales R. and Fuerte-Esquivel C.R. A new unified approach for the state estimation and bad data analysis of electric power transmission systems with multi-terminal VSC-based HVDC networks. *Electric Power Systems Research*, vol. 160, 2018, pp. 251–260.

[15] Romay O., Martínez-Parrales R., Fuerte-Esquivel C.R., Acha E. and Zamora-Cárdenas E.A. A new dynamic state estimation approach including hard limits on control device. *IEEE Transactions on Power Systems*, vol. 38, no. 1, 2023, pp. 463–474.

[16] CIGRE Task Force. Impact of interactions among power system controls. Power Systems Engineering Research Center, Tech. Rep. 16-02-38, May 2000.

[17] Electric Power System Research Institute (EPRI). Analysis of control interactions on facts-assisted power systems. Electric Power System Research Institute (EPRI), Tech. Rep. TR-109969, February 1998.

[18] CIGRE Working Group 14.29. Coordination of controls of multiple facts/HVDC links in the same system. Power Systems Engineering Research Center, Tech. Rep. 149, December 1999.

[19] Ullah K, Basit A., Ullah Z., Aslam S. and Herodotou H. Automatic generation control strategies in conventional and modern power systems: A comprehensive overview. *Energies*, vol. 14, no. 9, 2021, pp. 1–43.

[20] Asano H., Yajima K. and Kaya Y. Influence of photovoltaic power generation on required capacity for load frequency control. *IEEE Transactions on Energy Conversion*, vol. 11, no. 1, 1996, pp. 188–193.

[21] Simpson-Porco J.W. On area control errors, area injection errors, and textbook automatic generation control. *IEEE Transactions on Power Systems*, vol. 36, no. 1, 2021, pp. 557–560.

[22] Ojeda-Sesma R. Frequency stability in low inertia systems (in Spanish), *M.Sc. thesis*, National Autonomous University of Mexico, Mexico, 2024.

Index

www.ingramcontent.com/pod-product-compliance
Lightning Source LLC
LaVergne TN
LVHW011833200926
848484LV00002B/351